Bird Relics

Bird Relics

GRIEF AND VITALISM IN THOREAU

Branka Arsić

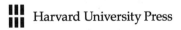

Harvard University Press

Cambridge, Massachusetts
London, England 2016

Printed in the United States of America

First printing

Library of Congress Cataloging-in-Publication Data

Arsić, Branka.
Bird relics : grief and vitalism in Thoreau / Branka Arsić.
pages cm
Includes bibliographical references and index.
ISBN 978-0-674-08847-4 (alk. paper)
1. Thoreau, Henry David, 1817–1862. 2. Grief. 3. Life.
I. Title.
B931.T44A77 2015
818'.309—dc23 2015011245

This book is for my daughter Emma,

in gratitude for making my life superlative

Contents

Abbreviations ix

Introduction: On Affirmative Reading,
or The Lesson of the Chickadees 1

Part I Dyonisia, 467 BC: The Mythology of Mourning

Introduction: Perpetual Grief and the
Example of the Fish Hawks 29

Literal Sounds 41

Homer's Music Box 49

Aeschylus's Paean 58

Cemeteries of the Brain 71

Pindar's Doubles 77

Odysseus, the Lyric, and the Call of the Dead 86

The Sound Dream 106

Coda: Melville's Seafowls 109

Part II Cambridge, Massachusetts, circa 1837: The Science of Life

Introduction: Harvard Vitalism and
the Way of the Loon 117

Birds 143

Fossils 163

Stones 187

Photographs 213

Swamps, Leaves, Galls: A Treatise on Decay 223
Coda: Antigone's Birds 244

Part III Walden Pond, Concord, Massachusetts, 1845: Epistemology of Change
Introduction: On Embodied Knowledge and the Deliberation of the Crow 251
Toward Things as They Are 258
Thinking with Geological Velocity 262
How to Greet a Tree? 266
Contemplating Matter 272
The Noontime of the Mind 277
Thinking with the Body 1: Walking 281
Thinking with the Body 2: Sitting 293
Personhood: Who or What? 311
Coda: Benjamin's Seagulls 318

Part IV Ossossané Village, Ontario, 1636: Acts of Recollecting
Introduction: The Ethics of Communal Mourning and the Flight of the Turtle Dove 325
Deathways of Things 330
Ragged Fragments and Vital Relics 338
The Huron-Wendat Feast of the Dead 357
Coda: Pythagoras's Birds 363

Appendix I Freud and Benjamin on Nature in Mourning 369
Appendix II On Thoreau's Grave 385

Notes 389
Acknowledgments 443
Index 447

Abbreviations

The abbreviations are used in the text and notes. For single-volume abbreviations the citation form is (xx, oo), where xx is the abbreviation and subsequent oos are the page numbers being cited. For multivolume works the form is (xx, oo, oo), where xx is the abbreviation, the first oo is the volume number, and subsequent oos are page numbers being cited.

C Henry David Thoreau, *The Correspondence of Henry D. Thoreau*, vol. 1, *1834–1848*, ed. Robert N. Hudspeth, The Writings of Henry D. Thoreau (Princeton, NJ: Princeton University Press, 2013).

CC Henry David Thoreau, *Cape Cod*, ed. Joseph J. Moldenhauer, The Writings of Henry D. Thoreau (Princeton, NJ: Princeton University Press, 1988).

CO Henry David Thoreau, *The Correspondence of Henry David Thoreau*, ed. Walter Harding and Carl Bode (New York: New York University Press, 1958).

J Henry David Thoreau, *Journal*, vols. 1–8, *1837–1854*, editor-in-chief, Elizabeth Witherell, The Writings of Henry D. Thoreau (Princeton, NJ: Princeton University Press, 1981–2002).

Jn Henry David Thoreau, *The Journal of Henry D. Thoreau*, vols. 1–14, ed. Bradford Torey and Francis H. Allen, Foreword by Walter Harding (New York: Dover Publications, 1962).

ST Aeschylus. "The Seven against Thebes," trans. Henry David Thoreau, in *Translations*, ed. K. P. Van Anglen, The Writings of Henry D. Thoreau (Princeton, NJ: Princeton University Press, 1986).

W Henry David Thoreau, *A Week on the Concord and Merrimack Rivers*, ed. Carl F. Hovde, William L. Howarth, and Elizabeth Hall

Witherell, The Writings of Henry D. Thoreau (Princeton, NJ: Princeton University Press, 1980).

Wa Henry David Thoreau, *Walden*, ed. J. Lyndon Shanley, The Writings of Henry D. Thoreau (Princeton, NJ: Princeton University Press, 1971).

Wk Henry David Thoreau, "Walking," in *Collected Essays and Poems*, ed. Elizabeth Hall Witherell (New York: Library of America, 2001).

On the death of a friend, we should consider that the fates through confidence have devolved on us the task of a double living-that we have henceforth to fulfill the promise of our friend's life also, in our own, to the world.

Henry David Thoreau,
Journal, February 28, 1840

Introduction: On Affirmative Reading, or The Lesson of the Chickadees

Strange things happen in Thoreau: sand starts moving like water, and stones vibrate with life; extinct species return; pine trees cry; fish become trees; men grow grass out of their brains; men, not gods, walk on water; like animals and with them, they also walk on four legs; they talk to fish and birds; birds migrate back to life after they have been seen dead; humans migrate into birds; birds migrate into other birds; humans migrate into other humans; two persons come to inhabit one body; two bodies come to be inhabited by one person. How are we to understand such strangeness? We can't treat it as fiction for, strictly speaking, Thoreau is not a fiction writer. The generic characteristics of all of his writings—*A Week* is a memoir, *Walden* is autobiography, the *Journal* is a record of perceptions and thoughts, while the natural history essays are structured according to the logic of scientific writing of the day—require that we treat their content not as fiction but as truth, and their utterances not figuratively but declaratively, as testimonies. Yet, his declarations are sometimes so eccentric, they so radically blur the distinction between what is possible and what is not, between miraculous and natural, that one must raise the question of whether to take them seriously. Does Thoreau really mean what he says when he asserts the possibility of what reason would call fantastic, such as, for instance, when he suggests that persons can shift their shape or, even more strangely, when he claims that life overcomes death, thus inverting the laws of causality and, as Colin Dayan puts it regarding Poe, implying that he does not inhabit a "privileged position in relation to the supposedly 'dead' or unthinking matter surrounding [him]"?[1]

Thoreau attached clippings to the back cover of one of his late commonplace books. Here is a sampling:

> *Boston Journal*, September 27, 1858 (excerpts); *Donati's comet seen in Cincinnati*
>
> On the evening of the 25th of September, the appearance of the comet, in the great refractor of the Cincinnati Observatory, was especially interesting. . . . The most wonderful physical feature presented was a portion of a nearly circular, nebulous ring, . . . the bright nucleus being in the centre, while the imperfect ring swept more than half way round the luminous centre. . . . No one can gaze on this gigantic object, in all its misty splendor, without a deep impression that the eye is resting on a mass of nebulous matter, precisely such as the nebular theory of La Place supposes to have been the primordial condition of our sun and all its attendant planets, and from which chaotic condition this beautiful system of revolving worlds has been evoked by the action of a single law. . . .
>
> O. M. Mitchell.
> Cincinnati Observatory, Sept. 27, 1858

> *Boston Journal*, March 2, 1860, *A TOAD STORY*
>
> A gentleman who witnessed the sight informs us that, about ten days ago, along one of the main roads near Forge village, in Westford, he observed the most marvelous collection of toads he ever witnessed or heard of. In the road for as many as a hundred rods the ground was so covered with them that one could not put his hand down without putting it upon a toad. An estimate was made, and it was determined that there were at least

> as many as twelve toads to the square foot! The sides of the road and fields were not examined, but for the distance we have named there were toads innumerable. Another fact not a little singular is that they were all apparently the same size—being about half an inch high, or in length, and in color and appearance seemed to be precisely alike, and all were sprightly and seemed as if very much at home. The question is where did they come from? There was a smart shower the night before they were discovered, but is it possible that they rained down? And if so, where did the clouds come in possession of such a multitude of juvenile toads? The fields around may have been as thickly populated, for aught we know to the contrary, as the road; and if so, who can explain their presence?

Reported in newspapers as events observed by reliable witnesses, these examples of the miraculous—vibrant and nebulous matter observed in the moment of creating new life, toads raining down from populated clouds—assume the status of the factual. More generally, the articles demonstrate that to an antebellum American the divide between fantastic and real was less distinct that it is to us postmoderns, which imposes the requirement that the faithful historian of ideas respect this blur. More specifically, to Thoreau, who collected them systematically, these reports were perhaps proof of his lifelong belief that far from being something surreal, which could at best function as a metaphor of something real, the fictional or even irrational is part and parcel of the real. In fact, that was a guiding conviction in *Walden*, whose opening famously declares that hardly any fiction can match the strangeness of ordinary reality:

> I have travelled a good deal in Concord; and every where, in shops, and offices, and fields, the inhabitants have appeared to me to be doing penance in a thousand remarkable ways. What I have heard of Brahmins sitting exposed to four fires and looking in the face of the sun; or hanging suspended, with their heads downward, over flames; or looking at the heavens over their shoulders . . . ; or dwelling, chained for life, at the foot of a tree; or measuring with their bodies, like caterpillars, the

> breadth of vast empires; or standing on one leg on the tops of pillars,—even these forms of conscious penance are hardly more incredible and astonishing than the scenes which I daily witness. (Wa, 4)

Nothing more incredible, then, than the lives of ordinary folks, whom Thoreau sees performing acts more fabulous than Brahmins involved in incredible corporeal and incorporeal transformations. But if the fantastic is so embedded in the common as to constitute it, how is it that we, so many ordinary people, can't see what Thoreau sees? What have we done to alter the real into what is coherent, explicable, and knowable, expelling the wondrous into an elsewhere that is only imagined?

On numerous occasions—to which I return in the sections that follow—Thoreau suggested that this filtering out of the fantastic from the real is generated by the dogmatic and critical epistemologies of the West, expounded on from Plato to Leibniz and Kant. Those epistemologies were predicated on the idealistic understanding of truth as noncontradictory. And since the incredible couldn't be deposited in the real in a noncontradictory manner—both because it is in itself often contradictory and because it would render reality simultaneously credible and incredible and consequently cancel the conceptual divides that generate noncontradictory, truthful thinking about the real—our thought is disciplined by mainstream epistemologies that produce a kind of magical transubstantiation: thinking is made to dematerialize what is really incredible into what is only imagined to be so. However, as Thoreau will suggest, the image formed through this dematerialization will not function as a representation of the material and real, but instead—and herein lies the perversity of the imagination and its ideations—as the merely metaphorical representations of the illicit and irrational longings of our mind. Thus, to reference only a few examples of this operation that are crucial for understanding Thoreau's thought, metamorphosis will be understood not as a representation of real, corporeal transformations, but as a metaphor of the mind's desire to change what is; the idea of a hybrid will refer not to something real that is generated by interspecies splicing, but to a metaphor of the mind's desire to experience its own otherness; the possibility of vitalized and agitated matter—which Thoreau will engage in a serious way—will be understood not as proposing a realistic ontology but as a metaphor for the mind's dream of immortality. The real is thus not only dematerialized into its own ideal representation; more radically, it is altered into a metaphor of what it is not, of the spiritual. An ideation of massive proportions is thus generated whereby everything becomes topsy-turvy, as matter is transubstantiated into an only more-or-less realistic representation of the mind's doings, an idea that de facto blocks access to embodied reality.

A passage from *Walden* additionally suggests that this transubstantiation (into a metaphor of our spiritual inclinations) of the material world that Thoreau calls "nature," exposes us to what he senses to be the greatest danger of all, the danger of forgetting the literal, the suchness of things: "But while we are confined to books, though the most select and classic . . . we are in danger of forgetting the language which all things and events speak without metaphor, which alone is copious and standard. Much is published, but little printed" (Wa, 111). Just as for the idea of matter as not a representation of anything corporeal but of the mind's operations, so all our words, Thoreau here suggests, react not to things but merely to other words. What is published now speaks to what was published before, the language we speak at present answers to classic languages; words are not representations of things, as Aristotle believed, but of words. Language is self-referential, its metaphors nothing but reservoirs of images deposited in them by the history of language, not things. In suggesting that all language is only "indirect speech" that converses with itself about reality in a complete absence of reality (or, as he phrases it, that words are not shaped by a direct "imprint" of things and literal events onto them), Thoreau gestured toward the discovery that, as Paul Patton puts it in another context, "it is not the representation of a nonlinguistic reality that is the primary function of language but repetition and therefore transmission of something already said."[2] As Thoreau specifies in *Walden*, that distance from the real turns all language into mere chatter:

> It would seem as if the very language of our parlors would lose all its nerve and degenerate into *palaver* wholly, our lives pass at such remoteness from its symbols, and its metaphors and tropes are necessarily so far fetched, through slides and dumb-waiters, as it were; in other words, the parlor is so far from the kitchen and workshop. The dinner even is only the parable of a dinner, commonly. As if only the savage dwelt near enough to Nature and Truth to borrow a trope from them. (Wa, 244–245)

Living in a parlor distanced from things we believe that we are talking about—distanced from the kitchen and the workshop where life is in the making, where it is busy changing—we end up living among "far fetched" metaphors. Our epistemologies have filtered the wondrous out of the real to reach a truth that has in fact relocated us in an imaginary real. Paradoxically, we have ended up living in a fantasized real from which the fantastic has been expelled. Our lives are highly ideated and "meaned up" ("the dinner even is only the parable of a dinner, commonly"), while ideated objects pass us by without affecting us. Thoreau's central discovery can thus be summed up in the insight that, as Cavell puts it, "an object named does not exist for us in a name."[3]

Thoreau's claim that human language means forgetting the very "language which all things . . . speak without metaphor" points to a major premise that will guide his investigations. It signals that for him—similar to what Walter Benjamin will later propose—in addition to the language of man, there is also "language as such," the language of things and other, nonhuman, beings. Everything is generative of language, then. As Benjamin explicitly puts it, echoing almost verbatim Thoreau: "There is no event or thing in either animate or inanimate nature that does not in some way partake of language, for it is in the nature of each one to communicate its mental contents. This use of the word 'language' is in no way metaphorical."[4] Just as Benjamin suggests that not only animals or plants speak but also things such as desks and lamps—as he puts it, "the language of lamp" expresses the "mental being of the lamp"[5]—so Thoreau proposes that nothing is abandoned to silent numbness, for there are meanings that things exchange among themselves. Things and beings bypass the human obsession with ideation and metaphorization to generate meaning by affecting other beings, or, as Thoreau has it, by imprinting themselves on other beings, thus literally or materially altering them, leaving their trace in them for them to "read."

Thoreau's recovered nonideational language of humans participates in this semiosis not by sacrificing the specificity of its word-making capability, but by requiring words to alter substantially. They have to find their way back to things: emerging out of imprints—a novel phenomenon generated by encounters with beings and things—they themselves have to become "events"; they have to be the repository of unprecedented meanings that can further act "literally," as "things," by generating alterations (imprints) in other beings and things. This semiosis of imprints evades the dualisms of mind and matter, literal and metaphorical. For when Thoreau says that nonideational language—in which "dinner even is only the parable of a dinner, commonly"—will have to be exchanged for another, in which words will be derived from bodies directly ("borrowing a trope from nature"), he is precisely gesturing toward a nondualistic ontology in which the difference between words and things, ideas and bodies, is not considered insurmountable. Instead, all those different phenomena will be understood to relate to one another on the same ontological plane, affecting one another in the same being. He therefore gestures toward the understanding of the sign that Peirce will later articulate, as something that can, but doesn't have to be linguistic, allowing linguistic and nonlinguistic signs to alter one another. If we can borrow a sign from a pine tree or a toad it is precisely because for Thoreau a sign is generated by any dynamic in which, as Peirce has it, "something . . . stands to somebody, for something in some respect or capacity." However, as Eduardo Kohn adds, "somebody . . . is not necessarily human and it need not involve symbolic reference or the awareness often associated with representation."[6] A "somebody" or a "self" is rather any locus of animation or motion in the embodied world that, in affecting another such "somebody," forces it to react to it, and thus to interpret it as a sign.

In fact, Thoreau will go further than gesturing toward the nondualistic ontology and the new language it generates. In a remarkable discussion of the reform of senses at the closing of *A Week*—to which I return, attempting to do justice to its complexity, in my discussion of his epistemology in the third part of the book—he will explicitly declare that his efforts are invested in reversing the ideational direction of our language: "Are we to be put off and amused in this life, as it were with a mere allegory? Is not Nature, rightly read, that of which she is commonly taken to be the symbol merely?" (W, 382). Instead of reading the natural—not just its beauty but its capacity to change, recover, or endure—as a metaphor of what in the spiritual is not yet accessible to the mind, Thoreau understands our mind and its thinking practices as mere metaphors of the extraordinary capacity for change in the material, a capacity that we have difficulty in understanding precisely because of our penchant for metaphorization: "We are still being born, and have as yet but a dim vision of sea and land, sun, moon and stars" (W, 385). If for Thoreau, as for Spinoza, we don't yet know what a body can do, what corporeality is, and what it is capable of, it is precisely because we don't have access to it, having obscured it with categories and representations as so many metaphors of the material. Thoreau thus wants to "read the Nature right" by despiritualizing it, that is, by passing through the curtain of metaphors to reach the material itself, and he tries to do that through a process I call "literalization," consisting of the twofold gesture I have discussed thus far: turning the word into some sort of thing, capable of affecting bodies; and bringing words closer to objects, recovering the presence of objects in names. In his terms, he wants to move the parlor into the kitchen, where words are still "cooking," still being concocted from fresh ingredients.

The literalization of language wasn't just one among many interests that Thoreau cherished. It was his central interest; as Barbara Johnson argued, it was what the experiment at Walden Pond was all about. On her understanding, *Walden* is not simply an elaborate metaphor of self-transformation but instead *is* that transformation, simultaneously the act and its description. *Walden* doesn't connote, but denotes:

> What Thoreau has done in moving to Walden Pond is to move *himself*, literally, into the world of his own figurative language. The literal woods, pond, and bean field still assume the same classical rhetorical guises in which they have always appeared, but they are suddenly readable in addition as the nonfigurative ground of a naturalist's account of life in the woods. The ground has shifted.[7]

The ground has shifted into figures, the bodies have moved into words, which is why, as Cavell puts it, "we don't know what *Walden* means unless we know what Walden is."[8]

One need not dwell on sentences insisting on literalization—as when Thoreau claims that nature is "that of which she is commonly taken to be the symbol of"—to realize how correct Johnson and Cavell were in claiming that, for Thoreau, to understand what something means is always to experience what it is. Some instances of that epistemological faith, which frustrates the drive to metaphorize, are famous. For instance, *A Week* is premised on the injunction that to understand life as a journey one must start traveling. Similarly for *Walden:* to understand that living means learning how to dwell, one must turn one's life into building one's dwelling; one can't understand what a house is unless one builds it. But there are many other, less famous examples of the same epistemological credo, asking us to somehow leave the images and concepts of our mind to access literally the life those images and concepts represent or symbolize. One finds such examples everywhere in Thoreau's writings as well as in his practice. They organize his thinking as well as his perception and everyday behavior.

Donald Worster identifies the behaviors generated by Thoreau's literalism—his desire actually to inhabit the perceived—as an "intense empiricism." By that he refers to a series of sometimes painful, rather than enchanted, practices whereby the mind is reduced to perceptions so intense that they lead the perceiver out of his self into the perceived. The perceiver's passages from a thought or perception into its object are identified by Worster as "becomings," that is, real or literal transformations, and he offers a series of examples to explicate the concept. Sometimes Thoreau becomes a muskrat: "[He] must allow himself to be engulfed to his very ears in the odors and textures of sensible reality. He must become, like the muskrat, a limpid eyeball peering out of the sedges of a flooded meadow."[9] On another occasion this "muskrat" turns into the earth's crust and becomes "terrene": "I felt as if I could eat the very crust of the earth; I never felt so terrene." If the "I" feeling terrene still points to a distance between perceiver and perceived, that distance is annulled on yet another occasion cited by Worster, when Thoreau performs literal behaviors that cause him to become a fox, a pine tree, or a frog: "Nineteenth-century Concord was home to many unusual individuals. But only one local citizen was likely to be seen snorting and galloping with glee after a fox on a snowy hillside. Or sitting in the top of a pine tree, swaying with the wind, or crawling about on his hands and knees endeavoring to communicate with a reluctant wood-frog."[10] Strange if not downright naive, such behavior nevertheless constitutes Thoreau's painstaking efforts to reach the real literally. His wager is that he will start experiencing differently thanks precisely to his (even if always only wished for) emancipation from abstractions embedded in our categories and tropes. He hopes that once his senses are entrusted to an unknown—conceptually unmediated—reality, they will allow him finally to experience how entities, whether corporeal or mental, are not fixed, as our traditional epistemologies would have it, but rather change, and so cancel the generic divides not only among beings (such as when Thoreau becomes a pine tree or a muskrat) but also, as I suggest in what follows, among the living and the dead.

Because Thoreau is so obsessed with testing the limits of the metaphorical, I address what literalization means to him throughout the book and from a series of different points of view. For now—and to clarify the method I have adopted in addressing Thoreau's work—the following general remarks concerning what counts as an act of literalization will suffice.

Literalization features the a-conceptual. The escape from "built-up" connotations that Thoreau envisions for a reformed language is a process that involves "stripping bare" our thought and language (our "walls must be stripped, and our lives must be stripped" (Wa, 38), producing an utterance or word so tied to its object that it only denotes. Literalization weakens the connotations we have deposited in the word to "mean up" our lives, but which are in fact, as Cavell points out in discussing literalization in Beckett, "curses under which the world is held," for they precisely distance us from the world. Literalization weakens connotations by unfixing clichés, making us acutely aware that we drown in empty meanings, that "our language [has] worked too little, because it worked too much."[11] Additionally, in trying to cancel metaphors, literalization also seeks to cancel concepts, because concepts are metaphors par excellence. As Paul Patton puts it, a "concept exists only when there is a distinction between what falls under the concept and what does not."[12] But such a distinction is always imaginary, always metaphorical only, not just because "no two particular objects or occurrences are identical"[13] and therefore can't be presumed to have the same essence that the concept confers on them, but also because singular occurrences always vary in the passage from class to class, from concept to concept, generating a process that can be called a "zone of connection."[14] If all concepts are metaphors, then, it is because they are the outcome of fictionalization. A concept emerges as a result of the cancellation of the real difference among singular cases under its jurisdiction, of their being merged into an imagined identity, which, by the same token, is a cancellation of real connections, transgressions, and mobility in a contrived stability. In undoing concepts, literalization seeks to achieve two things. First, to recover the particular, as when Thoreau insists that "pickerel" is a meaningless word, obscuring the fact that the "steel-colored pickerel" must be differentiated from a "bright golden kind," a "bright-golden kind with greenish reflections," and a "golden-colored" one (Wa, 184). The compound "bright-steel-golden with greenish-bluish-yellowish reflections" signifies less than, like a sensation, it affects. This particularization of words, turning them into hyphenated compounds on the verge of becoming a percept or a thing that merges into the singular it reflects, can be understood as a desire to reach the literality of what Thoreau calls the "this," so that we find ourselves where "this is and no mistake" (Wa, 98). The recovery of the "this" tells us that literalization really dreams about exiting the word, even the word that merely denotes, in favor of the percept. What it really wants is to enable the speaker's or viewer's immersion in the flow of ongoing perception. Hence, the second aim of literalization: in canceling categorical divides among the occurrences that concepts generate, literalization also recovers the fluid, the intermediate, and the variable. This explains Thoreau's obsession with transition, his dedication to detecting the connectedness of phenomena, emblematized, most notably, by his central discovery that Walden Pond is not isolated in its circular form but connected to Flint's Pond through many "smaller intermediate ponds" (Wa, 181). If Thoreau's literalizations render inoperative the abstractions that concepts are, it is because they want to recover not just the particular but also the process, the particular in the moment of its becoming otherwise, the particular that has already started differing from itself, a singularity that is already two, a pair that is already four. If, as Roland Barthes's formulaic phrasing has it, "nuance = difference (diaphora),"[15] then in recovering the nuance Thoreau's literalizations want to enable us to finally enter the diaphoric world, the plethora of differences.

Literalization recovers slow time. Because concepts are abstracts (that is, summaries of nuances), they enable thought to move quickly through varieties of phenomena. In rendering it unnecessary to dwell on the particular, they speed up thinking. They are economical, manufactured for those who are busy, exchanging lost nuances for gained time. In contrast, by refusing to conceptualize, literal thought is tied to nuance and dwells on it, following its very rhythm. It allows phenomena to take their own time, while teaching the mind patience.

Literalization is the critique of the literary. In its commitment to what is diaphoric, literalization can be understood as a critique of literary as well as conceptual thinking. For while literature indeed works with nuances—so much so that Barthes defines it as "a codex of nuances"[16] and contrasts it with philosophy, which operates through concepts—it does so only to the extent that it is able to restrict them by means of form. To maintain the very being of the literary, its own generic specificity, literature must interrupt the flow of the particular. Literary forms thus do for literature what concepts do for philosophy: they classify, segregate, bind, and regulate. In so doing those forms act like concepts, as forces of ideation and generalization, which means finally that they act as metaphors ("metaphor, hence, a generality" is how Barthes phrases it).[17] Thoreau's effort at literalization should therefore be registered as his resistance to the literary, as a subversion of literary forms that is by no means restricted to loosening the connections between paragraphs in *Walden*, for instance, or by constructing hybrid forms (*A Week* is at the least a memoir, a travelogue, and a history book). More radically, it involves working toward a complete dissolution of the form, such as he achieves in the *Journal*, where nuances flow unchecked, proliferating into formlessness. As Barthes again puts it, "nuance—if not kept in check—is Life," not literature, which is precisely where Thoreau wants literalization to take him: to life, to a becoming alive of the word.[18]

Literalization risks insanity. As Thoreau suggests in *Walden*, stripping our lives bare of connotations requires not only that we undo clichés and idioms (to leave "the mud and slush of opinion or prejudice"), but also that we abandon everything we thought we knew about the world, about "our poetry and philosophy and religion" (Wa, 98). It requires us to unfetter the systems and methods of thinking that we have designed to make sense of the world. However, in canceling the categorical and conceptual, as well as the methodical and ordered, literalization risks their opposites, and therein lies its greatest danger. It risks disorder, incoherence, and obscurity, if not insanity. Indeed, as Cavell reminds us, again in the context of literalization in Beckett, the disregard of categorical divides produces an incoherence that brings words so close to things that it even confuses them for things, which is "the mode which some forms of madness assume."[19] That is why, as Cavell puts it in *The Senses of Walden*, the "question of insanity . . . or at any rate, the extremity and precariousness of mood in which [Thoreau] writes is so recurrent in *Walden*."[20] Moreover, that extremity becomes its central theme, embodied throughout the book by the loon, the bird that risks or suffers insanity time and again by allowing itself to become disoriented, lost in its diving ("he would dive and be completely lost" [Wa, 234]). The loon reemerges on the surface as a crazed "demon," uttering an "unearthly laugh" [Wa, 235]), or metamorphosed into a wolf (howling "probably more like . . . a wolf than . . . bird" [Wa, 236]); indeed, becoming the wolf-bird that Thoreau himself would then strive to become ("While he [the loon] was thinking one thing in his brain, I was endeavoring to divine his thought in mine" [W, 235]). The loon not only merges the questions of insanity and metamorphosis—insanity functions as a type of molting—but also confuses corporeal and incorporeal transformation. And it is here that we finally encounter the most disturbing consequence of literalization and are required to deal with the crucial question of how to address Thoreau's work. For much of what Thoreau is saying can indeed be understood literally: for instance, the injunction "to think about the house requires building it" might change the way we think about dwelling, or change our habits, but not necessarily drive us to madness.

But how are we to treat metamorphosis, a word and experience so central to Thoreau? There is no reading of Thoreau that doesn't emphasize his preoccupation with "self-transformation," self-renewal and change. But if those assessments of his project are by now empty clichés, I would maintain that it is because we don't know what they really mean. Is Thoreau seriously talking about self-transformation? Of the mind or of the body as well? As Cavell puts it, "it is hardly necessary to insist on the concept of moulting, and metamorphosis generally, as central to *Walden;* but as elsewhere, it is hard to believe how thoroughly it is meant."[21] If metamorphosis is not meant in the spiritual sense only, representing a change of heart or mind; if it is not meant metaphorically, referencing corporeal reshaping to signal a change of mind; if, in contrast, it is meant really, literally, or "thoroughly," then we can no longer read it as poetic fancy only, as a mere wish for the impossible. Instead, we must come to terms with the *literal reality* of a change of heart that also generates a change in our bodies.

My readings are fashioned by the decision to take Thoreau's utterances—even when they seem most fanciful—as serious, nontrivial, and literal. In terms of Cavell's remarks from "Knowing and

Acknowledging" regarding how a non-skeptic is to treat the paradoxes of the skeptic, and appropriating them for the purpose of articulating a methodology of reading a literary text, in what follows I have approached Thoreau's work from the vantage of what, for want of a better word, I call "affirmative reading."[22] And while I will not deny that I find Thoreau's strange ideas deeply seductive, that doesn't mean that I have to believe in what he said, nor does my affirmative reading mean that I have to offer support or further evidence for his words as if I were defending their truthfulness. Instead, my reading obeys the following sets of presumptions.

Thoreau means every word he says, in the exact way that he says them; he means it literally. This is not to say that he didn't realize that he is doomed to language that is distant from things; that he didn't realize that swift transportation of things into words—which he calls for when he asks us to borrow directly from nature—is incredible; or that he wasn't always painfully aware that he is failing in his effort to bring tropes back to bodies. The promise of recalling actual bodies when we call their name was never actually fulfilled. If I then say that I treat Thoreau's words as if such a promise were fulfilled, it is because his desire to move words in the vicinity of things—to make them literal or to even turn them into what they mean—remains for him a regulative epistemological wish. Even if it is always failing, it is thus nonetheless always consequential in fashioning even his most ordinary acts and perceptions.

Because literalism must always—at least partially—fail, readings that take Thoreau's words as if he didn't mean them literally remain indispensable. We have gained fabulous insights into Thoreau's thinking through commentaries that propose, for instance: that his discussion of fish in *A Week* in fact refers to relationships among humans, or even among the thoughts in the mind of a single human; that his obsession with birds is really an obsession with what is ethereal and spiritual; that a little green bittern mentioned in *A Week* stands for platonic archetypes; that in talking about autumnal leaves, he has in mind the advanced modernity of the nineteenth century, or even projecting ideas concerning late capitalism; that his preoccupation with nocturnal walks registers his romantic devotion to what is dark and dim or, alternatively, his devotion to the woman to whom he proposed marriage earlier in his life; or, that when he talks about an apple tree he is pointing to Yankee resilience.

While acknowledging the complexity of such readings, I am interested instead in investigating what kind of Thoreau is revealed if one takes him to be saying exactly what he writes. What kind of philosophy appears if, when he says "fish" we take it to mean fish, or when he talks about matter (sand, mud, dust, stones) we take him to mean matter; or when he talks about the healing capacities of moonlight he means that moonlight literally heals bodies; or when, in a famous passage, he says that it was no longer he who "hoed beans" he indeed believes he wasn't hoeing the beans any longer; or when, as on so many occasions, he talks about corporeal metamorphoses, we understand him to mean that the bodies really metamorphose? To claim that Thoreau can't intend real metamorphosis when he says "metamorphosis," as he must know that corporeal transformations aren't really possible and so must be speaking metaphorically; to claim, additionally, that the desire to avoid metaphor must be relegated to the domain of epistemological phantasmagoria, as it amounts to an impossible desire to generate meaning without words, all of which are instances of tropes—such claims would, it seems to me, represent nothing less than a form of abuse, comparable to the way the nonskeptic abuses the skeptic by telling him, for instance, that he can't seriously doubt the existence of his own body since he obviously sees himself walking.[23]

Consequently, I tried to treat all of Thoreau's utterances as if they were meant literally, instead of choosing which ones might and which ones might not be so intended. For if we suppose that Thoreau meant only some of his words literally and others metaphorically (he really means that we must build a house in order to think it, but doesn't really mean that there is such a thing as corporeal transformation), we are immediately confronted with the necessity, and impossibility, of deciding the literal or nonliteral status of each statement. That would generate only arbitrary and ad hoc readings. Consequently, when encountering strange or even fantastic propositions in Thoreau, I refrained from normalizing them as allegories. Instead, I tried to respect and follow their strangeness by treating them as philosophical propositions that formulate a different ontology.

Thoreau knows at least as much as his reader. Following this rule I assumed that to tell Thoreau that he often writes strange things would not come as a surprise to him at all. In other words, I was convinced that he, like me, knew that the merging of persons, for instance, or calling the dead back to life are not quite credible events. Following the same rule I similarly assumed that Thoreau knew his statements to be often contradictory, for he revises his books carefully yet lets the contradictions remain. My belief that Thoreau was aware of the strangeness and contradictions of his statements also absolved me of the supercilious task of disclosing or denouncing them as contradictory or voicing my own disagreement with their strangeness. Instead of expressing my frustrations, my affirmative reading tries to bracket my beliefs (even if I realize that the task is impossible, it nevertheless regulates my readings here) to come nearer to the strangeness and difference produced by reading Thoreau, just as Thoreau so often did when faced with what he found incredible. That is an attitude akin to what Jonathan Lear terms that of a "bird-philosopher," by which he means a thinker or reader who listens instead of judging, somebody who is, as he specifies, wise only in the manner of a chickadee, for "the wisdom of a chickadee consists" of "learning to listen," "sharpening ears by constant use."[24] In Thoreau too, chickadees are extraordinary beings ready to listen to and follow what is different. In fact, in the later Thoreau, chickadees became an emblem of an exceptional capacity for self-transformation enacted through an effort to hear the beings that populate a non-chickadee world, as when a chickadee listens to a twitter that, as if to "attract a [twitter] companion" releases a "distinct gnah," whose meaning should be incomprehensible to a nontwittering creature, such as a chickadee, since its refrain, according to Thoreau's transliteration, is "tche de de de" (Jn, 12, 87). And yet, after patiently listening to this strangeness, the chickadee was observed to understand the call and "unfailingly followed," perhaps not quite becoming a twitter but nevertheless accompanying the twitter into its twitter world, forming an anomalous twitter-chickadee couple (Jn, 11, 391). Adopting the way of a chickadee, neither supporting nor disproving Thoreau's thinking—finally released from having to evaluate the "rightness" of an author's position—I have tried to follow him, not necessarily to accept but to "learn the particular ground" that his thinking occupies; that is, I tried to learn what his ideas could possibly mean, which, as Cavell reminds us, is "not the same as providing an evidence for them" but is instead the "matter of making them evident. And my philosophical interest in making them evident is the same as my interest in making evident the beliefs of another man, or another philosophy."[25] It is an interest not in reconciling difference into unity but in cherishing it, allowing it to stay.

To make Thoreau's thought evident I follow it as far as it leads. To follow Thoreau's thinking to its first cause or extreme consequences, rather than interrupting him every once in a while to critique him, doesn't mean that I am at the mercy of what I disagree with, since in the affirmative reading I imagine, to use Cavell's phrasing once again, the "critic and his opposition [don't] have to come to *agree* about certain propositions which until now they had disagreed about."[26] Instead, following Thoreau to his extreme consequences means reaching his final complication, experiencing the impasse he creates and, through the experience of this final boundary of what can be thought, formulating a new problem (instead of answering an existing question). Questions are formulated on the basis of already existing solutions (whether Thoreau's thought answers the criteria of truthful thinking, whether it has recourse to illicit forces of the fantastic, whether his call for literalization is a fantasy, or whether his books are well written—these are all questions one can raise and answer only because one already knows that truthful thinking must be conceptual, hence metaphorical, noncontradictory, and nonoccult, and because one already has a set of aesthetic standards outlining what a "good book" is). In contrast, as Gilles Deleuze, another advocate of affirmative thinking puts it, "the art of constructing a problem," the "invention of the problem-position" occurs "before finding a solution."[27] Problems are formed in an extreme precariousness of thinking, when it is unprotected by questions and faces the un-thought, what threatens to devastate it, what puts it at risk or has it confront an abyss. The Thoreau I encountered is a thinker who manages to avoid such protective questions; always ready to risk exposing thought to the awe of its own cancellation, he formulates unprecedented problems. I have tried, as humbly as possible, to follow his thinking to the very core of the problems it creates, where its precariousness threatens it with dissolution.

Following Thoreau's ideas is a task that is additionally complicated by the ritualistic if not obsessive nature of his thinking, which makes him seem to work through the same problem of transformation time and again. Respectful of nuance, he always formulates the problem of change from a slightly different vantage, which, especially in the *Journal*, often gives an impression of repetitiveness. This "slightness of difference," a difference that verges on repetition, confronts the critic with the difficult question of how radically to summarize his insights: does it suffice simply to suggest that transformation overwhelms his thinking, or should the critic, rather, follow the "slightness of difference," to see what it generates? Faithful to my effort to respect the general orientation of Thoreau's work, I try to solve this quandary by obeying his slow pace, letting Thoreau teach me how to respect his nuances, and realizing that he is anyway not a thinker for the impatient ones. As a result, in different sections of the book I return to seemingly similar problems but in order to draw, as Thoreau does, different conclusions from them.

While my Thoreau is influenced by the great ecological readings of Lawrence Buell above all, but also of Jane Bennett and Laura Dassow Walls, Thoreau is nevertheless less an ecologist than a thinker obsessed with the problem of life in a properly ontological sense. By this I mean not only that everything in his world—from stones to humans—is alive, but also that in his philosophy life is afforded the status of a force that precedes and generates all individuations and into which individual forms dissolve. Consequently, death is considered a process of deformation but not of cessation. Differently put, in Thoreau's world death does not have the power to interrupt life but instead functions as the force of its transformation, enabling us to experience finitude while ushering us into what remains animated. My book thus tells the story of how this central claim—which I have termed "vitalism" and whose meaning I develop in the Introduction to Part II of the book—came to be and how it fashions and complicates Thoreau's epistemology, science, poetics, and politics.

Because each part of the book is contextualized by its own introduction, here I offer only a very general outline of the whole. I argue that vitalism emerged as a central issue for Thoreau in the wake of his brother's death. The intense grief that remained following John's departure prompted him to ask sometimes disconcerting questions about what, and even whether, death was, leading him ultimately toward a stunning theory of grief as well as a novel epistemology and the outlines of a science of life. The first part of the book thus explores the theory of grieving that Thoreau formulated in response to John's death. That theory was based on a form of "unforgetting," which, by changing how we understand personhood, evolved into a philosophical proposition concerning life and an ethics of the treatment of living beings. In analyzing Thoreau's response to his own grief, I suggest that the ideas he will begin to formulate from 1842 on, relating to "perpetual grief," are explicitly predicated on the archaic Greek practice of perpetual mourning *(álaston pénthos)*. I suggest that Thoreau's perpetual grief—the ban on forgetting the loss—is closer to an ontological operation of restoration of the loss than the modern psychological commitment to protecting the interest of the mourner. Because his theory of grief is formulated under the influence of Greek sources—Hesiod's cosmology, Homer's epics, Greek tragedy, and Pindar's poetry—and because, following those sources it sometimes offers fantastic propositions and entertains ideas of magical transformations, I have titled the first part of the book a "mythology" of mourning.

As Thoreau's mourning leads him to question the existence of death and as this question leads him in turn deeper into an engagement with contemporary sciences of life, Thoreau comes to realize that many of the versions of vitalism he had been investigating through Greek sources—from Thales to Aristotle and Lucretius—had been revived and given scientific status. The second part of my book thus charts Thoreau's investigation of and contribution to the pre-

Darwinian sciences of life, at the same time reconstructing how he was influenced by the rarely discussed theories of life formulated by a group of scientists working in Boston and Cambridge, Massachusetts, and related to Harvard University, whom I came to call the "Harvard vitalists." That influence was profound, and the work of Bigelow, Felton, Guyot, Holmes, Nuttall, Tuckerman, Ware, and Waterhouse led, I argue, to Thoreau's formulating a vitalistic philosophy that would lead in later years to a more complex homeopathic proposal regarding the physiology and pathology of living beings. I investigate Thoreau's obsessive interest in vegetal tumors—galls and other abnormal plant outgrowths—as well as his lifelong preoccupation with vegetal decay, to propose that his research into life forms that generate through self-multiplication leads him to a larger philosophical claim about what constitutes life.

It is those theories of life, as I argue in the third part of this book, that enabled Thoreau to formulate a complex materialist epistemology, redefining not just the dualistic divide between matter and mind, body and memory, but by extension the Western understanding of subjectivity as well. In articulating such an epistemology Thoreau was guided by Eastern philosophical traditions. Thus, charting the epistemological consequences of Thoreau's study of Hinduism—from the *Gita* to Mahayana Hinduism and the *Sānkhya Kárika* school of thought—I investigate how that work helps him articulate, in *A Week* and in *Walden*, an image of new thought, a thinking predicated on radical dispossession and self-impoverishment verging on self-annihilation. But my reconstruction of Thoreau's epistemology of dispossession also makes an ethical claim. For if I am correct in suggesting that Thoreau worked toward weakening the self rather than—as is too often proposed—strengthening it, then certain political and ethical consequences necessarily derive from that. Perhaps nothing is more iconic in the history of American ideas than the image of Thoreau, sitting asocial and highly individualized in his cabin, distanced from a world that he has left to its own devices. His supposed strong version of individualism has typically been understood as emblematic of the brutal capitalism of Jacksonian America; or alternatively, his supposed resolutely isolated individual is taken as representing values that fit well with liberalism, which effectively weakens leftist efforts to enact collective social change. But if we understand that Thoreau was working not toward individualism but toward its opposite, toward a radical weakening of the self, advocating a precarious self that doesn't conform to any Jacksonian American value, we would be obliged to rethink our understanding of his politics and ethics. Indeed, since the self-cancellation that Thoreau proposes is so radical as to be almost unthinkable in the framework of Western logic and ontology, what kind of ethics could possibly be predicated on it?

To answer that question, in the last part of my book I discuss how Thoreau's understanding of mourning was mobilized as a means of gathering or recollecting community. I thus dwell on his practice of writing obituaries for people he didn't know and who didn't have anybody to bury and mourn them; I write about his habit of frequenting estate sales to recover personal effects of the dead, and I inquire into what kind of ethics or even politics might emerge from his idea that the loss I mourn doesn't have to be mine, that I can take over losses of others as if they were mine, and vice versa, that what I lost can be mourned and recovered by a community of others—whom Thoreau calls "travelers" he meets on the road—others who seek to recover my losses as if they were theirs. And I suggest that far from arresting the mourner in the stupor of grief, such practices of communal mourning in fact enable action and mobilize a community based on an ethics of caring, sharing, and participating.

Although my book is clearly a monograph dedicated to Thoreau's thinking, its reconstruction of the scientific and philosophical concerns of his America, with its religious and political turmoil and ethical quandaries, also makes this work a more general treatise on the antebellum cultural environment. Instead of addressing one topic through the work of many authors, the analysis moves through a range of discursive formations and offers a "feel" of antebellum culture based on the work of one author.

Birds fly throughout this book, because they fly throughout Thoreau's work. They fly through *A Week*, most notably in the discussion regarding the green bittern staring at two brothers as they are "rolling up" the Concord River; they fly through *Walden*, where a turtle dove embodies a loss taking its leave of Thoreau, and where the loon is summoned to emblematize capacity that all life has for metamorphosis. They also fly through Thoreau's Walden Pond cabin: "I sat in my sunny doorway . . . while the birds . . . flitted noiseless through the house" (Wa, 111). They are everywhere in his *Journal* and his walks, because they are always on his mind as he learns their different languages, caught in a genuine bird-becoming process. In the words of Frederick L. H. Willis, who visited him in his cabin in July 1847:

> [Thoreau] said: "keep very still and I will show you my family." Stepping quickly outside the cabin door, he gave a low curious whistle; immediately a woodchuck came running towards him. . . . With still another note several birds, including two crows, flew towards him, one of the crows nestling upon his shoulder. . . . He fed them all from his hand . . . and then dismissed them by a different whistling, always strange and low and short, each little wild thing departing instantly at hearing its special signal.[28]

If birds assume such a central role in Thoreau, it is because, as I argue, they are for him undying repositories of memory. Some readers have noted that his writing employs birds as metaphors of elegiac recollection, as when he addresses John in "Brother where dost do well?"—a poem probably written in 1842 and sent to Sophia in 1843—asking "what bird wilt thou employ / To bring me word of thee?"[29] In that question, on Sherman Paul's understanding, birds are employed in the same way as nature in "Lycidas," a poem whose parts Thoreau copied in one of his very early commonplace books.

Nature sympathetically records the poet's personal grief yet remains "barren and silent," failing to offer consolation. In Paul's account, birds are irrelevant as birds and assume meaning only as "prisms of [Thoreau's] own subjective idealism."[30] In contrast to Paul I argue that birds in Thoreau can become emblematic sites of recollection only because in a very materialistic manner he always afforded them the status of literal living relics, elevating them to immortal beings in perpetual change and capable of hosting what

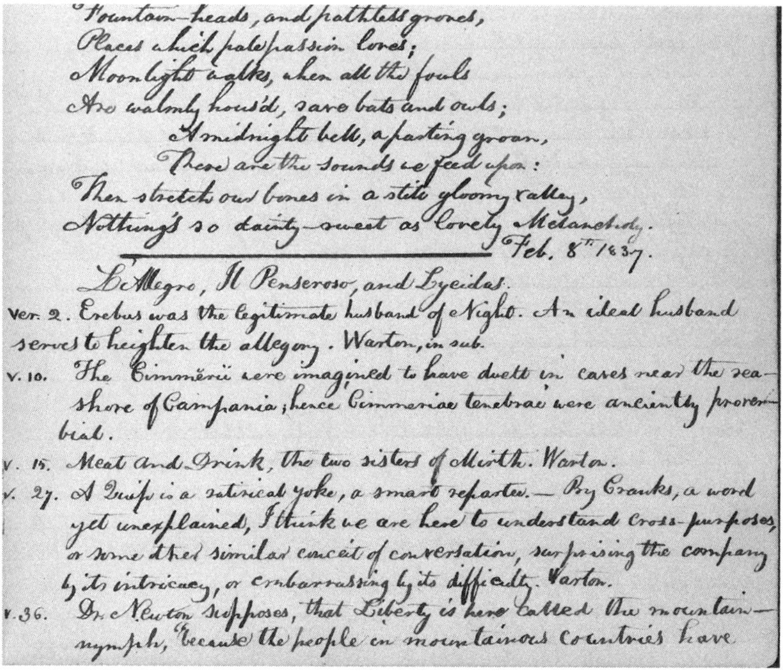
Fountain-heads, and pathless groves,
Places which pale passion loves;
Moonlight walks, when all the fowls
Are warmly housed, save bats and owls;
A midnight bell, a parting groan,
These are the sounds we feed upon:
Then stretch our bones in a still gloomy valley,
Nothing's so dainty-sweet as lovely Melancholy.
Feb. 8th 1837.

L'Allegro, Il Penseroso, and Lycidas.

ver. 2. Erebus was the legitimate husband of Night. An ideal husband serves to heighten the allegory. Warton, in sub.

v. 10. The Cimmerii were imagined to have dwelt in caves near the sea-shore of Campania; hence Cimmeriae tenebrae were anciently proverbial.

v. 15. Meat and Drink, the two sisters of Mirth. Warton.

v. 27. A Quip is a satirical joke, a smart repartee.— By Cranks, a word yet unexplained, I think we are here to understand cross-purposes, or some other similar conceit of conversation, surprising the company by its intricacy, or embarrassing by its difficulty. Warton.

v. 36. Dr. Newton supposes, that Liberty is here called the mountain-nymph, because the people in mountainous countries have

Lycidas, Thoreau's notes; thoughts on books. Autograph notebook signed: Cambridge, Mass., [ca. 1836–1839]. The Pierpont Morgan Library, New York. Purchased by Pierpont Morgan with the Wakeman Collection, 1909. MA 594.

has been. That is less strange than it might seem. For during the decades when Thoreau was writing, paleontology—itself a relatively novel science, the word *paléontologie* being coined by Cuvier's student Henry Marie Ducrotay de Blainville only in 1822—still hadn't discovered bird fossils as distinct from the widespread marine and reptile fossils that became the basis for Agassiz's work and his more general theory of the history of life. It was only in 1861 that German paleontologist Herman von Meyer discovered "the first remnant of a bird from pre-Tertiarty times," which he famously named *Archaeopteryx lithographica*.[31]

This discovery immediately generated the discussion that would enable Richard Owen, and later, Thomas Huxley, to speculate about the bird's "reptilian nature" and to suggest that birds flew from one period of earth's life to another, thereby maintaining its continuity

Archaeopteryx bavarica, Paläontologisches Museum, München. Solnhofen limestone (Plattenkalk). Photograph: Luidger, October 2, 2005.

while they themselves slowly underwent actual transformations.[32] Contemporary paleontologists know that the rarity of bird fossils is due to their small and fragile hollow-boned skeletons, which frustrate fossilization. But their absence from the paleontological archives in the 1830s and 1840s was understood by Thoreau as a lack of traces of death, which enabled him to imagine birds as an undying form of life capable of literal metamorphosis; hence his somewhat bizarre juxtaposition in *A Week* of human and avian bones prompted by the sight of reed-birds flying over "some graves of the aborigines" (W, 237). Both are metamorphosing; but while human bones are "mouldering elements preparing for . . . metamorphosis" into the plants they are going to feed, the reed-birds' bones undergo a different metamorphosis, rustling into new birds that render "the . . . race of reed-birds . . . undying" (W, 237). In the philosophical imagination of Thoreau's ornithology birds really are a form of life that cancels death by self-change, promising the fabulous renewals that Thoreau will extend to the whole of nature.

In the book that follows I summon birds in various ways. Birds fly here from mythology, the Bible, poetry, Greek tragedy, superstition, natural history, geology, and paleontology. Sometimes I invoke them as emblems of grief, at other times as the embodiment of lament. Sometimes they are clues into Thoreau's philosophy of life, at other times they are more specifically considered in the context of his ornithology. Sometimes they are omens of awesome events awaiting us, at other times their refrains voice the song of the dead. But whenever and however they appear, they are always avatars of the infinite life that is also its own—hence total—memory. They are always what Thoreau calls "living relics," embodying the central premise of his philosophy: that life commemorates itself.

Part I

Dyonisia, 467 BC: The Mythology of Mourning

Introduction: Perpetual Grief and the Example of the Fish Hawks

In discussing Emerson's understanding of grief, Sharon Cameron argues that grief is for him shallow, because it doesn't allow for "loss [to] injure the mourner's bodily integrity, although the primitiveness of supposing it could establishes the fantasy connection" between the griever's self and materiality of his loss.[1] Cameron doesn't specify why she considers the desire to have mourning alter the body to be primitive, but one might speculate that the primitiveness of such mixing of the self with what is material comes from its violating the presumably sophisticated Platonic-Christian paradigm where selves are ideal entities substantially different from bodies, never mixing with them. Contrary to such a paradigm the fantasy of a mourning that alters the body generates various illicit mixtures, such as that of a word that summons beings, or a mind that enters matter, blurring the boundaries between mourner and mourned, presence and loss—and perhaps even life and death.

But that primitive fantasy of grief is precisely what Thoreau took seriously. As early as 1839 he suggests in a *Journal* entry that grief is enlivening: "Make the most of your regrets-never smother your sorrow but tend and cherish it till it come to have a separate and integral interest. To regret deeply is to live afresh. By so doing you will be astonished to find yourself restored once more to all your emoluments" (J, 1, 85–86). Far from working on "economizing" his grief, as a modern mourner would do, Thoreau's mourner is here asked to dedicate himself completely to grief, to intensify it until it occupies him integrally, becoming identical with his life, which it will keep revitalizing ("to regret deeply is to live afresh"). This identification of the mourner's life with his grief will raise many disturbing questions, not the least of which is the following: if grief is not only a feeling of loss but also the mind's commitment to what is lost, isn't Thoreau's insistence that the mourner's grief coincide with his life also to be understood as his desire that the mourner's life somehow coincide with what has been lost?

Already here we can start to register the profound paradoxes of grief that Cameron calls primitive, which the chapters that follow will detail. For instance, if Thoreau imagines grief as the mourner's becoming who has been lost, what then remains of grief? Isn't the mourner's distance from the loss, his maintaining the boundaries of his grieving self, a necessary condition for grief to occur? Moreover, when Thoreau says that deep grief refreshes life, are we to understand this regeneration as merely figural, or as a profound psychological transformation, perhaps, even a sort of actual, as it were, material transmutation? And, if grief is something that brings, as Thoreau puts it, "astonishment" by restoring our once achieved but now lost gains (we will find ourselves "restored once more to all [our] emoluments"), then isn't it in fact closer to how we understand happiness? And what kind of grief could we imagine as identical with happiness? The *Journal* entry I quoted above may be too short and cryptic to even begin to answer these questions, and my questions may in turn be too logical—pedantically guided by a categorical belief in noncontradiction, for which Thoreau never had any particular respect—but they point to what will be the major feature of his understanding of grief: grief is uneconomical, committed to the lost rather than to those who survive, and in the end unending.

The most intense challenge to Thoreau's understanding of grief and commitment to loss came from personal experience. On January 11, 1842, his brother John died. Thoreau reacted to that death by means of the very grief Emerson thought impossible, by ravaging his body, developing symptoms of John's illness—a reaction to which I later return in detail—as if wanting to die his brother's death in an effort to defy the boundaries between survivor and the dead. The symptoms withdrew but the grief remained, mobilizing various interests that Thoreau already had—in natural history, ichthyology, epistemology, and Greek mythology—into a theory of grief that Thoreau will call "perpetual."

That such a theory was in the making can be seen in a letter Thoreau sent to Lucy Brown after John's death, where he suggested that not humans but "only nature has a right to grieve perpetually" (C, 102). Some critics have understood the phrase "perpetual mourning" metaphorically, while others think it meaningless or obscure. But in the Greek literature Thoreau had been reading both in the original and in translation ever since his Harvard years (some of which he even translated—Homer, Sophocles, Euripides, Pindar), the phrase "perpetual mourning" [*álaston pénthos*] is imbued with rigorous philosophical meaning. Tros uses it when he laments his son Ganymede in the *Homeric Hymn to Aphrodite.*[2] In the *Iliad* and *Odyssey*—arguably Thoreau's two most important literary texts—it is used numerous times. For instance, it is the phrase used by Achilles, whom Thoreau always praised as the greatest of heroes, in refusing Hector's desire to "exchange a reciprocal promise not to mutilate the corpse of the dead enemy."[3] Eumaeus uses it in conversing with Odysseus: Butler renders it as perpetual pain ("it always pains me"), Fagels as unstoppable grief ("How I grieve for him now, I can't stop"), Loraux as "grief without forgetting."[4] Similarly, in book four of the *Odyssey*, Helen uses a drug to "tear Telemachus and Menelaus away from Odysseus's *álaston pénthos.*"[5] Famously, at the very end, Eupithes uses the term to refer to his grief for his son Antinous, Odysseus's first victim ("Old lord Eupithes rose in their midst to speak out. / Unforgettable sorrow wrung his heart for his son, / Antinous, the first that great Odysseus killed").[6] "Unforgettable sorrow" or "unforgetting mourning"[7] is in each case *álaston pénthos*, the phrase by which the Greeks referred to a perpetually self-renewing grief *(pénthos)* abiding in nonoblivion *(álaston).*

Nicole Loraux, whose work on "amnesty and its opposite" is perhaps the most systematic engagement with the question of perpetual grief that is available to us, explains that like *alétheia* [truth], *álastos* "is built on a negation [*á*] of the root of oblivion." It is a negation that generates a radically positive existence in nonoblivion: "In fact, we may guess that, in nonoblivion, the negation must be understood in its performativeness: the 'unforgetting' establishes itself."[8] An unforgetting of the loss settles into the mourner, then, as if it were a kind of growing being that expands until the mourner—just as in Thoreau's description of deep regret—is completely occupied and so identified with it: "there is an obsessive component to *álaston*, a relentless presence that occupies, in the strong sense of the word, the subject and does not leave."[9] Such a grief can perpetually gesture toward the loss only if it never forgets to grieve, or, as Loraux puts it, only if it turns itself into "the oblivion of nonoblivion,[10] "occupying" the grieving subject "in a strong sense": "nonoblivion is all-powerful insofar as it has no limits—and especially not those of a subject's interiority."[11] Thus, when Loraux claims that we should think of Greek unforgetting mourning as a performative, she is in fact suggesting that far from resembling the emotional state or psychic pathos of the modern Freudian mourner, *álaston pénthos* literally changes the mood or modality of an existence. It is less a psychic than an ontological power, capable of creating illicit ontological mixtures between object and subject, mind and matter, living and dead. Above all, as Loraux insists, this grief is material: "concerning nonoblivion, I would prefer to insist on its *materiality*, indissociable from its psychic dimension. . . . Encompassing time and space completely, nonoblivion is everywhere. . . . It is there for the materiality of the *álaston* that silently keeps watch against oblivion."[12]

This nonoblivious grief, which enacts the presence of a loss on the way to completely possessing the survivor, is an obvious paradox. On the one hand, if the *álaston pénthos* perpetuates itself, it is because it wants to make what is lost constantly visible, drawing it into presence *(alétheia)*. In causing to appear what isn't or what is not apparent, which for the Greeks is the same thing, grief changes the mode of being—the ontology—of what is not into what is, but it can do so only as long as it perpetuates itself, imbuing what is lost with its own force. Thoreau could have found the example of this uncertain divide between grief and its object in Sophocles's *Antigone*, a tragedy that preoccupies him so much that he dedicates to it a series of mediations on burials in *A Week*. In that play, Creon, bearing the body of his son Haemon, is greeted by a messenger who tells him "master . . . you have come bearing these griefs in your arms"; or, similarly, after hearing of his son's death and on being allowed to see the dead body of his wife Eurydice, Creon exclaims "*Oimoi*, . . . I see a second grief! . . . I have just held my son pitifully in my arms, and I behold *her*, a corpse, before my eyes."[13] Or he could have registered it even more explicitly in Euripides's *Bacchae*, also the subject of commentary in *A Week*. There, Cadmus is seen carrying the remains of his grandson Pentheus and addressing the Theban citizens by saying: "I've brought the body back: I searched forever. It was in the folds of Cithaeron, torn to shreds, / scattered through the impenetrable forest, / no two parts of him in any single spot . . . I turned back again, up to the mountain, where I gathered the body of this boy. . . . I cannot watch this. This is grief that has no measure."[14] Cadmus's "extreme grief," as Loraux also calls it, is not only a state of mind but also the object grieved, for the name of his dead grandson precisely signifies mourning *(pénthos)*. Cadmus grieves the grandson whose dead body embodies mourning itself (when Cadmus later says to Agave "yes, now you know, I mourn Pentheus," it is as if he were saying "yes, I mourn mourning," where this second mourning is, in fact, the object that is lost).[15] And because his grief transcends the internal condition of his mind, being both an emotional condition and an external object causing that condition, it never reduces to a psychic operation or to the mental labor of recollecting, which it will later become for the Christian mind. As Loraux puts it, it is "much more than a simple inner state. At the same time outside and inside, sinister reality and psychic experience."[16]

In Greek tragedy this material mourning expresses itself by means of the interjection *aei*. Thoreau's translation of Sophocles renders it as "aie" or "alas, O," or "Oh, Oh," but it has the sense of "always." As Loraux explains, " 'always' is the standard, though imperfect, translation given for the adverb *aei*. If, however, we wished to examine the meaning of this term more closely, Emile Benveniste's essay "Expression indo-européenne de l'immortalité" on the notion of *aiōn* sheds vivid light on the subject."[17] Benveniste's seminal argument is twofold. On the one hand—and this elucidates why Thoreau correctly understood *álaston pénthos* as perpetual grief—Benveniste explains how it is that "always" (what is perpetual and, as it were, atemporal) is reconciled with a human temporality (birth, growth, and aging) that is incapable of eternity: "This 'always' indicates what perpetually recommences before being a permanent and immobile "always."[18] Benveniste suggests, then, that "always" was not previously understood to mean "stationary"; instead, for many centuries Eastern and Greek thought conceived of eternity not as stagnant but closer to the logic of human time, as incessantly agitated. As Benveniste explains, that agitation interminably perpetuates itself, constantly returning to itself; it is repetitive, ritualistic, and inexhaustible, hence enduring "forever." On the other hand, which also explains why mourning expressed by *aei* should be understood as "material," Benveniste determines a "convergence between *aiōn*, defined as designating primarily the 'vital force' [that is, as ontological force that materially creates beings], and adverbs derived from related forms that in the neuter mean 'always.'"[19] After discussing examples from Eastern thought, the Homeric hymn to Hermes, numerous instances from the *Iliad* and the *Odyssey*, as well as Pindar's fragment (111, 5; one should remember here that Thoreau translated Pindar's fragments for *The Dial* in 1843) Benveniste concludes:

> It is through a vital and immediate experience that the first thinkers of India and Greece conceived of eternity. The history of *aei* and especially of *aiōn* teaches us that this concept derives from a human and almost physical representation; the force that animates being and makes it live. This force is both one and double, transitory and permanent, being exhausted and reborn throughout generations, annulling itself in its renewal and subsisting forever by means of its always recommenced finitude. When *aiōn* becomes the name of "eternity" and therefore the universal mode of becoming, it is understood that the cosmic *aiōn* reproduces the structure of the human *aiōn*. There will be a necessary conformity between the two, because the force of life, implying the incessant recreation of the principle that nourishes it, suggests to thinking the most instant image of what is maintained endlessly in the freshness of what is always new.[20]

In a manner similar to Thoreau's understanding of the concept, Benveniste's analysis points to what we modern latecomers disqualify as magical thought, namely, the metaphysical slippage of word into matter and of promise into corporeal regeneration. For when the mournful "always" *(aei)* is cried out, it doesn't only promise that the loss will be forever remembered, but also expresses the desire to access loss as a vital force, a force with the capacity to gather bodies into forms. Each time *aei* is uttered it enacts what it means: it materializes itself into a sound that traverses matter as the "excessive vitality" Benveniste refers to, thus as mourning that turns itself into a force of life, mourning that regenerates or grief that recreates.

In what follows I seek to reconstruct Thoreau's theory of perpetual grief, whose historical and philosophical background I have just outlined. I start by arguing that Thoreau's empiricism abandons the traditional philosophical belief that human access to nature must always be mediated by concepts. Thoreau's effort to reduce thoughts to what is empirically, as if corporeally lived, thus leads him to sometimes astonishing revisions of what constitutes sensuous life and what the relationship is between mind and matter. As I read it, Thoreau will come to propose the recovery of a materiality and objectivity lost since Descartes and Kant, a loss that modern humanity never properly mourned, preferring instead, in a complete denial of that loss, to sustain itself on what it had philosophically renounced. To investigate how Thoreau imagined the recovery of that materiality, I reconstruct his practices of perception—seeing and listening—designed to reform the senses and enable the mind to settle into the empirical. However, I am here not interested in epistemological aspects of Thoreau's interest in the restoration of objectivity; instead I am primarily concerned with how that interest was mobilized by the grief that followed John's death and how it escalated into a remarkable theory of perpetual mourning.

I refer to Thoreau's diverse practices of performing such a mourning as "mythological" for at least three reasons.

1. Most obviously because, following John's death, Thoreau intensified his longstanding interest in Greek philosophy and tragedy. He translated Aeschylus and Pindar, wrote on Sophocles and Euripides, but in each case chose texts related to the mythology of brothers who died for one another, mourning each other or even

switching their personalities to substitute one for the other (Eteocles and Polynices, Prometheus and Epimetheus, Castor and Pollux).

2. In articulating his own philosophy of grief through elaborations of Greek mythology, Thoreau, like other mythologists, encounters what, for a modern philosophical mind, might be called insurmountable logical incongruities: beings and minds are found to fuse and switch, substances mix, and everything is on the move, *in becoming*, as if substances were itinerant and everything existed in a zone of process. As numerous twentieth-century classicists have suggested, *becoming* is precisely the form of mythological thought. Thus, I understand myth to be neither a literary contrivance rooted in the fantastic nor a primitively "absurd" theogony. Instead, following Vernant, I understand it as a "veritable system of thought not, at every level, immediately accessible to the habitual working of our minds."[21] If this style of thinking is not "immediately accessible" to our philosophical habits, it is precisely because philosophy thinks of beings as stable identities, as Plato explained in the *Republic*, whereas myth thinks of becomings. As Vernant puts it, "Plato grants an important place in his writing to myth, as a means of expressing both those things that lie beyond and those that fall short of strictly philosophical language. . . . How would it be possible to speak philosophically of Becoming, which, in its constant change, is subject to the blind causality of necessity? Becoming is too much a part of the irrational for any rigorous argument to be applied to it."[22] Obeying Plato's definition, then, I call Thoreau's thinking mythological in order to suggest that it is a thought committed to becoming, always positioned on the threshold of transformation, seeking out the secret of generation. That Thoreau imagines the experience of grief to lead to real becomings is signaled already in his notes for *A Week*, a book in which, as so many readers agree, Thoreau performs his own mourning for John. On the first leaf of an early commonplace book, he drew a picture of two hands clasped in a handshake, both of the hands, as in a visual riddle, also resembling birds whose fusion—by means of the handshake—metamorphosed them into another bird, a single new being. Thoreau circled that drawing with titles of *A Week's* chapters, visually suggesting that metamorphosis would be its central concern.

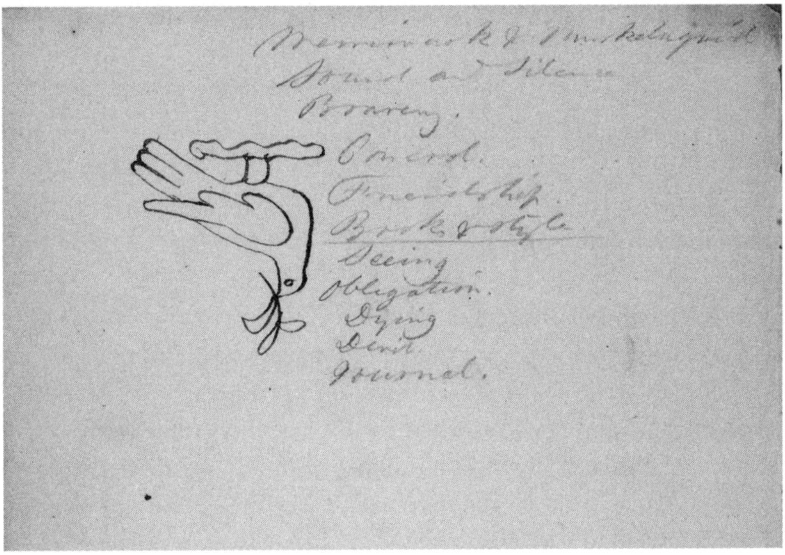

Drawing of a "bird-handshake," surrounded by the following headings: Merrimack & . . . : Sound and Silence; . . . Concord; Friendship; Brook . . . Seeing; Obligation; Dying; David; Journal. Commonplace Book of Henry David Thoreau [HM 945, Front Left Page, "Walden," (Index), 1840]. The Huntington Library. Reproduced by permission of The Huntington Library, San Marino, California.

But he didn't leave that concern on the level of a pure idea. The challenge of Thoreau's thought resides is in the fact that, thanks to what Worster called his "intense empiricism," he always seeks embodied meanings for the ideas that preoccupy him. Hence the Greek suggestion that beings can fuse is also a lesson he starts learning from hawks, a couple of which he observed, two months after John died, playing "one above the other in endless alternation" (J, 1, 393). Intrigued by the possibility of an "endless alternations" of beings, Thoreau continues observing hawks, until on October 26, 1857, yet another encounter with fish hawks allows his realization concerning the fusion of beings to crystalize:

> A storm is a new, and . . . more active, life in nature. Larger migrating birds make their appearance. They . . . sympathize with the movements of the watery element . . . I see two great

> fish hawks . . . slowly beating northeast against the storm, by what a curious tie circling ever near each other. . . . Two long undulating wings conveying a feathered body through the misty atmosphere, and this inseparably associated with another planet of the same species. . . . Damon and Pythis they must be. The waves beneath, which are of kindred form, are still more social, multidinous, ανηριθμον. (Jn, 10, 126–127)

The wings of two discreet entities touch, enacting their fusion into a single feathered body that the perceiver sees as a continuous mobile line similar to the continuity of waves on the water, always connected through the liquid medium into oneness, yet always appearing on the surface as different, "multidinous, anêrithmon." Because *anêrithmon* is connected to what is "multidinous," it is safe to presume that what Thoreau refers to here is *anêrithmon gelasma*, a famous phrase used by Aeschylus in *Prometheus*. Alternatively translated as "innumerable laugher of the sea-waves," or "many-twinkling smiles of Ocean," the phrase is understood by the classical tradition to be a formulaic expression drawn from Aeschylus's ontology of becoming, in which everything emerges to continue into something else, thus making the universe both one and multiforous.[23] In Thoreau's dense passage, this ontology of becoming includes humans too, for it is translated into the story of Damon and Pythias, two friends who, according to the legend, could die one instead of another as if their discreet personhoods, like the bodies of fish hawks or the forms of Aeschylus's waves, were substitutable and continuous.[24] In Thoreau's world, then, mythology is more than a fantasy, for in that world fish hawks actually perform, as it were, onto-pictorially, the truthfulness of how everything moves into something else, how beings pass through one another, fuse, and get revised. From his position in the place of mutation Thoreau will be able to investigate whether the inanimate can be transformed into life and whether a word can alter bodies: two questions, as I will argue, that are central for understanding his theory of perpetual grief.

3. Thoreau's thought can be called "mythological" also because of its openess to contradictions. In Plato's words, the becomings with which myth concerns itself can't be argued logically, because they

are neither stable nor generalizable; they cannot be named for thought or abstracted into a concept. Mythological thinking understands each becoming as an irreducibly singular process, varying in unpredictable ways and often contradicting or erasing its previous shapes or conditions: "Plato often appears to reject *muthos* utterly as, for instance, when in the *Philebus* (14a) he writes of an argument, *logos*, which, being undermined by its own internal contradictions, destroys itself as if it were a *muthos*."[25] This is to say that mythological thought thinks processual existences in a nonlinear way and, following their unpredictible changes, remains unconcerned with what it has already claimed; it thinks in the same way as the existences it conceives, contradicting itself and so destroying itself. Anne Carson defines the processual existences with which mythology is preoccupied as "existences that crash into other lives or brush against gods."[26] But if such unstable existences crash into other forms and lives, they do so only to renew the process called life; for, having once moved into something else, they don't vanish but affect and alter the new environment they inhabit—as it were, moving it on. Becomings are therefore to be understood as inexhaustible attempts at surviving and going on. Like the "*aie*" of grieving they are the promise of perpetual renewals; and like *aiōn* they are never permanent even though, because they are incessantly renewed, they are eternal. In underlining the structural coincidence in Thoreau's thinking between the thought of becoming and that of perpetual grieving, I also seek to emphasize that far from simply being thematically concerned with what grief is, his ideas interminably perform grief. In pursuing the question of loss or death and what it means to grieve perpetually, Thoreau's thought leads to extreme consequences; it encounters insurmountable contradictions and, in Plato's sense, destroys itself. But it begins again, trying to answer the same question, each time in a slightly different way. Its crashing fragments it, but it tirelessly renews itself, as if it were inexhaustibly evoking the action of perpetual grieving.

Faithful to that mournful thinking, which often falters or falls and contradicts itself, the chapters that follow are written in a discontinuous but thematically linked way. As Thoreau was with his losses, so I attempt to be with his thoughts: as if performing an obsessive ritual of mourning, I collect its fragments into a provisional collage, so that, once gathered and juxtaposed, they are seen or heard to echo one another. The fragments can be rearranged, but regardless of their position in one chapter or another, my readings all tell the story of extreme grief and its ethics, one so dedicated to the loss that it is reluctant to be prevented by sophisticated ontological divides from pursuing its effort to recover what it grieves for.

Literal Sounds

It would be stating the obvious to say that the question of perception preoccupies Thoreau. There is no major critical engagement with Thoreau's thinking that doesn't analyze his effort to generate techniques of perceiving designed to alter the way humans experience the world. Thoreau's proposed reform of the senses sought to enact what Jane Bennett understands as "enmeshing oneself with Nature," or what Lawrence Buell identifies as defying the "myth of human apartness" from the environment.[1] For those influential readers, Thoreau's perceptual techniques would "enmesh" the self by generating a vision of nature as something so enchanting that it would suspend the perceiver's mind—the source of his considering himself apart—bringing about a wondrous feeling of unity with the perceived. Thus for Bennett, Thoreau's perceptual practices were meant to transform "nature" into something "beautiful, sublime, and Wild"; whereas for Buell, they were similarly a "studious exercise in romantic literary sublimity."[2]

But even though it seems unquestionable that Thoreau wanted new sensing to defy the self's separation from its surroundings, it is less certain that such defiance was to be achieved through the sublimation of nature. In contrast to readings that understand Thoreau's effort to vitiate the isolation of selfhood by generating sublime percepts, I will thus be arguing that such defiance was to be achieved through "desublimation," which, as Ross Posnock puts it, breaks the "insulating habits of idealism imbibed from Plato."[3] I will argue that it is through disenchantment that Thoreau wanted to access what surrounds us literally, as opposed to sublimely. For it is by means of that literal access, identified by Donald Worster as Thoreau's "intense empiricism,"[4] that the self will reconnect with the world as proposed in a famous passage from *A Week:*

> We need pray for no higher heaven than the pure senses can furnish, a *purely* sensuous life. Our present senses are but the rudiments of what they are destined to become. We are comparatively deaf and dumb and blind, and without smell or taste or feeling. . . . The eyes were not made for such grovelling uses as they are now put to and worn out by, but to behold beauty now invisible. . . . When the common man looks into the sky, which he has not so much profaned, he thinks it less gross than the earth, and with reverence speaks of "the Heavens," but the seer will in the same sense speak of "the Earths," and his Father who is in them. . . . What is it, then, to educate but to develope these divine germs called the senses? (W, 382)

Thoreau's senses are here called on to change the direction of perception: not to look upward, toward the skies—toward what is ideal or divine—but downward, into the slimy swamps and mud of the earth. If there is anything divine or heavenly, Thoreau suggests, it resides only in earthly matter, or more precisely, it *is* matter itself. But neither is matter dumb and inert; to the contrary, if perceived with reformed senses it will show itself as a site of miniscule exchanges and processes, processes that our imagination transposes to the spiritual "Heavens" only because our current senses are incapable of registering them. By discovering in the material life what is currently conferred on the spiritual, Thoreau's new perception would, by the same token, radically transform the idealist hierarchy between matter and mind. In contrast to idealistic claims that nature is merely a "symbol" of the spirit—as in Emerson's famous formulation, according to which "particular natural facts are symbols of particular spiritual facts," and "the whole of nature is a metaphor of the human mind"[5]—Thoreau's reformed senses reveal it to actually be the heaven it is presumed only to emblematize. For Thoreau's new senses, then, the "Father is in the Earths," God is corporeal.

Not just in *A Week*, but also in *Walden* and throughout the *Journal* Thoreau offers a critique of "our present senses"—their practices of figuration and sublimation—while simultaneously suggesting how are they to be remade to hear "new sounds" and see what is "now invisible." For Thoreau, such a remake requires that the senses be trained to register sensorial experiences beyond any figuration or as he puts it, "ideation." In a short *Journal* entry, Thoreau calls "ideation" the operation of abstracting figures out of mobile and unformed perceptual diversity: "It [the eye] is the oldest servant in the soul's household-it images what it imagines-it ideates what it idealizes. Through it idolatry crept in-which is a kind of *religion*" (J, 1, 182). Ideation is the eye's habit through which the specificity of what is perceived now is sacrificed by being incorporated into an idea of it that the mind has formed on the basis of previous perceptions; the singularity of the present is thus lost in the ideality of an image. And because the objects of those images are nonexistent, Thoreau can say that our sensorial habits generate a *cultic* perceptual field. He explains this process of ideation more elaborately in a long *Journal* entry written on December 15, 1840:

> In the woods one bough relieves another, and we look into them, not with strained, but relaxed, eyes. . . . -But as soon as man comes into nature, by running counter to her, and cutting her off where she was continuous, he makes her angular and formal . . . I saw today where some pines had been felled at various angles with the rest of the wood, and on that side nature offended me, as a diagram. . . . I saw squares and triangles only. (J, 1, 203–204)

By confronting the visible from a viewpoint external to it, the eye objectivizes it into a "picture," a kind of a photograph that freezes the motion of the visible into a form: it "cuts [it] off where [it] was continuous," into what is "angular and formal." These ideated forms that Thoreau calls "diagrams" represent the visible *as it is not*—the visible escaping the eye being the processual continuum of the unformed—and so amount to its fictionalization. What is commonly called the visible world is thus already an effect of mimesis: it is a fiction or a fable.

In contrast, in an entry from September 13, 1852, Thoreau describes the eye freed from the habit of ideation as desiring to perceive translation of one color into another at a speed that transcends forms. To perceive a transformation of colors not attached to forms results in a demand for the a-figural:

> I wish to see the earth translated-the green passing in to blue . . . I must walk more with free senses- . . . I have the habit of attention to such excess that my senses get no rest-but suffer from a constant strain. Be not preoccupied with looking. Go not to the object let it come to you. . . . -What I need is not to look at all-but a true sauntering of the eye. (J, 5, 343–344)

"Free" looking sees things with eyes emptied of the capacity to stabilize the mobile empirical into an image of it the mind has already formed; this is to say that it frees the eye of the demand to recognize the seen as something known and classifiable. Because it has suspended or forgotten what it had previously seen, the eye freed from "the constant strain" of recognizing encounters the object as something completely novel. As a passage written in October 1859 additionally specifies, in encountering phenomena in their unprecedented novelty, a free eye will not be able to say what they are, but it will be able to see that they are, and that will be its biggest success: "If you would make acquaintance with the ferns, you must forget your botany. . . . you must approach the object totally unprejudiced. . . . Your greatest success will be simply to perceive that such things are" (Jn, 12, 371). Emancipated from the figural, the unprejudiced eye finally becomes capable of perceiving the forces and elements in the very moment of their formation; it finally comes to see "the earth translated-the green passing in to blue." And to perceive passings and translations is to perceive the truth, because, as an entry from December 15, 1840, specifies, nature is precisely a motion of forces that defy any figural or geometrical spatiality: "Every where [nature] preaches not abstract but practical truth. . . . The moss grows over her triangles" (J, 1, 204). The free eye sees this "practical truth," the fact that life, empirically, escapes forms, that out of each form there grows moss in all directions, rendering form formless. The free eye, then, sees the incessant process of generation that undoes shapes, the very practice of life revealing the empirical in the plethora of its mobile differences.

As many *Journal* entries suggest, Thoreau conceived the whole of nature not only as the incessant translation of colors that frustrated figuration but also as matter through which sonorous motions roam. Thus in Thoreau's ontology vibrant sound becomes the force of rendering corporeality and is elevated to something having the status of a being. All nature is for him acoustic and rhythmic: "*all nature is tight-drawn and sonorous* like seasoned wood" (J, 1, 439–440, italics mine); or: "All the elements strive to *naturalize* the sound" (J, 1, 347); or again: "All sounds, and more than all silence, do fife and drum for us. The least creaking doth whet all our senses" (J, 1, 96). Sounds contrive a pervasive ambience in which all entities bathe; the whole universe resonates with an "acoustics of space" (J, 1, 61) as if all matter were traversed by sonorous vibrations and so animated. Moreover, a *Journal* entry from July 7, 1845 explicitly identifies sonorous vibrations with the motion of life: "Sound was made not so much for convenience, that we might hear when called, as to regale the sense-and fill one of the avenues of life" (J, 2, 158). That sound can "fill" us additionally explains why hearing rather than seeing becomes for Thoreau something of an exemplary perceptual experience: "The five senses are but so many modified ears" (J, 1, 27). For it suggests that more than the eye, the ear allows an object to inscribe itself into the perceiver. In the case of listening, then, external materiality—actual sonorous vibration—literally or physically enters the perceiver's interiority. In listening—more obviously than in seeing—the perceiver hosts in himself a sound external to him, which is why listening can inherently subdue the perceiver to the point of turning him into the perceived, annulling the difference between external and internal.

Yet, despite this extraordinary capacity of listening to allow a reality external to the mind to inhabit that mind, Thoreau discovers that tuning the mind to a sonorous reality still requires hollowing out its habitual sensorial patterns. He finds that the aural field no less than the visual is susceptible to figuration that ideates what is empirical. To experience non-figural sonority—to hear what, as early as 1840, Thoreau called a "stately march to an unheard music"[6]—requires undoing the habit of recognizing sounds that the ears have contrived. Ears are normally "cultivated" so as to register certain sounds as melodious or arranged, and to disqualify silence as the absence of sound, or to dismiss all discordant sonorous phenomena as meaningless noise. Thus, whereas the ear is not intentional in the sense that the eye is—it doesn't have the power to focus or close itself by will—it has nevertheless been conditioned to identify certain combinations of sounds as harmonious. Harmony is to listening what geometry is to seeing; just as geometry negates differences among bodies by reducing them to their presumed, but in fact only imagined, proportionate shapes, so harmony cancels dissonance by generating a concord understood to be the perfect essence hiding behind all discord. That is what Thoreau's notebook for "The Service" suggests in quoting a passage from Coleridge's *Specimens of the Table-Talk*, where harmony functions precisely as acoustic geometry: "Take a metallic plate and strew sand on it; sound a harmonic chord over the sand, and the grains will whirl about in circles, and other geometrical figures, all, as it were, depending on some point relatively at rest. Sound a discord, and every grain will whisk about without any order at all, in no figures, and with no points of rest."[7] Coleridge's overlapping of aural and visual—for in this passage harmonic sound generates geometrical images—discards discord as a multiplication of meanings that fragments reality instead of totalizing it as a unity ("every grain" whisks its own "message"). For Coleridge this reality made of many contradictory messages amounts to a falsity, since he equates truth with a coherence and unity of meaning gathered into one form, which is itself generated according to a preexisting ideal of perfection.

In contrast to Coleridge, Thoreau's early essay "The Service" (1840), imagines a man with "new" ears that, deaf to harmony, are capable of "nicer" listening:

> Let not the faithful sorrow that he has no ear for the more fickle and subtle harmonies of creation. . . . If his pulse does not beat

> in unison with the musician's quips and turns, it accords with the pulse beat of the ages. A man's life should be a stately march to an unheard music, and when to his fellows it seems irregular and inharmonious, he will be stepping to a livelier measure, which only his nicer ear can detect.[8]

Unlike musical harmonies, the "pulse beat of the ages"—the time of humans and beings in the world that one could perhaps call history—is irregular and discordant. It cannot be arranged according to a unifying understanding of form and truth ("does not beat in unison with the musician's quips and turns") but embodies precisely what Coleridge dismissed as discord. To register this arrhythmia—which the ear cultivated on harmony can't perceive ("unheard music")—a "nicer" ear is required; an ear that is—like the unprejudiced eye—emptied of its habits to become an indifferent receiver capable of hosting whatever aural phenomenon traverses it. In the same year that "The Service" was published, Thoreau also formulates the requirement of a nicer ear as a theoretical proposition regarding new perception in general: "to discover a gleam in the trenches, and hear a music in the rattling of the tool we work with- is to *have* an *eye* and an *ear*" (J, 1, 213). When Thoreau thus equates "hav[ing] an ear" with "hear[ing] a music" he doesn't mean that the new ear he has in mind will be the same as the old only more "capable" than it, discerning harmony behind the dissonant rattling of a hammer; rather he suggests that the new ear will listen to and follow the disorder of rattling dissonance, keeping it divorced from aesthetic production. To "have an ear" is thus to hear as novel and unrecognizable—hence to "discover"—any sound an ear prejudiced by its aesthetic cultivation would render repetitive and familiar; it doesn't dismiss what we ordinarily identify as the musical but instead seeks to expand it by proposing a listening that is attuned to cacophony and discord. And while an entry from August 31, 1851, will only briefly confirm that when by "musical" Thoreau meant any discordant sound ("Every sound is music now" [J, 4, 23]), in June 1852, the *Journal* explicitly states that music is not only what is harmonious but also what transcends the harmonious to include the atonal and discordant:

> A child loves to strike on a tin pan or other ringing vessel with a stick, because its ears being fresh sound attentive & percipient it detects the finest music in the sound at which all Nature assists. . . . So clear and unprejudiced ears hear the sweetest & most soul stirring melody in thinking cow bells & the like (dogs baying the moon) not to be referred to association-but intrinsic in the sound itself. Those cheap & simple sounds which men despise because their ears are dull & debauched. Ah that I were so much a child that I could unfailingly draw music from a quart pot[.] Its little ears tingle with the melody. To it there is music in sound alone. (J, 5, 82–83)

Thoreau's "unprejudiced" ears do not hear sounds as material representations of ideal essences or spiritualized souls (for instance, the dog's baying at the moon is not heard as an acoustic metaphor representing human solitude in nature); moreover, they do not *associate* at all ("not to be referred to association"). Instead, they are traversed only by what Thoreau says is "intrinsic" to sound itself, the very material vibrations that the prejudicial ear, aestheticized by the habit of judging what is harmonious, dismisses as dull. They hear sound stripped down to sound: "But the music is not in the tune-it is in the sound" (J, 5, 146). On this understanding, then, any acoustic event is music if heard clearly or intrinsically, that is, not by adjusting the sonorous according to subjective preferences but by releasing the subjective altogether. By fusing the senses with the sensed, and by refusing all idealizing figuration, clear ears finally access the materiality of the aural sensible, as if recovering it.[9]

Homer's Music Box

THE EXTENT TO which undoing perceptual habits results hearing sounds as other than repetitive can help explain Thoreau's love of music boxes. He listened to them obsessively, never tiring of the same tune they released. At least three music boxes circulated in his household: one he listens to throughout January 1842 in all likelihood belonged to a family member, quite possibly to John; another belonged to Nathaniel Hawthorne and in Walter Harding's words "charmed" Thoreau after July 1842;[1] a third was, according to Jeffrey S. Cramer, also given to Thoreau in the winter of 1842 by Richard Fuller. That was the one that Thoreau left with Hawthorne in the spring of 1843. On April 8, 1843, Hawthorne praised it in his *American Notebooks*, noting that he was "occasionally refreshing" himself with a "tune from Mr. Thoreau's musical-box, which he had left in my keeping." But the freshness of its sound was limited by its repetitiveness, for the next day, after listening to the tune some more, Hawthorne adds: "Many times I wound and re-wound Mr. Thoreau's little musical-box; but certainly its peculiar sweetness had evaporated, and I am pretty sure that I should throw it out the window were I doomed to hear it long and often."[2] If Hawthorne's reaction is that of common sense, it only confirms what Thoreau suggests: common sense is already irrecoverably aestheticized. For Hawthorne, repetition degrades sound into meaningless noise, replacing the affect of "sweetness" he wants to enjoy with irritation. Thoreau's listening is in opposition to Hawthorne's cultivated taste; if he never tires of listening to music boxes, it is because he follows dissonant—

hence unpredictable—sounds, his ears astonishingly emptied of expectation.

Thoreau is listening to one of those music boxes in early January 1842, in the room with his dying brother John.

According to Walter Harding's reconstruction, the symptoms of John's illness appeared only several days after the injury:

> On New Year's Day John Jr. was stropping his razor when it slipped and cut off a little piece from the end of the ring finger of his left hand. It was a very slight cut, just deep enough to draw blood. Replacing the skin, he put on a bandage and paid no more attention to it for several days. . . . by mid-week [January 5, 1842] he noticed pains in the finger and when he removed his bandage on the 8th, he found part of the replaced skin had "mortified." That evening he called on Dr. Bartlett, who dressed the finger but thought nothing unusual of it. On the way home, John had strange sensations and acute pain in various parts of his body, and was barely able to reach the house. He woke the next morning complaining of a stiffness of the jaws. That night he was seized with violent spasms and lockjaw set in.[3]

It is between John's convulsions of pain that Thoreau sits in his room, listening with him to the music box. In a stunning meditation on sound prompted by these nights of listening, which has never been adequately explained, Thoreau begins to ask a series of disconcerting, albeit logical questions regarding how this undoing of the ears relates to the continuity and unity of persons.

His meditation begins on the night of the 8th. At that point, perhaps sensing that time is what John is running out of, Thoreau will be concerned with the temporality of sound:

> One music seems to differ from another chiefly in its more perfect time- In the steadiness and equanimity of music lies its divinity. It is the only assured tone . . . Because of the perfect time of this music box-its harmony with itself-is its greater dig-

> nity and stateliness. This music is more nobly related for its more exact measure-so simple a difference as this more even pace raises it to the higher dignity. . . .
>
> Of what manner of stuff is the web of time wove-when these consecutive sounds called a strain of music can be wafted down through the centuries from Homer to me-And Homer have been conversant with that same unfathomable mystery and charm, which so newly tingles my ears.
>
> Am I so like thee my brother that the cadence of two notes affects us alike?
>
> Shall I not sometime have an opportunity to thank him who made music? I feel very sad when I hear these lofty strains because these must be something in me as lofty that hears- Doest it not rather hear me? . . . God must be very rich who for the turning of a pivot can pour out such melody on me.-It is a little prophet-it tells me the secrets of futurity where are its secrets wound up but in this box? So much hope had slumbered- There are in music such strains as far surpass any faith in the loftiness of man's destiny—He must be very sad before he can comprehend them—The clear liquid note, from the morning fields beyond seems to come through a vale of sadness to man which gives all music a plaintive air. (J, 1, 361–362)

What is it that Thoreau calls "the perfect time" of the music box? Thoreau hears the difference among sounds as minimal acoustic variations (the music he hears, he says, is "steady," generating a "simple" difference, moves at an "even pace," and is "exact"), as if what he hears is a chant rather than a complex composition. Because the tonal variations are minimal, their duration is always "exact." That is to say that a just-released tone, taking the same amount of time to reach the listener, will *always* have an element of simultaneity with the one the box releases immediately after. Simply put, the listener's present—what Thoreau hears in any given now—always reaches him from the past, because sound needs some time, however minimal, to traverse space. But as he is hearing this past in his present, the box generates a new tone about to become his future; thus, what the listener hears now—his present arriving to him from the past—

becomes contemporaneous with what will be his future. This coincidence of past and future in a present is what Thoreau calls "perfect time."

In Thoreau's terms the perfection of time experienced through the emission of sound in a small room is emblematic of how time is generated in cosmic space. That is what he describes in the second paragraph of the entry, where the cosmos replaces his room and where Homer's poetry assumes the role of the music box. "Consecutive sounds," he suggests, counter the progression of time that would keep Homer at an insurmountable distance from him. Time becomes an embodied acoustic strain that, starting from Homer, "wafts" into Thoreau's now, to put them in touch; it is as if, in sharing the same acoustic space, their existences were rendered simultaneous. The sound that reverberates through the room of the cosmos and enters Thoreau in his "now" is thus assimilated to the same sound heard by Homer, and their respective presents encounter each other in the vast simultaneity of the universe. But when Thoreau's past (Homer's Greece) and Homer's future (Thoreau's Concord) come to coincide in Thoreau's present, the two ends of time—what was and what is—are no longer separated in a linear manner but instead touch, and time folds into a circle. Thus, Thoreau's evocation of Homer and an archaic sense of time is to be understood in the specific context of John's impending death, as enacting an existential coincidence of past and present that will place that death in a very different conceptual context.

The circle that enfolds time is not, however, the one regulating seasonal and generational renewal, so often claimed to dominate Thoreau's understanding of time.[4] Instead—because Thoreau is talking about time whose beginning moves into its end and because this time is generated by the corporeality of sound that arrives at a present from a past, as if materially encircling space—the time emblematized by the acoustics of the music box must be the time of Chronos. In the Orphic theogonies that preoccupied Thoreau throughout his life, Chronos is the time that encircles the cosmos materially, by means of its own body.[5] As Jean-Pierre Vernant summarizes, in those theogonies Chronos is not an abstract notion but "a living being," "a snake whose body forms a circle, a cycle that, by enfolding the world and binding it together, makes the cosmos a single eternal sphere,"[6] eternal because it doesn't perish, not because it is immobile. For in fact, events and causes incessantly move along it as it eternally circles. That is what Thoreau specifies in March 1842, two months after John dies, when he remarks that "all things indeed are subjected to a rotary motion" (J, 1, 376), suggesting that the circle of chronic time coincides with the mobility of material life. Embodied yet eternal chronic time differentiates between past and present, which occupy different sites on the sphere of its body, while being contemporaneous on the sphere of its eternal duration. Such time is chronic because past and present never pass, yet beings pass through them, making them more or less acute and thereby making eternity mobile. As a result of the simultaneity it affords past and future, chronic time is able to contain contraries generated by different causes at different times; for instance, it can simultaneously contain effects generated by death and birth. On this circle—and that is obviously what commands Thoreau's attention as he listens to the music box—the death of a being can coexist with its birth or any point of its life.[7]

The question of what happens to the coherency of human persons on this circle is wrongly posed. For the circle is a simultaneous totality of all historical instants, whereas persons are effects of linear temporality characterized by a finite sum of never-coinciding instants.[8] Mortals live tied to a finitude in which past is separated from future; only rarely do some of them exit the linearity of human time to enter the chronic sphere along which they can then transit into the past, and—because to move the circle backward is to move it forward—so discern the future. The Orphic traditions known to Thoreau identified such persons as seers or, as Thoreau prefers, poets.[9] In 1838, discussing spheric time in Homer, Thoreau describes the nature of this visionary poetic existence as follows:

> Homer
>
> The poet does not leap-even in imagination, from Asia to Greece through mid air, neglectful of the fair sea and still fairer land beneath him, but jogs on humanly observant over the intervening segment of a sphere—

—έπειή μάλα αολλά
Ουρεά τε σκιόεντα, Θάλασσα τε ήχήεσσα.—
—for there are very many
Shady mountains, and resounding seas between—. (J, 1, 33)

Quoting Achilles's words from the *Iliad*, Thoreau suggests that poet doesn't live by leaps from mental to material (translating into words the visions of imagination), past to the future, site to site (Asia to Greece). Instead, the poet is the creature of a (by definition) infinite continuity that connects differences between what is and what isn't, here and there, mind and body, differences that the ontology of finite mortals renders insurmountable. That is what Thoreau means when he says that the poet doesn't skip anything, but walks the whole circle, going through what is between entities ("the intervening segment of a sphere"); to reach a continent he moves through water; to reach the day he goes through the night; and, more generally, to reach what is formed he moves through what is disarticulated: shadows and echoes ("shady mountains, resounding seas"). And because for the archaic Greek mind the world of shadows is the world of the dead (Hades), to say that the poet moves from daylight to shadows is to say that by walking on the sphere of chronic time, he precisely passes through the world of dead, coming back to the living. In so doing he realizes that those hidden from us—dead to us—still are found on some other site of the sphere, in the past or future of those who are living. The poet thus comes to understand that everything is always alive, albeit not revealing itself to him in his own present. As Thoreau explicitly puts it: "There seem to be two sides to this world presented us at different times-as we see things in growth or dissolution in life or death For seen with the eye of a poet-as God sees them, all are alive" (J, 1, 372). Walking on a circle of time Thoreau's poet sees the simultaneity of what to us is presented at different times; what we understand as succession—of life and death, for instance—he sees as co-present. He thus sees that what we fancy as lost in the past can, in chronic time, be encountered in the future.[10] He sees those who are gone as forthcoming.[11]

It is that type of walk on the circle of chronic time that Thoreau calls for in the third paragraph of the meditation on the music box.

The music box, whose sound arrives from a past, is now explicitly identified as what resounds the future: "God must be very rich who for the turning of a pivot can pour out such melody on me.-It is a little prophet-it tells me the secrets of futurity where are its secrets wound up but in this box?" As the pivot generating the sound turns, Thoreau is poised to unmoor his mind from the confines of personhood and to step into the wheel of time. As he starts walking on the circle he will, in the first place, become the sound he hears and so come to inhabit with John the same sonorous present, the difference between them shrinking into identity as they vibrate into the same acoustic being: "Am I so like thee my brother that the cadence of two notes affects us alike?" But, in the second place—and like Homer from the earlier *Journal* entry—he too, without making leaps, will be able to walk the circle and so partake of the time that is the totality of the world (hence, two months after John's death, a remark: "I am time and the world. I assert no independence" [J, 1, 392]). Only, as he suggests, he will enter it from the opposite end: "There are in music such strains as far surpass any faith in the loftiness of man's destiny—He must be very sad before he can comprehend them—The clear liquid note, from the morning fields beyond seems to come through a vale of sadness." Following the wheel of the sound he hears, Thoreau will thus start his walk not from the happiness of morning fields, where it originated in Homer, but from the deep sadness that is settling in as he writes. Walking through the valley of sorrow, he will reach the origin of daylight. Thoreau will walk through the vale of death—where John is going—to come out of it on the other side, where life and happiness are, not avoiding the horror of death but defying it in encountering it.

As if taking literally Homer's and his own words about going to the vale of death, when John died Thoreau notoriously developed symptoms of his brother's illness, trying to take upon himself John's death. As Walter Harding recounts, "at the time of John's death . . . [Thoreau] . . . became completely passive—sitting, pondering, saying nothing. . . . Then . . . on January 22 he too became ill, exhibiting all the symptoms of his brother's disease, lockjaw."[12] Two *Journal* entries written after John's death register Thoreau's disappointment at his own recovery. On February 21, 1842, he records the failure of his transformative listening supposed to enact the transformation of personhood, suggesting that personhood is more firm—more resilient to change—that he believed: "I was always conscious of sounds in nature which my ears could never hear-that I caught but the prelude to a strain-She always retreats as I advance- . . . I never saw to the end nor heard to the end-but the best part was unseen-and unheard" (J, 1, 365). A sentence transcribed later in 1842 is even more explicit, stating that "we have no divination or prospective memory" (J, 1, 429),[13] thus admitting that mourning stunned him into the stasis of the present, immobilizing his person in its historical, finite form, dimming the past and thus the future.

Yet, despite that realization, his experimentation with chronic time continued. On June 14, 1849, two weeks after *A Week on the Concord and Merrimack Rivers* was published, Thoreau's sister Helen died. Walter Harding reconstructs the event of the funeral:

> The funeral was held in the home on the 18th with both the Unitarian and the Trinitarian ministers in attendance. Thoreau sat seemingly unmoved with his family through the service, but as the pallbearers prepared to remove the bier, he arose and, taking a music box from the table, wound it and set it to playing a melody in a minor key that seemed to the listeners "like no earthly tune." All sat quietly until the music was over. She was buried in the burying ground next to her brother John.[14]

In what is perhaps the only critical assessment of this scene, Richard Lebeaux argues that Thoreau's listening to the music box at the moment the bier was about to be removed signals his feeling of guilt:

> Now the music box reappeared in connection with Helen's death, and, especially considering her burial site next to John's, it is impossible not to conclude that Helen's death was linked with his brother's. Her death could not have failed to rekindle many of the agonizing emotions associated with John. Coming as it did, when he was seemingly on the brink of a literary triumph, Thoreau may have felt with renewed intensity that he

> deserved to be thwarted or even punished for his pursuit of success.[15]

Even if Lebeaux's interpretation is psychoanalytically plausible, it is also credible that Helen's death intensified an ontological, rather than just a psychological possibility that first articulated in the context of John's death: the possibility for a person or a seer walking on the circle of time embodied by the transition of sound to become the sonorous motion he listens to; the listener would be inhabited by the sounds that once also traversed the dead and would thus come to partake not just of a being they all used to share (it wouldn't simply be a question of reminiscence), but, because on the circle of time all temporal modes are contemporaneous, the being they still share (the question of a different manner of existence). This strange ontological phantasy is precisely what Thoreau articulates in a text that was perhaps intended to be an essay, perhaps to be included in a larger project, titled "A Sister," which he starts writing any time after May 26, 1849, when it became obvious that Helen would die.[16] In that meditation a "sister" is not just an entity with which one can fuse thanks to the mediation of sound (as for instance, when listening to the music box while John is dying, John and Thoreau are confused into a single affect: "the cadence of two notes affects us alike"). More strangely, in that meditation the sister is a being transfigured into a flow of sound—also identified as the flow of life—invading the listener: "My sister, it is glorious to me that you live. Thou art a hushed music to me- a thousand melodies commingled and filling the air. Thou art transfigured to me, and I see a perfect being. . . . Into whom I flow Who is not separated from me" (J, 3, 18). From this point of view, when Thoreau opens his music box at Helen's funeral and releases sounds in a minor key, it may be less to convey his feeling of remorse than to bring the sound that traversed John's and Helen's life back into the room where he and Sophia are found, placing them all—a dead brother and sister and a living brother and sister—once again on the same circle of time. It is indeed, therefore, about summoning the dead back to life; or—for on the circle it amounts to the same thing—it is about going to visit the dead.

Aeschylus's Paean

John died on January 11, 1842. Following that event, Thoreau wrote three letters in which he explicitly commented on his grief—to Mrs. Lucy Brown, on March 2; to Emerson, on March 11; and to Isaiah T. Williams, on March 14, 1842—formulating, albeit cryptically, the core of what was to become, as I will be suggesting, the most complex nineteenth-century theory of mourning apart from that which would later be developed by Freud. In the letter to Brown he hinted at how his interest in the operation of sound, time, and life would become indistinguishable from the question of loss:

> Soon after John's death I listened to a music-box, and if, at any time, that event had seemed inconsistent with the beauty and harmony of the universe, it was then gently constrained into the placid course of nature by those steady notes, in mild and unoffended tone. . . . But I find these things more strange than sad to me. What right have I to grieve, who have not ceased to wonder?
>
> We feel at first as if some opportunities of kindness and sympathy were lost, but learn afterward that any *pure grief* is ample recompense for all. That is, if we are faithful;-for a just grief is but sympathy with the soul that disposes events, and is as natural as the resin on Arabian trees.—Only nature has a right to grieve perpetually, for she only is innocent. (C, 101–102)

As the letter specifies, there is something like a pure grief—which Thoreau seems to be using synonymously with "just" and "perpetual" grieving—that requires "faithfulness" to the loss, sympathy with it, of which only nature is capable. From pure and perpetual grief, then, humans are excluded. And because Thoreau doesn't specify here what would be characteristic of human grief, to understand it one can only use contraries: if natural grief is perpetual, then human is limited in duration; precisely because it is perpetual, hence unforgetting, natural grief is said to be faithful to the loss, whereas human grief, by opposition, is limited in duration and reaches the point when it forgoes the loss, becoming unfaithful to it. And if faithfulness to the loss of natural grief is just, then the economy of human grief, where one forgoes loss to attend to the mourner, is suggested to be unjust. Thoreau confirms that understanding of human grief in his letter to Emerson of March 11, 1842. There the difference between how humans mourn and how nature mourns is again referred to, only this time Thoreau identifies human grief as "private" and its interest in self-restoration as "selfish":

> How plain that death is only the phenomenon of the individual or class—Nature does not recognise it, She finds her own again under new forms without loss. . . . We are partial and selfish when we lament the death of the individual, unless our plaint be a paean to the departed soul, and we sigh as the wind sighs over the fields, which no shrub interprets into its private grief. (C, 104–105)

What is so astonishing about Thoreau's distinction between partial and natural or impartial grief is less his suggestion that nonhumans could mourn, that there could be a grief that is not personal, or that there is mourning where there is no self to experience it. Rather, what astonishes is the idea that *because* there is no self to experience it and enclose it ("no shrub" to interpret it into its privacy), selfless grief is more profound, even more ethically desirable. Human grief is less ethically appreciated, because it is guided by the interest of the mourner's self, its sorrow, its recovery. Human grief doesn't infinitely abide with the lost but instead figures ways to self-regeneration, the "egotism" of which renders it ethically invalid ("*we* are partial . . . when we lament"). In contrast, and by extension, pure grief is a "better" way of grieving, because it is not motivated by partial interests. Its purity derives from its not having to end, being able, with the same intensity, to perpetuate itself and thus to remain faithful to the loss.

Such unbound natural mourning is for Thoreau embodied in the motion of the wind, whose sound he identifies as nature's sighing ("sigh as the wind sighs over the fields"). Thoreau's phrasing makes clear that this sigh is not a personification of human mourning, as in English romanticism, for instance in Shelley's "Ode to the West Wind," where the wind becomes subjectivized "thou." Rather, the wind is here referenced in its objectivity, a concrete nonpersonified phenomenon, the sound of which *is* lament. This sheer sound operates as a natural or ontological force that traverses everything, passing through other natural bodies—in the letter to Emerson, Thoreau lists grass, foliage, and flowers in decay—making them vibrate and thus animating them, or, as he puts it, regenerating them "*without loss.*" By identifying this incessant animation with pure grief, Thoreau not only points to the ontological indiscernability of life and grief but also signals what grief should be all about; he imagines deep or extreme grief to be intense enough to recall the dead.

However, as if to rethink the human exclusion from the performance of perpetual grief, Thoreau also adds in the same letter that perpetual grieving could be accessed by humans were they somehow to turn the partiality of their grief into a "paean to the departed soul," the operation of which would have effects identical to the sound of the wind. But what, one might ask, are we to make of these analogies that conflate a natural phenomenon (the motion of the wind) with a highly structured artifact, the paean? What are we to make of a paean that transcends the finitude of human song to follow the souls of the departed and recall them back to life?

It is not immediately evident that the paean can be the expression of grief that Thoreau wants it to be, because from archaic times to Pindar, Plato, Aristotle, and Euripides, the paean was, as Nicole Loraux explains, addressed not to the dead but to the living. It is typically considered to be an "eminently Apollonian form of song": it " 'shines' *(lampei)* . . . like every 'luminous' *logos*" and so is "clear as daylight."[1] In addressing the living, it reflects the "basic movements . . . of a life that is orderly and brave," celebrating its divine goodness. In the tradition of Greek tragedy that Loraux identifies as orthodox, the joyous paean is accordingly imagined as clearly opposed to lament *(thrēnos)* in all its forms:

> In a perfectly orthodox way . . . [tragedy] associates hymn and paean with Apollo. . . . Paeans are endlessly associated with the one often invoked as *Paian*, in simple address or in solemn appeal. . . . In these moments of theological and musical orthodoxy, tragedy seems to distinguish carefully between the orders, for example, opposing the *thrēnos* sung over a tomb with a joyful paean, or insisting that the Muse of funeral *thrēnoi* remain separate from paeans because hers is the song of Hades.[2]

But if the paean is so strictly delineated both from formal lament and from excessive grief expressed in the cry of inarticulate moaning—alternatively, if its solemnity is for the living and by the living—why is it that Thoreau understands it oppositely, as what is addressed to the dead, and why is it that he identifies its sound with the sound of infinite lament? How can the lament be joyous, how can the logos coincide with the confusion of acategorial night? How is it that the paean of life can reach the dead?

Yet, it is precisely in pairing the balanced or metrical happiness of the paean with a cry of unbounded lament—thus impossibly morphing logos into what cancels it—that Thoreau shows just how precisely he understands the Greek tragic tradition. For as Loraux explains, several important tragedies will respond to the orthodox distinction between the "luminosity of the solemn song" and the nocturnal lugubrious cry by blurring that distinction. In that "counter-tradition" the rhythmic and harmonious will be submerged by the inharmonious but interminable sonority of a howling voice: "The respite [of orthodox distinctions] is short-lived, however: numerous counterexamples come to the fore at once, in *Oedipus the King*, where we underscore the simultaneity *(homou)* of the paean and the moans that rise in one voice toward Oedipus. . . . Is Aeschylus recanting when—after stating that there is no paean for honoring the god *Thanatos* (Death), or that there is no defense against this god *(oude paionizetai)*—he identifies death as the only Savior *(hō thanate paian)?*"[3] Certain tragedies, then, will complicate the paean as an oxymoronic "grieving happiness" and employ it precisely as a "modifier" of measured mourning. As a "*coincidentia oppositorum*" the paean will be intoned to sound the fusion of categories cancelling each other's boundaries—those separating measure from anger, love from hate, happiness from sorrow as well as life from death—and producing an unorthodox ambivalence. Even though this crucial modification of grief enacted by the change in the operation of the paean can be registered in a series of tragedies, in Loraux's estimate it is most prominent in *Seven against Thebes*, Aeschylus's tragedy produced at the City Dionysia in 467 BC: "In every case, from *Seven against Thebes* and *Agamemnon* to *The Trojan Women* and from *Choephori* to *Alcestis*, the paean sounds lugubrious, whether it denotes the music of flutes or is addressed to the dead."[4] Thus, *Seven against Thebes* inaugurates something of a new tradition of the modified paean, so that even "*thrēnos*-paean" in *Helen* "may be a Euripidean variation on a passage in *Seven Against Thebes* recalling 'The ill-sounding Furies' hymn, / And Hades' hateful paean'" (Aeschylus, *Seven against Thebes*, 869–870).[5]

Seven against Thebes is precisely the tragedy that Thoreau began translating in 1842, either immediately or soon after John died.[6] In it Aeschylus rethinks what it means to mourn dead brothers by suggesting a complex understanding of the paean that blurs the divide between dead and living, the measure of a composed mind and the infinitude of perpetual lament.[7]

Standing above the bodies of Eteocles and Polynices—two brothers who killed each other in the battle for Thebes—the chorus, made of Theban maidens, registers the entry of Antigone and Ismene (Eteocles's and Polynices's sisters) by referring, in Thoreau's translation, to the paean in a way that reshapes the boundaries of personal grief:

> But these come to a sad affair
> Antigone and Ismene,
> The Mourning of their brothers, not
> dubiously

I think that they from their loving
  deep-bosomed
Breasts will send forth worthy grief.
But it is just that we howl before
Relating, the sad sounding hymn of
Erinnus, and to Hades
Chant the hostile paean. (ST, 101)

Thoreau's translation differs from most English translations of the tragedy, which points to his accurate understanding of the shift in the operation of grief that Aeschylus brings about. Following Tauchnitz's Greek edition of Aeschylus, Thoreau's translation reflects the deep ontological ambivalences signaled by the chorus, registered not just in the overflowing of Antigone's and Ismene's private grief into the public sphere (their "deep-bosomed" love will "send forth [the] grief"); not just in the fact that anger (Erinnus) will be voiced in a "hymn," confusing solemnity with black hate, as indicated by the oxymoron of "tranquil fury"; and not just in how balanced rhythmical song coincides with unsustained howling. The ambivalence is expressed most of all in the chorus's prediction that the contradictory lamenting-paean will be chanted not to the living—as the orthodoxy would have it—but to the dead. The paean will be addressed to Hades, or, as Aeschylus says elsewhere in the tragedy, in Thoreau's translation, to the "sunless, All-receiving, invisible waste."[8] In Thoreau, the paean is a sound of life sent to the dead to recall them, just as in the letter to Emerson, where it is represented as a plaint of the living to the departed.[9]

Thoreau's translation systematically negates the "formalism" of the chorus's lament, repeatedly affording the chorus the power to voice an oxymoronic paean that enlivens the dead, as if magically, through the lament. Thus, as Argive's army is encircling Thebes, in Thoreau's rendering, Eteocles doesn't ask the chorus (as is the case in Tauchnitz's edition that he is translating) to stop lamenting and start singing the ululation (ὀλολυγμός) that is to voice a more adequate form of prayer that would satisfy Apollo, who abhorred mourning.[10] Instead, in Thoreau's translation, Eteocles tasks the chorus to come up with a lament that would simultaneously be a paean: "Pray the gods to be our allies unto / better things; / Hearing my wishes, then do thou / For thy lament sing the sacred and / enlivening paean" (ST, 75). The chorus meets this demand for a different lament—which Eteocles imagines as an ontological intervention into reality, for it will quicken or "enliven" what is inanimate—at the highly disputed end of the play, after Eteocles and Polynices die and Antigone and Ismene appear:[11]

> Some evil chill falls around my heart;
> I composed a song for the tombs
> As Priestess of Bacchus, blood-dropping
> Dead weeping unfortunate
> Having died; Surely ill omened this
> Concert of the spear. (ST, 100)

Here again—as was the case with the paean that in Thoreau's translation is voiced *to* Hades—the sad melody that the living composed *for* the tombs (τύμβω μέλος) becomes by the fourth line the weeping *of* the dead in their tombs, and the "I" referencing the choral voices fuses with the multiple "weeping" dead. If the dead are fused with the sound of living grief it is, then, because its sound transcends the person voicing it, materially reaching the other realms of the real.[12] This is to say that whether voiced by Antigone and Ismene, the chorus, or, as happens toward the end of the play, by semi-choruses (to which the ambiguity of the Greek text sometimes accords simultaneous, instead of alternative, laments),[13] grief in *Seven against Thebes* is not simply a personal emotion but also an ontological force, for it transcends persons and travels to the world of shadows while still being a "*materiality*, indissociable from [their] psychic dimension." It is thus much like the grief that Loraux identifies as "nonoblivious" in the *Illiad* or in Sophocles's *Electra:* "at the same time outside and inside, sinister reality and psychic experience."[14] Most explicitly, after the brothers die in *Seven against Thebes*, the chorus will come to insist explicitly on the material power of the grief to rearrange reality:

Some others labor surely may labor
Sitting on the hearths of the houses.
But, friends, with a fair wind of
  lamentations
Row around the head a conveying
  motion
Of the hands, which ever conveys
  through Acheron
The groanless, black-sailed
Navigated bark,
To the untracked by Apollo, the sunless,
All-receiving, invisible waste. (ST, 101)

The common gesture of mourning in ancient Greece, head-beating, is here transformed into a dance that animates through lament: attuned to the voice of its mourning, the lamenting body dances, moving in circles around itself, its hands above its head, physically creating the wind by its "rowing motion."[15] It is this wind, matter in motion manufactured by the sound of grief—which is why in the letter to Emerson, Thoreau identifies perpetual grief as the coincidence of lamenting paean and wind—that will function as the sonorous whirlpool of life, capable of catching the "black-sailed bark" that ferries the dead down the river Acheron, on its way to the underworld, and move it against the river's current, back into the land where the dead can be seen again.[16] This then is the precise ontological operation of grief in Thoreau's Aeschylus: it transforms the dead from invisible waste into the daylight of creaturely life.

Grief understood as a force of life helps explain why Thoreau's translation of *Seven against Thebes* renders death less final than the Greek text he translates. For where Tauchnitz's Greek edition suggests that the strife among brothers caused by Oedipus's curse will end in death *(kai Thanatos telos)*, Thoreau's version proposes that it is death itself that ends:

And a groan has gone through
The city, the towers groan,

Groans the man-loving plain;
And possessions remain for the
  decendants,
On account of which strife arose,
And the end of death. (ST, 103)[17]

Not, then, death as the end *(telos)* of strife, but strife as the end of death. But how are we to understand Thoreau's decision to declare that strife bans death? In *Seven against Thebes* strife signifies not only a condition generated by persons (disagreement in any common-sense meaning of the word) but is also the meaning of the dead brother's name: Polynices means "much strife." Thoreau's decision to translate the arising of strife as the ending of death could thus be read as his suggestion that a brother is, precisely, what doesn't end, but is instead the end of death. However, according to the mythology reconstructed by Thoreau's translation, for the end of death to occur, one would have to suppose that the wind of lament did indeed catch the "black-sailed bark" that ferries the brother to the all-receiving world of shadows, diverting it back to the shore of life. That doesn't happen in Aeschylus, because Polynices dies; but it does happen in Thoreau.

In 1849 Thoreau published *A Week on the Concord and the Merrimack Rivers*, registering in it, as scholars agree, much of the labor of mourning for John that he had undertaken during the preceding seven years. The time of that labor that the book puts in motion is, however, conditioned by space. The book recollects the voyage that Thoreau and his brother took temporally, unfolding the chapters—named after the days of the week—in successive order. But those chapters follow an introduction titled "Concord River," reminding us that there is no journey in time that is not contained within the materiality of a site. In *A Week* that site is the port of Concord, "for Concord, too, lies under the sun, a port of entry and departure for the bodies as well as the souls of men" (W, 15). As they are crossing this passage, moving away from the shore, the brothers catch a glimpse of friends gathered to wave them a "last farewell," as if at a funeral, but the brothers can no longer speak back to them, having already departed, as if dead: "Some village friends stood upon a promontory lower down the stream to wave us a last farewell; but we, having already performed these shore rites, with excusable reserve, as befits those who are embarked on unusual enterprises, who behold but speak not, silently glided past the firm lands of Concord" (W, 16). If the brothers can't reply to the valedictorian address of friends, it is because they are already "glid[ing] noiselessly" in their bark (W, 20)—like the "groanless, black-sailed, Navigated bark" sailing on Acheron, in Thoreau's translation of *Seven against Thebes*—down the stream of death.

For, like Acheron, the Concord River is dead. Its current is "stationary" (W, 27) and "still" (W, 19); it is a "sluggish artery . . . without . . . a pulse-beat" (W, 11), so that, quite explicitly, in the midst of discussion of Greek mythology, Thoreau describes it as "a dead stream" (W, 61).

But this death that is the Concord "empties into the Merrimack" (W, 6), a stream that lives by contrast, described as "restless all" (W, 84): "unlike the Concord, the Merrimack is not a dead but a living stream" (W, 88).

What empties the pulseless stream into the living Merrimack, inspiring death to become life, is the force of animation, provided by the wind: "Many waves are there agitated by the wind, keeping nature fresh, the spray blowing in your face, reeds and rushes waving; . . . such a healthy natural tumult as proves the last day is not yet at hand" (W, 7). The wind can agitate matter into life because, as we are immediately told and in more ontological terms, the wind is a living breath that materially traverses nature with its acoustic waves, thus generating locomotion: "As yesterday and the historical ages are past, as the work of to-day is present, so some flitting perspectives, and demi-experiences of the life that is in nature are, in time, veritably future, or rather outside to time, perennial, young, divine, in the wind and rain which never die" (W, 8). If the wind (and water) are timeless, it is not because they don't change—for their modalities constantly shift, from calm to storm, from rain to ice, for instance—but because they are inexhaustible: wind always is. It embodies unpassing yet processual life; it is something that moves, and just like the sound of Homer's poetry or John's music box, always arrives

The dead stream of the Concord. Courtesy David Wills.

The living stream of the Merrimack. Courtesy David Wills.

from another space. It comes from "ages past" into "the work of to-day" while simultaneously being in the time "veritably future." Wind is the breath sliding along the circle of chronic time and in so doing, by extension, it moves among and into different ontological conditions: from death to life, illness to health, and back again, just as Thoreau suggests in the letter to Emerson or in Aeschylus's *Seven against Thebes.* Thus, as Thoreau specifies elsewhere in *A Week,* only those who are recovering from a deadly illness—those who have already begun sailing to the shore of shadows—know the saving, animating force of the wind: "Only the convalescent raise the veil of nature. An immortality in his life would confer immortality on his abode. The winds should be his breath" (W, 379).

Despite the infusion of life that the Merrimack provides, however, the brothers experience difficulty on their way back along the Concord River. For the wind on that dead stream is not absent but stilled, seeming to withdraw from them. Thus, "we endeavored in vain to persuade the wind to blow through the long corridor of the canal. . . . When we reached the Concord, we were forced to row once more in good earnest, with neither wind nor current in our favor" (W, 366). "Aspir[ing] to an enduring existence" (W, 379)—aspiring to "immortality in this life"—the brothers create the breath of wind by rowing. Their hands and oars generate a wind strong enough to bring their boat back to the shore of the living, just as, at the end of Thoreau's translation of *Seven against Thebes,* the wind of lamentation helps the mourners to "row around the head a conveying motion of the hands," to generate a countercurrent that renavigates the bark against the stream of Acheron to the shores of Thebes.

A Week thus operates in a similar way to the lamenting paean that Thoreau talks about in the letter to Emerson, which, by coinciding with the literal acoustics of the revivifying wind, forces the lost back into life. Additionally, as in the archaic mind of the Greek poets, the evocation of the departed in *A Week* is circular. Thoreau's dead brother returns to life on the circle of chronic time that the book constructs by ending where it started, at the "entry of living souls." In that book, then, John returns, and the brother is, as is the case in Thoreau's Aeschylus, what doesn't end.

Cemeteries of the Brain

A STUNNING, albeit cryptic passage from *A Week*—combining sentences written as early as November 1837 with meditations from the mid-1840s—will radically complicate the claims concerning partial and impartial grief that Thoreau hinted at in the letters to Brown and Emerson. It argues that the partiality of human grief is intrinsic to the structure of human subjectivity but in the end entertains the possibility, also hinted at in the letter to Emerson, that under reformed epistemological conditions, perpetual grief might become accessible to humans. Here is the passage:

> When a shadow flits across the landscape of the soul, where is the substance? . . . Every man casts a shadow; not his body only, but his imperfectly mingled spirit; this is his grief; let him turn which way he will, it falls opposite to the sun; short at noon, long at eve. Did you never see it?-But, referred to the sun, it is widest at its base, which is no greater than his own opacity. The divine light is diffused almost entirely around us, and by means of the refraction of light, or else by a certain self-luminousness, or, as some will have it, transparency, if we preserve ourselves untarnished, we are able to enlighten our shaded side. At any rate, our darkest grief has that bronze color of the moon eclipsed. There is no ill which may not be dissipated, like the dark, if you let in a stronger light upon it. Shadows, referred to the source of light, are pyramids whose bases are never greater than those of the substances which cast them, but light is a spherical congeries of pyramids, whose very apexes are the sun itself, and hence the system shines with uninterrupted light. (W, 352–353)

This image of the mind's landscape divided by shadows flitting across it, as well as the conception of shadows as pyramids—hence as enclosed burial sites or monuments—depicts the mind's scenery as very much like a graveyard. The "I" of this mind can perhaps take a walk in this cemetery, among the monuments, but it cannot go into what is deposited beneath them. For although there are monumental signposts on the terrain of the mind, what they point to remains shadowy. A certain portion of what contrives the mind—its feelings, thoughts, or memories—is thus enshrined in it, as if the mind housed a crypt that remains unavailable to it. Such a mind is constitutionally opaque to itself ("not greater than his own opacity"). No matter how much it tries to come closer to the stuff of which it is made ("let him turn which way he will"), it remains "imperfectly mingled"; it manages only to circle around its shadows, unable to coincide with what produces it (hence man's shadow, in contrast to the sun, "short at noon, long at eve"). That the mind sees itself only as a shadow it can't know—as if it were lost to itself—is for Thoreau the fundamental impairment that generates grief in every man: "Every man casts a shadow; not his body only, but his imperfectly mingled spirit; this is his grief." The mind grieves its inability to come closer to what it harbors.

We might understand this grief less as an affective state than as an epistemological condition, and indeed, the word can suggest a more general misfortune than that related to a specific loss. It might simply mean that man's greatest problem is his inaccessibility to himself; but toward the end of the paragraph another grief appears. To explain how our feelings can remain distant from the mind, Thoreau's passage gives an example of grief, this time clearly more understood as an affect of sorrow ("our darkest grief"). As Thoreau's analogy has it, just as the eclipsed moon is rendered invisible when caught in the space between the "opaque body of the earth" and its shadow—a space in which the moon, however, continues to move, casting back on the earth its own shadowy light, which is detected by Thoreau as "bronze colored"—so our grief is entombed by the mind and rendered shadowy, as if inessential (hence it too "has that bronze color of the moon eclipsed").[1] Yet, according to the logic of the analogy, even though this shadowy grief is expelled to sites inaccessible to the mind, it nevertheless continues to circulate, just as the eclipsed moon moves between the earth and its shadow. That, then, is how the mind disposes mournful events in the process of "partial" grief: it less disposes of them than "eclipses" them, excluding them from its explicit content. The affect of grief is thus not extinguished, but neither does it function as a mental representation, with which the mind can coincide. The mind doesn't live grief, although grief lives in it, buried for it but circulating through the body, casting here and there its shadow and so rendering the mind lunar, darkening its mood. Grief thus becomes buried alive in a vault in the mind, from which it acts on that mind while remaining withdrawn from it. An answer to Thoreau's question about what happens "when a shadow flits across the landscape of the soul" could thus be this: when the mind sees a shadow in its landscape, a crypt is formed, whereby grief for the loss is deposited so as to remain living there but mentally inaccessible.[2]

Whereas the first part of the passage I am discussing clearly claims this structure of the mind and hence this economy of mourning to be constitutive (the mind remains imperfectly "mingled" no matter which way it turns), the second allows for a phantasmatic projection, suggesting that under certain circumstances the opaque shadows cast by the mind can be dispelled. The mental condition coinciding with the dispersion of shadows is something Thoreau calls "self-luminosity." In an almost ciphered sentence, composed of terminology taken from contemporary optics, Thoreau describes the self-luminosity of the mind as a natural phenomenon behaving, in certain circumstances, in the same way as material bodies. The self-luminosity of the mind is thus rendered identical to the luminosity of matter generated by the rapid "refraction of light" ("by means of the refraction of light, or else by a certain self-luminousness . . . we are able to enlighten our shaded side" [W, 352]). Thoreau would have found the description of the experiment producing luminous matter in Farrar's *Experimental Treatise on Optics*, a scientific treatise he studied at Harvard in 1836: "When we take two prisms of polished glass, and . . . rub the two prisms against each other . . . Then if we present the prisms to full daylight . . . we shall perceive a greater or less number of coloured rings . . . formed by the reflection of light."[3] Thus, when

Thoreau explains the operation of the mind in the passage I am discussing, by referencing the fusion of different substances—such as matter and light—he is relying on the possibility of their actual intersection that Farrar describes. Like matter (considered opaque) made transparent by light (considered ethereal and substantially different from matter), the mind can reach self-luminosity by fusing with what it has rendered obscure. The mind can dispel its shadows only if it manages to coincide with its body, through which grief—and by extension other buried feelings—circulate. The desired self-luminosity is thus the coincidence of mental and the sensuous, since one can inhabit buried emotional content only by living it.

A passage immediately preceding "the shadows of grief" describes the accomplishment of such self-luminosity precisely as a free circulation of the mind into sensations and bodies that dispels all shadows:

> When I visit again some haunt of my youth, I am glad to find that nature wears so well. . . . There is a pleasant tract on the bank of the Concord, called Conantum, which I have in my mind;-the old deserted farm-house, the desolate pasture with its bleak cliff, the open wood, . . . and the moss-grown wild-apple orchard. . . . It is a scene which I can not only remember, as I might a vision, but when I will can bodily revisit. . . . When my thoughts are sensible of change, I love to see and sit on rocks which I *have* known, and pry into their moss, and see unchangeableness so established. . . . There is something even in the lapse of time by which time recovers itself. (W, 350–351)

In the terrestrial landscape Thoreau describes here there are only "tomb stones": the old deserted house, the desolate pasture, the bleak cliff, the hollow tree, the moss-grown orchard, the moss-grown rocks. They stand for decay, abrasion, and loss, buried in the life that surrounds them, much as memories and grief are buried in the landscape of the mind. But they are reincorporated into the life they commemorate not through the mental practice of recollection—

they are not "visions" remembered by the mind, as Thoreau explicitly says. Instead, they are enlivened through the bodily practice of touching. Memory is embedded in the intimacy of present sensing, thus always embodied and hence something that can be physically encountered. Thoreau positions his body on the rocks to physically inhabit what once he knew sensuously, his hands lifting the moss as if to excavate what is covered by it. Prying into the moss, his hands reach the surface of the rock, and past life—sensations that have been—recurs in present touching. When Thoreau refers to this operation as turning the lapse of time into time's recovery, he thus means this: the recollection of a sensation is turned into sensation; the feeling of what had been lost is not expelled any longer into realms inaccessible to the mind but freely circulates through the mind and the body, fusing them. Thanks to touching, the mind is able to walk freely through different temporalities and to inhabit and revive what was. Through this walk, which begins with hands that recover past sensations, the mind finally restores and so gains access to its losses; it finally annuls the operation of partial grief, which is now released to circulate through the mind that is morphed into the body. In that way the "unchangeableness" of what was thought to have decayed is established.

Bizarre as it might appear, grief and loss thus become indistinguishable from life, which is what Thoreau describes in a letter to Isaiah Williams, the third letter in which he comments on John's death and his own grief. There, grief is released from privacy of the mind to act as an essence of life *(elixir vitae)* capable renewing life as it renews itself:

> Replace the . . . elasticity of youth with faithfulness in later years. . . . We do not *grow* old we *rust* old. Let us not consent to be old, but to die (live?) rather. Is Truth old? Or Virtue-or Faith? If we possess them they will be our *elixir vitae* and fount of Youth. It is at least good to remember our innocence; what we regret is not quite lost-Earth sends no sweeter strain to Heaven than this plaint. Could we not grieve perpetually, and by our grief discourage time's encroachments? (C, 108)

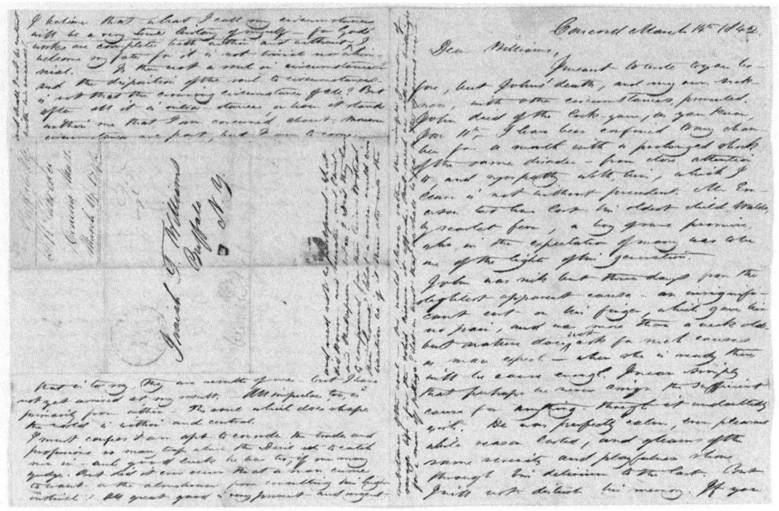

Autograph letter signed: Concord, to Isaiah Williams, 1842 Mar. 14. The Pierpont Morgan Library, New York. Bequest; Gordon N. Ray; 1987. MA 4500.

Here, grief explicitly acts as an ontological rather than psychological operation, one that reconfigures the landscape of what is. For in transforming the departed into a recurring existence ("what we regret is not quite lost"), regret annuls the succession of linear time ("discourages time's encroachments"); and precisely because it updates the present by planting in it what has been, grief is afforded the status of an ontological force that generates eternal life.

Pindar's Doubles

DESPITE THE fabulous quality of this ontology, the *Journal* entries written after John's death show that Thoreau thought seriously about how to enact a revision of human personhood that would weaken the "partiality" of human grieving. In one of the first entries written after John's death, on February 21, 1842, Thoreau thus accuses himself of living unethically (living "ill") because he lives too close to his "self":

> I have lived ill for the most part because too near to myself——I have tripped myself up-so that there was no progress for my own narrowness- . . . - And the soul dilutes the body and makes it passable- My soul and body have tottered along together of late tripping and hindering oneanother like unpractised Siamese twins- They two should walk as one that no obstacle may be nearer than the firmament-. (J, 1, 366)

Instead of resisting the corporeal by opposing it, while "standing pledged" to its identity, the mind is asked to diminish the limits of its selfhood. In the imagery of free walking and tripping that Thoreau evokes here, this shrinking is imagined to generate an openness in which the mental (the "soul") fuses with the corporeal so that the two move freely, without obstacles, not just across the boundaries separating each other ("they two should walk as one"), but also through and into any other body. Wai Chee Dimock is thus right to claim that in this entry Thoreau connects body and soul "against the grain of individuation," as if he imagined life to be a transversal cutting across the boundaries of persons.

On Dimock's understanding, in this entry body and soul are said to be

> Siamese twins because a single human being houses both. And they are Siamese twins in a different sense as well: not just within the compass of one person but, even more strikingly, across the relational fabric of more than one, dramatized Odyssey, the Lyric, and the Call of the Dead by Thoreau's peculiar spelling of "oneanother." With that spelling, body and soul are also united, but transversely, against the grain of individuation. The physical death of one person binds the stricken soul of someone else to an analogous state, dissolving the boundaries between them, in a faithful memorializing of the physical by the nonphysical. . . . Those twinnings have a further implication. . . . While it is true that the symptoms of the dead are reproduced among the living, the explosiveness of that reproduction also suggests that not all the symptoms are pathological or malignant. The spilling over of death is a form of "life" after all, however odd it might seem.[1]

This "binding" through grief, of the dead body of one person to the living soul of another, is understood to be an actual trespassing of a soul from its own body into the body of another, what Dimock calls "twinning." And while from any stance committed to categorical clarity, such twinnings are "pathological or malignant," Dimock is precise in suggesting that Thoreau wants them to be ontological: the spilling of death over into life to mobilize what is inanimate and to recreate, materially, just as nature does, with minimal loss. What Thoreau desires, then, are existential shiftings in reality, with no essences to block becomings but only attributes that could be variously assembled.

In fact, in a *Journal* entry written on March 2, 1842—the same day that he formulates for Lucy Brown the idea of perpetual grieving—Thoreau explicitly formulates the idea of existing through attributes instead of through essences as a preferable mode of being:

> The greatest impression of character is made by that person who consents to have no character. He who sympathizes with and

> runs through the whole circle of attributes can not afford to be an individual. Most men stand pledged to themselves. . . . They are like children who cannot walk in bad company and learn the lesson which even it teaches without their guardians for fear of contamination. He is [a] fortunate man who gets through the world without being burthened by a name and reputation. (J, 1, 367–368)

Ontologically, attributes issue from a cause (rather than having the capacity to cause) and consist of everything changeable (functions, conditions, or qualities) that gets allotted to something essential, thereby additionally qualifying it. So when Thoreau suggests the possibility of a nonindividuated existence—an existence that wouldn't be anything other than a series of attributes it passes through—he is pointing to a reality in which nothing is essential but identities are arranged temporarily through a provisional arrangement of attributes. In this flexible reality nothing belongs to something constitutionally, but what belonged to one may be conjured differently to constitute another. On March 11, the day Thoreau writes to Emerson against partial grief, a *Journal* entry describes his idea of such transformations in the following way: "I must receive my life as passively as the willow leaf that flutters over the brook. I must not be for myself. . . . I feel as if [I] could at any time resign my life . . . and become an innocent free from care as a plant or stone" (J, 1, 371). That is the life he wants, then, moving from one transformation to another, personal to impersonal, human to plant, and even inanimate. It is a life whose nameless existence is precisely its fortune, its happy chance: "he is fortunate man who gets through the world without being burthened by a name."

This fantastic ontology of "twinnings" will, I shall shortly suggest, generate the logic of *A Week* in a central way. It was something that Thoreau had found elaborated in yet another Greek text that greatly preoccupied him.

In late August 1843, after translating *Prometheus* and *Seven against Thebes*, Thoreau, encouraged by Emerson, began work on some of Pindar's Pythian, Olympian, and Nemean Odes. In January 1844 *The Dial* published a selection of his translations that included the

Tenth Nemean Ode, in which Pindar transformed the classical Discouri myth into a proper, albeit counterintuitive, ontology of life. In Pindar's version two brothers, Castor and Pollux, have different fathers who have given them two different sorts of life. Fathered by Zeus, Pollux is made immortal, whereas Castor, fathered by the mortal Tyndorus, is rendered mortal. What interests Thoreau the most in Pindar's ode is the moment when Pollux is about to be attacked, and presumably, killed by Idas, who has already fatally wounded Castor, the moment at which Zeus intervenes to save his son and so maintain his immortality. As *The Dial's* editorial explained, the part of the ode published begins from the moment when "Castor lies mortally wounded by Idas," while "Pollux prays to Zeus, either to restore his brother to life, or permit him to die with him." To Pollux, then, a shadowy "existence" in Hades is as good as one on Olympus, as long as it promises a being-together with his brother. But in negotiating with Pollux, Zeus agrees to meet him only halfway, as it were, dealing to Pollux a double deal, which Thoreau translates thus:

> Nevertheless, I give thee
> Thy choice of these; if indeed fleeing
> Death and odious age,
> Thou wilt dwell on Olympus,
> With Athene and black-speared Mars;
> Thou hast this lot.
> But if thou thinkest to fight
> For thy brother, and share
> All things with him,
> Half the time thou mayest breathe, being
>   beneath the earth,
> And half in the golden halls of heaven,
> The god thus having spoken, he did not
> Entertain a double wish in his mind.
> And he released first the eye, and then the
>   voice,
> Of brazen-mitred Castor.[2]

What Zeus offers Pollux is this: there will be only one life, hence, one death, but two persons. In Zeus's arithmetic, if Pollux so wishes, he can have his brother share his breath, making one life two-headed, as it were. For in Pindar's ode—and that is what makes Pindar's take on the Discouri myth of transformation astounding—those in Hades are not quite dead, but instead continue to breathe ("Half the time thou mayest breathe, being / beneath the earth").[3] Life is continuous, extending into the "other side," beneath the earth, so that, as Heidegger puts it in his interpretation of this Pindaric Ode, "unlimited and limitless are zones without stopping places, where every sojourn loses itself in instability."[4] What is called "death" is thus still a life, albeit a life in the zone of shadows that is Hades, so that the dead become not those who are not, but those who can't be seen. As Heidegger interprets it, in Pindar's ontology—which he renders central not just for understanding Pindar but also for understanding Aeschylus and Sophocles—"beyond the modes of dissemblance and distortion, there prevails a concealment appearing in the essence of death, of the night, and everything nocturnal."[5] The dead are concealed not distorted; and those who are concealed can become unconcealed, or, as Pindar puts it in Olympic Ode VII, "the signless cloud of concealment" can be "thoughtfully disclosed." What discloses those covered in signless (and hence, mute) shadows is precisely continuous life itself, moving to and fro through its own different zones—limited and unlimited, finite and infinite, formed and unformed—and in so doing shifting those in the shadow into the light of day. This shifting, this economy of concealment and unconcealment, is what Heidegger calls the "awe" of Pindar's ontology. Far removed from the sense of fear experienced by an individualized psyche, this ontological awe means that "being itself sustains awe, namely the awe over the 'to be.' "[6] concealing beings but revealing them again. That is the logic of the Ode that Thoreau translates, where Zeus plays the role of "being" that deals in concealing and disclosing beings, so that while one brother is visible, the other is concealed in the "signless cloud" of the shadow, yet still breathing, sharing in the same life.

And if Thoreau's *A Week* is an exercise in Aeschylus's way of lamenting—where lament is an ontological rather than psychological force of unconcealing the dead, by calling them back—it is also, I am suggesting, an exercise in Pindar's ontology of persons. For it systematically generates switchings, whereby a unity of life shared by two brothers transforms their separate selves into a compound signified by the first person plural. As Steven Fink has demonstrated, in *A Week* the brothers keep moving into one identity even if, or when coming out of it: "it is significant that Thoreau never uses the first person singular while narrating the actual enactment of the river journey, but always retains an identification with his brother by using the plural 'we.'"[7] This becoming one (plural) life of two persons has different modalities. Sometimes, in an almost explicit reference to Pindar, the brothers become interchangeable and never identified individually by name: "One . . . was visited in his dreams this night by the Evil Destinies, and all those powers that are hostile to human life . . . but the other happily passed a serene and even ambrosial or immortal night, and his sleep was dreamless" (W, 116). At other times the switching of identities comes into focus in a moment in which the brothers separate persons fuse into the same life. For instance, in a passage from "Monday," the brothers' bodies follow the same rhythm of activities and are attuned to the same phenomena in their environment, which ultimately moves their separate psyches into dreaming the same dream: "Then, when supper was done, and we had written the journal of our voyage, we wrapped our buffaloes about us . . . listening awhile to the distant baying of a dog, or the murmurs of the river, or to the wind, . . . or half awake and half asleep, dreaming of a star" (W, 172). There are still other modalities of this switching: the brothers share the same life by sharing the same breath, as when they fall asleep as one person ("we . . . fell asleep" [W, 333]); or they share the same past ("we naturally remembered" [W, 218]; "we remembered" [W, 332]); or when a shared imagination proleptically transports them in a future moment, suggesting they will continue to participate in the same life ("there we imagined" [W, 233]; "we thought that we should prefer" [W, 209]). But the fusion is never definitive, because the two brothers get separated time and again, like Pollux and Castor, as Thoreau's voice keeps coming out of the fusions into the first person singular. It is as if in those temporary fusions the two brothers take a glimpse of each other at the edge of the shadow as they are separated again. If Thoreau never tires of reenacting this ontology of awe—for Pindaric switchings and fusions traverse the whole text of *A Week*—it is because he was prepared to keep experiencing the loss of the brother (each time the fusion is dismantled) only in order to keep salvaging his brother's shadowy being.

Thoreau offered the most explicit articulation of functioning of these vital cohabitations (twinnings, switchings, becomings) in his correspondence with Otis Blake. Blake initiated a correspondence after having his attention attracted precisely by Thoreau's ideas about the weakening of the self leading to its partaking in other beings. In March 1848 Blake wrote:

> When I was last in Concord, you spoke of retiring farther from our civilization. I asked you if you would feel no longings for the society of your friends. Your reply was in substance, "No, I am nothing." That reply was memorable to me. It indicated a depth of resources, a completeness of renunciation, a poise and repose in the universe . . . ; which in you seemed domesticated, and to which I look up with veneration. I would know of that soul which can say "I am nothing" . . . I honor you. . . . It is noble to stand aside and say, "I will simply *be*." (C, 357–358)

The two claims by Thoreau that Blake paraphrases—"I am nothing" and "I will simply be"—do not contradict each other. Rather, read together they explain what is meant by the relinquishment of the self and Thoreau's desire to open up the space of becomings. While signaling the abatement of the "I," the "I am nothing" claim simultaneously confirms the existence of a sheer, nonindividualized, being. For "I am nothing" is not a negation; instead, it affirms being without attaching an identity to that being; it affirms the existence of a delimited life running through all attributes—such as Thoreau sought in the first entries written after John's death—animating them. In a letter from April 3, 1850, Thoreau explains that such an existence results from an attitude of utmost passivity, which enacts endless transformations: "If for a moment we . . . cease to be but as the crystal which reflects a ray,-what shall we not reflect! What a universe will appear crystallized and radiant around us!" (CO, 257). A mind so emptied and passive that it can't impose on itself or endure any form is nothing but a pure reception through which it becomes whatever it receives, and through those becomings moves through “the whole circle of attributes.” Still clarifying the same point, on March 13, 1856, Thoreau sends Blake a diagram of this “run” through minds and bodies, now called, as it was in 1842 after John’s death, “sympathy”: “As you suggest, we would fain value one another for what we are absolutely, rather than relatively. How would this do for a symbol of sympathy?

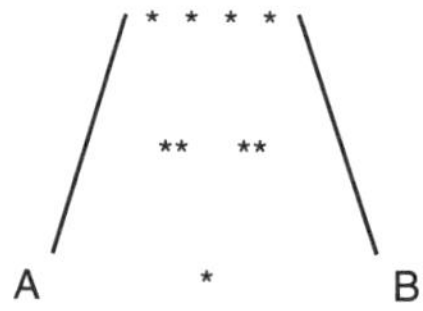

” (CO, 420)

As shown by the distribution of asterisks, the attributes contriving the two individuals (A, B) do not vanish, but migrate, forming a new composite (AB). We might imagine a new AB individual as a composite formed of attributes—thoughts, dreams, sensations—belonging to both A and B, like the ones Thoreau depicts in *A Week*, where two brothers become one. But we can also imagine it as an individual on its way to yet another compound, with a C for instance, forming a new temporary individual ABC, each of those compounds being only an “arena” of provisionally related mental phenomena. That is in fact how Thoreau described them to Blake in a letter from April 10, 1853, where he claimed a mind so provisionally assembled as to be the effect of a “singular kind of spiritual foot-ball,—really nameless, handless, homeless, like myself,—a mere arena for thoughts and feelings; definite enough outwardly, indefinite more than enough inwardly” (CO, 302). However, inward indefiniteness or the absence of a formed person—“nameless, handless, homeless”—is also said in the same letter to weaken the “outward definiteness” of corporeal existence, transforming it into a being so fragile, so lacking in form that it leaves no material trace in the world, verging instead on the imperceptible:

> I am so faint an entity, and make so slight an impression, that nobody can find the traces of me. I try to hunt myself up, and find the little of me that is discoverable is falling asleep, and then

> I assist and tuck it up. . . . How can *I* starve or feed? Can *I* be said to sleep? There is not enough of me even for that. If you hear a noise,—'t aint I,—'t aint I. (CO, 302–303)

The entity without identity that Thoreau describes here, an imperceptible, undiscoverable, yet existing being (leaving no impression, making no noise, neither hungry nor awake), adds up to life emancipated of all figures. An existence ready to run through all the attributes on the circle of time.

Odysseus, the Lyric, and the Call of the Dead

During the summer following John's death Thoreau wrote in his *Journal* the first eight lines of what was later to become the poem titled "The Funeral Bell," here quoted in its entirety:

One more is gone
Out of the busy throng
That tread these paths;
The church bell tolls,
Its sad knell rolls
To many hearths.

Flower bells toll not,
Their echoes roll not
Unto my ear;—
There still perchance,
That gentle spirit haunts
A fragrant bier.

Low lies the pall,
Lowly the mourners all
Their passage grope;—
No sable hue
Mars the serene blue
Of heaven's cope

In distant dell
Faint sounds the funeral bell,

A heavenly chime;
Some poet there
Weaves the light burthened air
Into sweet rhyme.[1]

Everything here seems to be consistent with the idea of two manners of grieving that Thoreau began to articulate two months earlier. To the grief of nature, inaccessible or barely accessible to human perception (Flower bells toll not / Their echoes roll not / Unto my ear), the poem juxtaposes human mourning, represented by a series of rituals (the tolling of church bells, the procession of mourners, the burial) that formalize grieving according to a communally determined pattern. And while human mourning fades away ("Faint sounds the funeral bell"), the muted grief of nature continues, haunted by the gentle spirit of the dead. Yet, the poem doesn't simply repeat Thoreau's juxtaposition of two types of mourning, for now the figure of a poet appears to suggest that poetry is a human operation capable of performing the perpetual grieving previously reserved for nature. For, as the last stanza suggests, the sound of human grief can be prolonged only if ending in a poem, as if poetry could somehow save the lost ("Some poet there / Weaves the light burthened air / Into sweet rhyme"). Indeed, what is in this poem only a faint suggestion—that a true poet could be the artist of perpetual mourning—was to become, I will argue, the basis of Thoreau's poetics. In developing his early counterintuitive ideas of poetry as resistance to the art of versification, Thoreau gradually came to see how the work of depersonalization resisted figuration in favor of "passages through attributes" that enabled constant grieving as the highest poetic act. In contrast to readers who argue that Thoreau's poetics was informed by Emerson's, expressing the self's interiority, such that the success of the artistic form depends on how accurately it reflects the self, I propose that Thoreau elaborated his own idiosyncratic poetics, one intensely opposed to self-expressionism. That poetics is predicated on the systematic erosion of the self—similar to freeing the senses of all perceptual habits to empty the mind, only this time it is performed through the dismantling of logical and conceptual patterns—

to the point where there would be nothing left of the mind to express.[2]

As early as 1836, in a letter to Charles Rice, Thoreau advocates resistance to composition, whether of thoughts or words, in order to replace generic forms with a less articulated assemblage of words. Explaining at the end of his letter why it violated rhetorical demands for the epistolary genre formulated by rhetorician Charles Lamb (for whom letters were to be a crafted combination of news, sentiments, puns, and topics), Thoreau writes: "Now this [letter] is neither one thing nor the other, but a kind of hodge-podge, put together in much the same style that mince pies are fabled to have been made, ie by opening the oven door, and from the further end of the room, casting in the various ingredients—a little lard here, a little flour there —now a round of beef, and then a cargo of spices—helter skelter" (C, 13–14). As Sanborn has noted, already here Thoreau was indicating his desire for disconnection, a desire generated by his "unwillingness willfully to change the flow of his rapidly moving thoughts."[3] The idea of a mind as container (oven) that opens to various ingredients without in any way organizing them so that they remain helter skelter reappears in a *Journal* entry titled "Composition," from March 1838. There Thoreau proposes that the art of versification must be a reflection, rather than a figuration or representation, of what is external to the poet's mind:

> We should not endeavor coolly to analyze our thoughts, but keeping the pen even and parallel with the current, make an accurate transcript of them. . . . The nearer we approach to a complete but simple transcript of our thought-the more tolerable will be the piece, for we can endure to consider ourselves in a state of passivity-or in involuntary action. (J, 1, 35)

The complete transcript Thoreau refers to is not the mind's representation of perceptions or an idea generated by their interpretation, hence, an intended meaning; the mind, in other words, is not asked to translate and hence to reformulate into language what it has experienced. The "complete transcript" Thoreau wants is thus not disoriented by the time needed for a linguistic formulation to occur but is rather a coincidence of event and word. Regardless of whether any such thing is possible, what Thoreau desires is a transcript in a literal sense: a language that doesn't interpret or figure things but simply relocates them in writing, as writing; a language that utters *things as they are* (I use the word "things" to include perceptions and thoughts too, as they are not registered or attended to by the mind but remain as external to it as any external thing). And, finally, because these perception-words are not gathered into the transcript he has in mind according to any willed order—which would require the "compositional self" that Thoreau refuses in favor of passive involuntariness—one could claim that Thoreau's "composition" also cancels rhythm and meter. Thoreau's use of the term "composition" therefore means something commonsense composition is not: the absence of structure and arrangement, something closer to a helter skelter notebook than a crafted form. Formless, arhythmical, and disconnected, the text Thoreau imagines as poetic is more a heap of words than a versified art, coming closer to a syntagmatic lump than a language working with meter to enact closure and so generate form. Thoreau's "composition," then, enacts a scattering of the compositional limitations that poetics imposes to bind the mind into coherent meaning. In so doing, it brings what Thoreau understands as poetic composition closer to literality, letting an unwrapped poem risk its coincidence with a cry. In fact, on May 9, 1841, Thoreau insists that the poet's words deteriorate into the literality of a cry, having his speech coincide with the motion of the wind. In a wording soon to be used almost verbatim in the letter to Lucy Brown that describes the motion of perpetual grief as the motion of the paean or the wind, he specifies that "the poet speaks only those thoughts that come unbidden like the wind," and that "the breath that inspires the poet has traversed a whole campagna" (J, 1, 309). What is here called the "utterance" is made to coincide with natural acoustic phenomena, arriving from the landscape and passing through poet's brain as if it too were a leaf of grass in the field, moved by the wind.

Although Thoreau never questioned that early idea of the poet as mental emptiness coincident with what is external to it, after John's death that externality is explicitly identified as life. Thus, in a long *Journal* entry written on March 25, 1842, composition is understood as a scalp, as what remains when life is over: "very dangerous is the talent of composition-the striking out the heart of life at a blow-as the Indian takes off a scalp. . . . The artist must work with indifferency-too great interest vitiates his work" (J, 1, 390–391). In contrast, in an entry written on September 24, 1843, the uncomposed mind pulsates with life, and the poet becomes an existence experiencing "pure life": "He [the poet] hibernates in this world, till spring breaks. He records a moment of pure life" (J, 1, 465). More radically still, in 1852 the poet doesn't only record pure life but also becomes such life, his thoughts like lichen, perpetually propagating without having to stop for the recreation that requires death: "Poeta nascitur non fit, but under what conditions is the poet born?— Perchance there is such a thing as a perpetual propagation or reproduction of the human without any recreation—as all botanists assert respecting plants and as Meyer in particular concerning lichens" (J, 4, 360–361). That the poet is "poor within," that he is nothing but the succession of embodied external phenomena will perhaps be claimed most succinctly in *A Week:*

> in comparison with his task, the poet is the least talented of any; the writer of prose has more skill. See what talent the smith has. His material is pliant in his hands. When the poet is most inspired . . . then his talent is all gone, and he is no longer a poet. . . . Great prose, of equal elevation, commands our respect more than great verse, since it implies . . . a life more pervaded with the grandeur of the thought. The poet often only makes an irruption, like a Parthian, and is off again, shooting while he retreats. (W, 342)

In contrast to the craftsman, the poet is completely nonartisanal, and in contrast to the (prose) writer, he is untalented, emptied of all "elevated" thoughts (such as "pervade" prose artists), possessing a personhood so fragile that it constantly fails to exist (his talent all gone, and he no longer a poet), but "retreats" into what he receives and, leaving no trace of himself, saying ",—'t aint I,—'t aint I.'" Far from expressing his fulfilled self, Thoreau's poet is the least expressive being of all, having no interiority to express apart from what his self has eroded into. But if that is the case, can we still think of Thoreau's enterprise as literary? As Lawrence Buell phrases the concern in a slightly different context: "what sort of literature remains possible if we relinquish the myth of human apartness?"[4]

Following Plato, who dismissed the cry on the basis of its nonfigured corporeality, musicologists understand the sound as that of a passion emitted by an unhinged mind addressing nobody in particular. A song is also capable of being addressed to no specific addressee, but unlike the cry, it is a figural articulation of the singer's affect and even if that affect is mournful, the song always expresses, in addition to what it sings, the satisfaction of successful articulation.[5]

S. E. Rena testified in the *Boston Evening Transcript* of February 15, 1896, that following John's death Thoreau stopped singing but continued to play the flute, which Harding corroborates.[6] Louisa May Alcott's "Thoreau's Flute" even identifies Thoreau's life more generally with flute playing, suggesting that his surviving friends will experience his death as the silencing of his flute.[7] Sanborn also speaks of Thoreau's habit of playing the flute on the "edge of the wood or stream" identifying the sound as a "pastoral note," bringing "pleasure in its strains." However, Thoreau's own account of his and John's flute playing makes it sound more strange than pastoral. In a *Journal* entry of May 1850, where Thoreau remembers the effects of John's flute, the sound of the instrument coincides with the sound of natural acoustic phenomena:

> I have heard my brother playing on his flute at evening half a mile off through the houses of the village, every note with perfect distinctness. It seemed a more beautiful communication with me than the sending up of a rocket would have been. So, if I mistake not, the sound of blasting rocks has been heard from down the river as far as Lowell,—some twenty miles by its course,—where they were making a deep cut for the railroad. (Jn, 2, 12)

Far from generating aestheticized sound, the sound of the flute is here evoked as identical in its effects to the sound rocks release after being blasted. That the sound to which Thoreau compares the sound of John's flute is the sound of exploding stones—the nonarticulated, unfigured cry of the injured earth—suggests that on his understanding the flute's sound is able to be what it is not commonly taken to be: the literality of the unfigured cry. The sound of the flute emerges out of, and merges back into, the natural sounds—identifiable as cries—in a *Journal* entry from August 1841, where Thoreau describes his own music:

> I sailed on the north river last night with my flute—and music was a tinkling stream which meandered with the river—and fell from note to note as a brook from rock to rock. I did not hear the strains after they had issued from the flute, but before they were breathed into it—for the original strain precedes the sound—by as much as the echo follows after—and the rest is the perquisite of the rocks—and trees—and beasts. Unpremeditated music is the true gage which measures the current of our thought-the very undertow of our life's stream. (J, 1, 320–321)

Rather than being premeditated by aesthetic intentions, the sound of the flute was simply breathed into the instrument by the rhythm of life itself ("life's stream")—which from a musical point of view can of course mean that it is completely arhythmical—and then released into the forest, as one among many sounds in it. When Thoreau first hears it—when he hears what in fact is its echo—he hears it as a natural sound supplementary to the sound emitted by rocks, trees, and animals. This *Journal* entry will be used, almost verbatim, in *A Week*, with this change: there the nonintentionality of Thoreau's flute playing and the naturalness of the flute's sounding—a cry that supplements the sounds of trees and animals—will be ascribed to Orpheus and his lyre: "Orpheus does not hear the strains which issue from his lyre, but only those which are breathed into it; for the original strain precedes the sound, by as much as the echo follows after; the rest is the perquisite of the rocks and trees and beasts" (W, 340).

This switch—having the lyre release a "natural sound," or a cry like a flute—is, I am suggesting, crucial, as it signals not only that Thoreau is intervening in a long dispute between lyricist and flautist but that, in so doing, he is also revising what constitutes poetry.

In the ancient tradition to which Thoreau's paragraph refers, Orpheus—the "antibarbarian [who] constitutes the civilization of the lyre"[8]—is in contrast with the flautist whom Plato famously banned from the city for resisting the logos and thus the "perfect" or lyric song that expresses it. The perfect song is "composed of three things, the words, the tune, and the rhythm," arranged in such a way that "the music and the rhythm must follow the speech."[9] The speech *(logos)* or the articulation of meaning in discourse thus comes to define song, whereas the music and rhythm are rendered accidental to the *logos*, which must keep them in check. This privilege given to logical speech will eventually transform song into poem, turning poetry into contrivance that both forgets and commemorates the singing body inhabited by rhythm.

If the lyre becomes "the most authentically Greek instrument,"[10] incomparably preferable to the flute *(aulos)*, it is precisely because it allows for the existence of Plato's "perfect song" of the *logos*. The lyre becomes the instrument of lyric poetry, because in simply following the poet's speech it allows for the consolidation of linguistic meaning into form. In contrast, the flute replaces the poet's speech with breath, transforming meaning into a sound that is closer to the literality of voice than the ideality of words. Because it sounds on the outskirts of the *logos*, the flute is said not to address reason but to "tame rats and charm snakes"[11] and is associated with the hubris (mania/mourning, *Appassionato/Lamento*) that is routinely suspected of deteriorating into disharmony. Plato, for example, explicitly identifies the flute as worse than "many-stringed and polyharmonic" instruments, because its sound dissolves into the incongruence of the "panharmonic."[12] The sound of the flute is thus deployed along a fragile line that delineates the coherence of composed speech from a meaningless sonorous materiality, which is why in the classical Greek world it is "constantly accused of blending too perfectly with wailing" or groaning, verging on the literality of a cry, becoming a vocalization of grief that refuses language as a substitute for pain, and resisting the lyric figuration that controls grief by manufacturing its closure.[13]

Why is it, then, that in the passage from *A Week* that I quoted, Thoreau first affirms Orpheus as a poet of the lyre but subsequently makes Orpheus's lyre sound like his "literal" flute, as if to turn it into a wind instrument played by the breath and thus to admit into the house of the poet the lyre-less cry that even Sappho thought had "no place" there?[14]

Orpheus, the "most famous poet who ever lived," received his lyre from Apollo, but unlike Apollo composed lugubrious songs to bring his wife back from the dead. When Eurydice dies of a serpent bite and is taken by Hades, Orpheus fetches her back by composing lyrical laments of such great beauty that they not only "charmed the ferryman Charon, the Dog Cerberus and the three Judges of the Dead,"[15] but also "soothed the savage heart of Hades," who gave him "leave to restore Eurydice to the upper world."[16] Thus, the myth of Orpheus is about lament composed into song that actually enlivens. But that is not all, for if it were, if Orpheus's lyre had succeeded in transforming Eurydice back into creaturely life—it would be hard to understand why Thoreau changed the powerful lyre into the nonlyric flute. True, Orpheus's lyric possesses an ontological power of enlivening inasmuch as it indeed affects death; it affects Hades's heart so that he lets Orpheus take Eurydice from the underworld under the condition "that Orpheus might not look behind him until [Eurydice] was safely back under the light of the sun."[17] Thus it is an enlivened Eurydice who "followed Orpheus up through the dark passage, guided by the sounds of his lyre."[18] When he reached the sunlight, Orpheus famously turned back—thus interrupting his playing— to make sure that Eurydice also had emerged into the light, "but he looked too soon, since, as she wasn't yet under the sunlight, Hades took her back and Orpheus lost her for ever."[19] In this turn through which Orpheus loses to death the woman he has already restored from the dead, at the precise moment when his lyric comes to a close, is found what Maurice Blanchot called the "essential ambiguity of the figure of Orpheus," an equivocation that will come to haunt the lyric while also determining it. For in looking back to verify that the work of his lyric is accomplished—if through its lament the lyric had finally awoken the dead—Orpheus in fact enacted a closure of his song that determined its form and so annulled the restorative work his lyrics had accomplished. As Eurydice's body disappears into formless darkness, we thus realize, together with Orpheus, that it is the closure of the lyric—something that will become its constitutive feature—that cancels what it is meant to generate and precisely keeps the dead forever lost.

When Thoreau makes Orpheus's lyre sound like a flute whose sound is identified with breath and wind (or with a cry entrusted to animals), or when he specifies that such a cry is "unpremeditated," something that Orpheus didn't intend or compose, he is suggesting something stranger than the idea that producing form is not the poet's task. He proposes that poetry might not have much to do with form at all, being instead related to the poet's capacity to enter what precisely is the dark realm of the unformed. In the same entry where he discusses the naturalness of his flute's sound, Thoreau identifies that realm as the west, the untamed, the non-civil, or the wild: "The best poets, after all, exhibit only a tame and civil side of nature- They have not seen the west side of any mountain. Day and night- mountain and wood are visible from the wilderness as well as the village- They have their primeval aspects-sterner savager-than any poet has sung. . . . Wordsworth is too tame for the Chippeway" (J, 1, 321). But if the Chippewa language—precisely the language about which Thoreau in "House-Warming" says that, strictly speaking, it has no tropes but borrows directly from beings and things, rather than from other words (Wa, 245)—is poetry; that is, if poetry is meant to enable words' access to the night of the unformed (the untamed, "west side of a mountain," not coerced into form), which Orpheus's lyre foreclosed for the Western literary tradition, but which Thoreau's flute wants to reopen, then what he imagines as poetry is, as Jane Bennett suggests, the "inexplicably foreign dimension of anything" that one nevertheless must try to explain.[20]

To abandon the tension between words and sound—as Thoreau does when he substitutes the literal cry of a wailing flute for lyric composition and when he turns that cry into a natural phenomenon that doesn't end—is to do away with what we presume to be necessary for a poem to exist (closure, frame, intentionality of the poetic speaker). The consequence is not just a disorientation of prosody, but of logic more broadly and hence the disturbance of a gathered, logical mind and the self-identity predicated on it. In arguing that the disturbance of prosody is a disturbance of self-identity of personhood, I am guided by Clive Scott's radical understanding of prosody as not only what relates the asemantic elements of meter and rhythm to the semantic units of speech, but as what is also concerned with paralinguistic phenomena (such as "loudness, tempo, degree of stress, tone, intonation"), whose motion generates what Scott calls the "fundamental 'I'" of personhood. On Scott's understanding, such paralinguistic phenomena inhabit meter—meter not being the "originary source" of rhythm—and so come to represent the "foundation of inner experience," an experience to which the mind doesn't have immediate access, instead encountering it only through a formed poem, and in so doing, encountering itself.[21] Prosody is thus a "mental synthesis" rather than a rhetorically quantifiable object. In allowing a self to experience a poem, prosody also enables that self to experience an intimacy with itself, an intimacy of paralinguistic phenomena that constitutes the inner sounding of the "I," which gives the self its existence.[22]

Many nineteenth- and twentieth-century readers of Thoreau's work presumed the same connection between prosody, regulative artistic principles, and the generation of the self and criticized him for failing to develop such a relation. What concerned them in the cases of both *A Week* and *Walden* was the absence of unity among chapters, the contingent order of paragraphs in chapters, the disconnection of sentences in paragraphs, and the multi-vocality in sentences. Thoreau's sentences, it was often claimed, were obscure, cryptic, indecipherable, hence inaccessible to the prosodic self. They were said to confuse subject with object, one with many, time with place, mental with material, in the end defying the reader's effort to detect a sense of taxonomy, order, and meaning. Thus, the *New York Times* reviewer of *Walden* found it a work of "erratic genius . . . apt to confuse rather than arrange the order of things, mental and physical." Elizabeth Barstow Stoddard similarly diagnosed the incapacity of Thoreau's mind to apprehend artistic norms, suggesting "his ideas of beauty" to be "limited," and "the world of art" "beyond his wisdom." Notoriously, James Russell Lowell explicitly claimed that Thoreau was a "sorry logician," incapable of "digesting and assimilating thoughts," confounding them with rhetoric and style, a confusion that resulted in "sophism."[23] Commenting on both *A Week* and *Walden* in 1946, Joseph Wood Krutch diagnosed Thoreau's tendency to fragmentation, his mind's resistance to "chronological order," his incapacity to organize a "united . . . tight system," his radically illogical thinking causing the reader to "logically topple."[24] In 1957 Shanley compared various versions of *Walden*, reconstructing the process of its writing, and opined that readers will "fail to appreciate the structure of *Walden*" as long as they understand it as "organized according to sharply defined and separated topics." No topical clarity is intended in *Walden*, he claimed to prove, since categorical overlapping is precisely what resulted from Thoreau's revisions, so that, to mention only three examples, from version one to version five winter came to represent not the end but the middle of seasonal change, remarks made about visitors became remarks about solitude, names shifted to refer to different experiences.[25] The revisions, Shanley argued, often made particular points—or phrases—clearer but that clarity came at the expense of a "rigorously constructed argument," which was something readers shouldn't seek in *Walden*,[26] for instead of categorical rigor Thoreau aimed to destabilize meanings, rendering them volatile and processual. In accord with this long tradition, Walter Benn Michaels's more recent reading of *Walden* argues that Thoreau's practice of writing that confuses categories and escapes prosodic regulations "'divides the individual,' that is, it divides the reader himself, confronting him repeatedly with an interpretive decision that calls into question his notion both of the coherence of the text and of himself." On Michaels's account, then, the incoherence of Thoreau's writing questions the coherence of the self and frustrates "the search for *Cogito*," which Michaels understands to be precisely what Thoreau wants to find.[27]

Thus, readers concerned about the clarity and coherence of Thoreau's writing—those nineteenth-century "guardians of sanity," as Perry Miller somewhat viciously called them[28]—tend to obscure that the weakening of prosodic features was precisely Thoreau's enterprise, in order not to strengthen the *Cogito* but, on the comtrary, to cancel it completely. For what Thoreau identifies as real or true thought derives from nothing like a unified *Cogito*, because in unifying thoughts according to categorical and logical rules, such a *Cogito* in fact cancels their truth. Thoughts in Thoreau are unintended phenomena, they are said to "visit a mind" (J, 3, 377). Like all other phenomena in the world they are heterogeneous, which allows him to say that "thoughts of different dates will not cohere" (J, 4, 336). Because he does not work from a basis of coherence, the heterogeneity of the real—which includes both beings and thoughts, both things and events—constitutes for Thoreau not contradiction but difference. Whereas contradiction is a logical category imposed by the mind on the real to eliminate difference and to render reality coherent, difference is reality in the process of self-generation or self-differentiation, hence a reality that resists the idealization of contradictions. That is why for Thoreau, even the mind that would try not to generate any logical or prosodic coherence but instead would simply record contraries and antagonisms without trying to dispel them, would still risk producing an abstraction that negates truth: "There is a littleness in him who is made aware of antagonism. The opposition I feel reproaches me" (J, 1, 438). This suggests that, to exist, prosodic coherence would have to be generated through the negation of reality, a reality in which coherence is nowhere to be found. And since prosodic coherence is nothing but an abstraction, a self enacted by prosody will is also be an abstraction, a formalistic contrivance generated by a preexisting idea of continuity and coherence. That is why for Thoreau there is no self that is not already "styled": "to study style in the utterance of our thoughts is as if we were to introduce Homer or Zoroaster to a literary club" (J, 1, 189). To the extent that a formed self is stylized by the prosodic idea of it, it is also always aestheticized. Persons are like so many styles of being, each reflecting a certain literary taste.

It is this abstracted, aesthetisized personhood that Thoreau wants to do away with in his "poetics," just as he sought to deactivate it by means of his critique of musical harmony. For in the same way that he considered any dissonant and contingent series of sounds to be music, so he believes that any disorganized assemblage of thoughts constitutes a poem: "The very thrills of genius are disorganizing" (J, 1, 109). The greatness of the poem that Thoreau has in mind lies in its disorganized heterogeneity, which means its objectivity, an objectivity that results from an annulled personhood. That is how we should understand the truth that is referred to in a long *Journal* entry written on February 5, 1854, from which a famous sentence from *Walden's* "Conclusion" is taken: "I fear only lest my expressions may not be extravagant enough-may not wander far enough beyond the narrow limits of our ordinary insight . . . so as to be adequate to the truth. . . . From a man in a waking moment-to men in their waking moments. Wandering toward the more distant boundaries of wider pastures" (J, 7, 268). Thoreau's truth is here conceived of not epistemologically—something resulting from conceptual mediation—but as an incoherency external to the mind, which that mind can enter ("wander toward") only by unbounding its "narrow limits," abandoning its prosodic form, going heterogeneous itself, becoming adequate to the nonabstracted real.

On Thoreau's account, *A Week* came close to the disorganized real. In contrast to critics for whom its incoherence amounted to its failure, Thoreau himself found it relatively successful on the same grounds. In June 1851 he suggests that a good thing about *A Week* is that it tries to overcome prosody—not to smell of the "poet's attic"—and allows itself to be heterogeneous and unregulated, or, as Thoreau puts it, open and hard to contain: "I thought that one peculiarity of my "Week" was its *hypaethral* character, to use an epithet applied to those Egyptian temples which are open to the heavens above—*under the ether* . . . I trust it does not smell of the study & library—even of the poet's attic . . . that it is an ahypaethral or unroofed book, lying open under the *ether*—& permeated by it. Open to all weathers—not easy to be kept on a shelf" (J, 3, 279). Yet, it seems that Thoreau finds its unboundedness still too prosodic, since he considered neither *A Week* nor *Walden* but rather his *Journal* to be closest to the thrilling disorganization of natural life that he wants poetry to be adequate to. He describes the *Journal*, in a famous entry, as follows:

> A faithful description as by a disinterested person of the thoughts which visited a certain mind in 3 score years & 10 as when one reports the number & character of the vehicles which pass a particular point. As travelers go around the world and report natural objects & phenomena—so faithfully let another stay at home & report the phenomena of his own life. Catalogue stars—those thoughts whose orbits are as rarely calculated as comets It matters not whether they visit my mind or yours—whether the meteor falls in my field or in yours—only that it came from heaven. . . . A meteorological journal of the mind. (J, 3, 377)

The disinterest described here is a way of referring to a self that has been cancelled out. For, as Deborah Elise White explains, disinterest doesn't refer to a lack of interest in certain objects or beings; it is not an aesthetic preference. Instead, disinterest is what "disarticulates—it *disinter-ests* the self from itself."[29] As she explicates, the disarticulation of the self enacted by disinterest results in the "eruption or interruption of sheer being" over the categorical coherence of the mind and cancels the mind in favor of the material externality of Being, allowing for Thoreau's "thrill of disorganization." In the terms of Thoreau's *Journal*, the disarticulation enacted by disinterest also interrupts the self's capacity to establish relations among beings and persons (they remain as unconnected as natural phenomena scattered around the world). That means nothing less than that Thoreau's disinterest interrupts the self's capacity to transform objects into subjective representations, which it would then cohere into its own noncontradictory unity. Such a disarticulation of the self, which cancels that self's capacity to transform what is objective into subjective—the mind doesn't add anything of itself to the phenomena but simply records them as they are—also annuls any capacity to transform what is conscious into something self-conscious. Perceptions and thoughts thus assume the same objectivity as any other being or thing. They are not ideas—images of things—but "natural objects & phenomena" themselves, like the wind blowing through the field, or a meteor falling to the earth. They therefore share the same ontological status with any other thing, closing the gap that separates minds from bodies maintaining them as different substances. As the wind passes through a field, so thoughts, in the same flow with other phenomena, pass through a mind (they "visit it" and then exit it, Thoreau says). And it is by cataloging this objective externality of natural phenomena that the *Journal* comes close to the great poem Thoreau had in mind: not a finely fashioned representation of the real, but instead a dissonant, arhythmical, and indeed, asyntactical real itself—precisely what natural life is in the absence of a self to idealize it into prosodic coherence and precisely what elevates the "natural," unbounded, and crazed *(alogoi)* cry of the flute into the instrument of the poet. Differently put, the *Journal* comes closest to Thoreau's highest poetic statement, because, as he specifies in *A Week*, "the divinest poem is . . . as impersonal as Nature herself, and like the sighs of her winds in the woods, which convey ever a slight reproof to the hearer" (W, 309). This coincidence between the poem and the impersonality of natural life explains why Thoreau's description of the divinest poem uses the same words that he employed in a letter to Emerson to describe perpetual grief. There, it is said that the person would achieve the right to "lament the death of the individual" only by managing to transcend the partiality of lament and to transform it into "a sigh as the wind sighs over the fields" (C, 65); here, "the divinest poem is . . . like the sighs of her winds in the woods." Thoreau thus arrives at a final identification: like the paean, the truest poem that is the *Journal*, coincides with the creativity of natural life in perpetual grief. But this coincidence between the poem and the perpetual grief of nature also affirms what was already suggested in "The Funeral Bell": it is to the poet that the task of enacting perpetual grief will be entrusted. The poet grieves by becoming a very different Orpheus; not the Orpheus of the stylized lyre but the Orpheus of the unhinged lamenting flute, whose sound is generated by the wind of the breath, and which, in Thoreau's phrasing, keeps echoing through the woods, perpetuating itself.

Finally, then, if the *Journal* embodies what counts as a poem for Thoreau, it is because—like the lament of the flute, which is associated in Plato with a hubris that resists closure—the *Journal* rethinks what a boundary is. In her commentary on Emily Dickinson's fascicles, Sharon Cameron suggests that they stage the question of

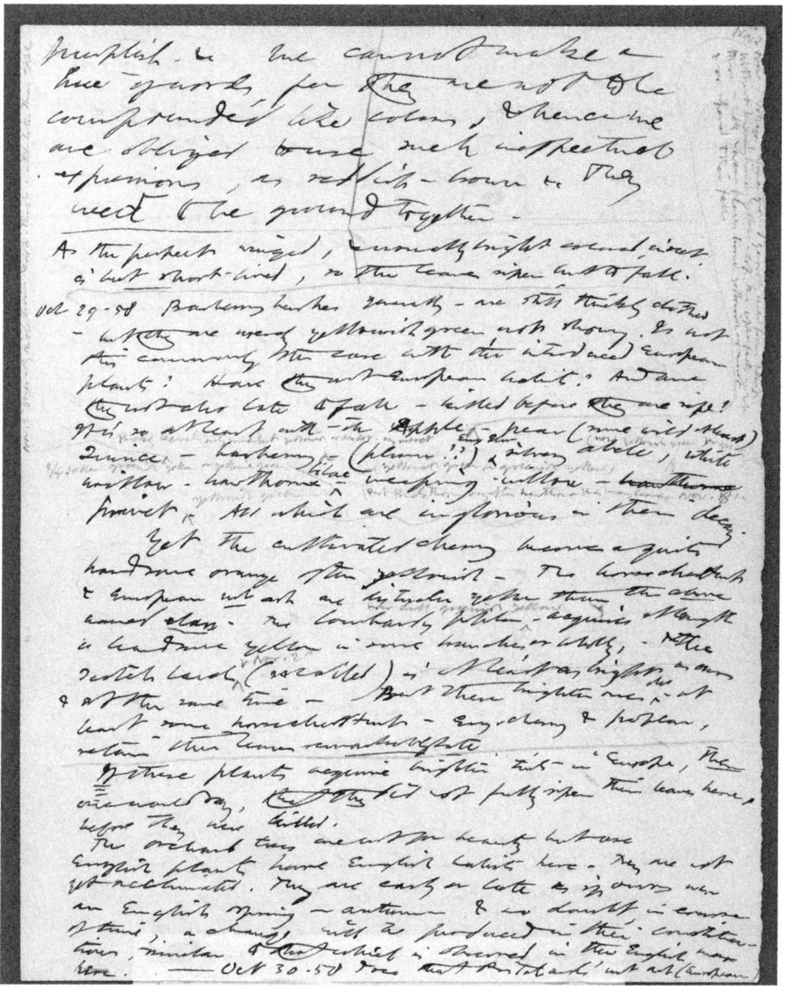

The growth of Journal sheets. Houghton Library Manuscript collection bMS Am 278.5 (15), section 15J. Courtesy Houghton Library, Harvard University.

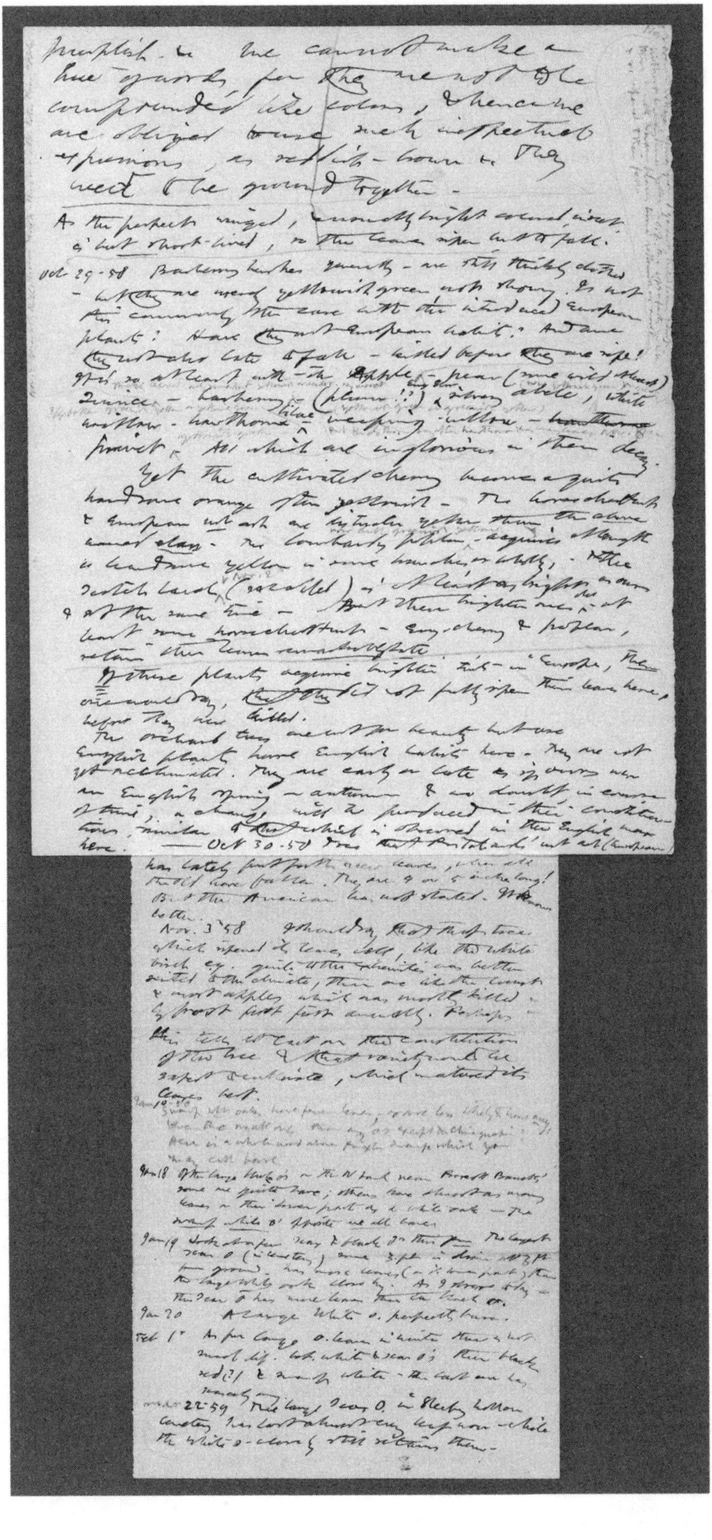
purplish — we cannot make a hue of words, for they are not to be compounded like colors, & hence we are obliged to use such ineffectual expressions, as reddish-brown &c. They need to be ground together —

As the perfect winged, & usually bright colored, insect is but short-lived, so the leaves ripen but to fall.

Oct 29-58 Barberry bushes generally — are still thickly clothed — but they are mixed yellowish green & [illegible]. Is not this commonly the case with the introduced European plants? Have they not European habits? And are they not also late to fall — killed before they are ripe? It is so at least with — the apple — pear (some wild [illegible]) Quince — barberry — (plums!?) silvery abele, white willow — hawthorne — lilac weeping willow — [illegible] privet. All which are inglorious in their decay.

Yet the cultivated cherry become a quite handsome orange & then yellowish — The horsechestnuts & European [illegible] are [illegible] yellow than the above named class. The Lombardy poplar — acquires at length a handsome yellow in some branches, or wholly, — & the Scotch larch (so called) is at least as bright as ours & at the same time — But these brighter ones, at least some horsechestnuts — [illegible] cherry & poplar, retain their leaves [illegible].

If these plants acquire brighter tints in Europe, then one would say, that they [illegible] not fully ripen their leaves here, before they are killed.

The orchard trees are not for beauty but use. English plants have English habits here. They are not yet acclimated. They are early or late as if [illegible] in an English spring or autumn & no doubt in course of time, a change will be produced in their constitutions, similar to that which is observed in the English man here. — Oct 30-58 Does not [illegible] not all (European) has lately put forth new leaves, when all the old have fallen. They are 4 or 5 inches long! But the American has not started. It knows better.

Nov. 3-58 I should say, that those trees which ripened their leaves well, like the white birch e.g. quite [illegible] were better suited to the climate, than are the [illegible] & most apples which are mostly killed by frost first [illegible] annually. Perhaps this tells [illegible] on the constitution of the tree & that variety would be [illegible], which matured its leaves best.

Jan 18 Of the large black o's on the N bank near Brooks [illegible] some are quite bare; others have almost as many leaves on their lower parts as a white oak — The [illegible] white & [illegible] are all bare.

Jan 19 [illegible] scarlet & black o's this year. The largest [illegible] O (in [illegible]) some 3 ft in diam [illegible] from ground — has more leaves (on its lower parts) than the large white oak close by. As I [illegible] — the [illegible] O has more leaves than the black os.

Jan 20 A large white o. perfectly bare.

Feb 1st As for large o. leaves in winter there is not much dif. bet. white & scar o's then black red (?) & [illegible] white — the last are by [illegible] any

March 22-59 The large scar O. in Sleepy Hollow [illegible] has lost almost every leaf now — while the white o. close by still retains them —

boundedness, which the form of a lyric poem thematizes. Dickinson, Cameron argues, had difficulty imposing a limit on her poems and instead opened them onto variants, which, in destabilizing the poem's frame, functioned as a kind of exteriority that was nevertheless constitutive of the poem's identity. But the variant—as the linguistic growth of a poem out of itself, or as the poem's semantic multiformation—is then tamed into a boundedness or entity by being organized in a fascicle governed by certain principles (hence, the preparation of a fascicle is, as Cameron puts it "the preparation of an entity").[30] Thoreau's *Journal*, however, closer to Dickinson, not only often cancels the prosodic coherence of a sentence or a passage, but also enacts its decomposition by means of a variety of diacritical marks, from single or double dashes to other, less obviously meaningful marks (+, ++) as well as small or large blank spaces between words. More radically, it even challenges the physical boundedness of the material notebook or sheet of paper, as if coercing it into transcending its form. In that context I am not referring to the trivial fact that Thoreau sometimes ripped apart the notebooks of the *Journal*, but rather to his more significant practice of using the pages so ripped out as a part of a new, "polymorphous" binding, bringing into proximity passages that are thematically incoherent. For instance, Thoreau glued cuttings from his other notebooks, on the same topic or on a different topic over the early *Journal* entries on autumnal tints. The entries thus form strata, tied into a strangely unformed physical heap that, unlike a Dickinson fascicle that obeys its own material dimensions, grows out of the *Journal's* interiority, turning two-dimensional writing into a three-dimensional body. In addition to the process of constructing a new notebook as a glued pile of sheets, a leaf from one *Journal* notebook could always be attached not *in* another notebook, but *to* the lower or upper parts of its leaves, mutating the formats of two leaves into a new body, producing a long strip, that potentially extended the word indefinitely into space.

In Thoreau, then, piles of words generate material growths, similar to the vegetal galls that will so preoccupy him throughout his life, and the bound form of the *Journal* was abandoned in favor of polymorphous "increase." That is precisely the geological vocabulary that Thoreau uses to describe the process of "journalizing" in an early entry, comparing it to the stratification of the earth that ends in the actual augmentation of matter: "Let this daily tide leave some deposit on these pages, as it leaves sand and shells on the shore. So much increase *of terra firma*" (J, 1, 151).[31]

If Thoreau's *Journal* is an exercise in "divinest" poetry, as I am suggesting, it is not only because, in their syntagmatic proliferation, his notes evade both the regulations that govern poetic form and sometimes even logical discourse, thus realizing his idea that the "highest condition of art is artlessness." It is also because in letting compiled words engender a growing corporeality, his *Journal* moves finally toward making language indiscernible from matter. In that sense it can indeed be called a "book of life." For it is neither an aesthetic representation of life's meaning nor a book of extraordinary experiences—it is not Whitman's book of specimens—but instead an endless network of "natural phenomena" (things, beings, thoughts, and sensations, which are all equally exterior to any internal mental experience), contingently proliferating, moving, and connecting to create anew. The *Journal* thus becomes not the *representation* but the *operation* of the natural; it does what nature does: it creates corporeality, thus achieving precisely what Thoreau wanted perpetual mourning to perform. The recording of the *Journal*—the fact and process of its keeping—thus becomes Thoreau's lifelong performance of the poetics of perpetual mourning.

The Sound Dream

On October 26, 1851, Thoreau had a dream:

> I awoke this morning to infinite regret. In my dream I had been riding-but the horses bit each other and occasioned endless trouble and anxiety & it was my employment to hold their heads apart. Next I sailed over the sea in a small vessel such as the Northmen used-as it were to the Bay of Funday & thence over land I sailed still over the shallows about the sources of rivers toward the deeper channel of a stream which emptied into the gulf beyond.
>
> Again I was in my own small pleasure boat-learning to sail on the sea-& I raised my sail before my anchor which I dragged far into the sea- I saw the buttons which had come off the coats of drowned men- and suddenly I saw my dog-when I knew not that I had one-standing in the sea up to his chin to warm his legs which had been wet-which the cool wind numbed. And then I was walking in a meadow-where the dry Season permitted me to walk further than usual-& there I met Mr Alcott-we fell to quoting & referring to grand & pleasing couplets & single lines which we had read in times past-and I quoted one which in my waking hours I have no knowledge of but in my dream it was familiar enough- I only know that those which I quoted expressed regret-and were like the following though they were not these-viz-
>
> "The short parenthesis of life was sweet"
>
> "The remembrance of youth is a sigh." &c

> It had the word memory in it!! And then again the instant that I awoke methought I was a musical instrument-from which I heard a strain die out-a bugle-or a clarionet-or a flute-my body was the organ and channel of melody as a flute is of the music that is breathed through it. My flesh sounded & vibrated still to the strain- & my nerves were the chords of the lyre. I awoke therefore to an infinite regret- to find myself not the thoroughfare of glorious & world-stirring inspirations-but a scuttle full of dirt-such a thoroughfare only as the street & the kennel-where perchance the wind may sometimes draw forth a strain of music from a straw. . . .
>
> I heard the last strain or flourish as I woke played on my body as the instrument. Such I knew I had been & might be again-and my regret arose from the consciousness how little like a musical instrument by body was now. (J, 4, 154–155)

The dead are summoned by Thoreau's dream out of the unstable element of water that hosts them. The dreamer's boat streams down rivers he once sailed on with his brother. This time, however, he goes farther than he went with his brother, toward the "deeper channel of a stream," leading him into the "gulf beyond." Sailing on the waters of life might be pleasurable ("again I was in my own small pleasure boat-learning to sail on the sea"), but it uncannily takes the dreamer "far into the sea" of drowned men, requiring him to navigate his boat through the manifold flotsam of the drowned. In the strange and fragmented logic of this dream, the material flotsam of the dead is related to the dog that appears to enact a fabulous enlivening, allowing the same possibility for the other dead too. For the dog that the dreamer accepts, but doesn't know as his, is either dead or on his way to being dead (his legs numbed, his body frozen). Its appearance in the waters of the dead is thus logical, yet it is in this water of the dead that its benumbed body is restored to the warmth of life ("I saw my dog . . . standing in the sea up to his chin to warm his legs"), which suggests to the dreamer that the revivifying heat of life is also found there where the dead are housed.

Such a magical vicissitude whereby a relic becomes a living body is repeated in the second part of the dream as commemorative sign is turned into living breath. The dreamer is now on dry land, walking in a meadow where he meets Alcott, with whom he starts a memory game. Alcott and the dreamer entertain themselves by reciting poetry they had read in "times past." And just as the "small boat" from the first part of the dream was at first pleasurable, but eventually sailed the dreamer into the water of the dead where he encountered a frozen dog he couldn't recognize, so the "couplets & single lines" the dreamer recites in the second part of the dream are "pleasing," until the pleasure is interrupted by the sudden appearance in the dreamer's mind of a quote he utters but doesn't recognize. Hence, the poetic line of the second part of the dream assumes the role performed by the dog in the first, which makes poetry Thoreau's dog of metamorphosis. The line the dreamer recites but doesn't recognize expresses an unknown person's regret over lost life ("the short parenthesis of life was sweet"). In giving voice to a dead other—assuming his personhood by saying "I,"—the dreamer enlivens him and once again, that takes place as if literally. For by being voiced in the dream, the word composing the line causes the dreamer's body to vibrate, channeling motion into it: "I was a musical instrument . . . a flute-my body was the organ and channel of melody as a flute is of the music that is breathed through it. My flesh sounded & vibrated still to the strain—& my nerves were the chords of the lyre." In this uncanny representation of Thoreau's theory of poetry, the dream turns poetic utterance into the cry of a flute diffusing its breath into dormant bodies, and the lyre becomes something whose fabulous immediacy moves flesh and nerves, enacting, as does a flute—indeed, indistinguishably from the flute—the corporeal vibrations of life. As the dreamer awakes—still inhabiting the demonic interval between dreaming and waking reality—he glimpses the truth of his dream: he feels his body still vibrating with the rhythm of the oceanic or acoustic waves that, in the dream, gave to both living and dead a new life.

Coda: Melville's Seafowls

The ocean on which Thoreau's dreamer finds himself, a place "beyond" where vitality restores itself, is not only an oneiric site. In Thoreau's thinking the ocean figures as an actual seat of life. Summoning sources as diverse as Homer and Darwin, a series of texts edited by Sophia Thoreau and posthumously published in 1865 as *Cape Cod* details how the ocean blossoms into living beings. If the ocean can't bury "the bones of many a shipwrecked man" (CC, 14) and doesn't preserve vestiges of wrecks, it is because it doesn't bury anything; it doesn't abandon things to a deadly isolation, but instead moves and mingles them into new connections. The ocean is, as the archaic Greeks believed, a creative chaos, generative of life: "before the land rose out of the ocean, and became *dry* land, chaos reigned; and between high and low water mark, where she is partially disrobed and rising, a sort of chaos reigns still" (CC, 54–55). When Thoreau writes that the creative oceanic chaos "still reigns," he has in mind modern scientific insights, which argue, in accordance with ancient mythology, that oceanic water is precisely what gathered together and sustained the first living being: the earth. As Thoreau details, quoting Edouard Desor's treatise on "The Ocean and Its Meaning in Nature," life first progressed from oceanic waters to generate dry land: "the dry land itself came through and out of the water." That is why the ocean is not "unfruitful" but should be "more truly called, the 'laboratory of continents'" (CC, 100). But the ocean doesn't generate only the earth. It brings all life into being; as Desor puts it in Thoreau's quotation, "modern investigations merely go to confirm that great idea that the Ocean is the origin of all things . . . the principle seat of life." As Darwin noted and as Thoreau also finds via

Desor, the ocean is inhabited by forests so thick, so abundant with life, that in comparison "our most thickly inhabited forests appear almost as deserts" (CC, 99–100). The ocean is the laboratory of continents, because it is the laboratory of life, a bricolage reassembling beings into new beginnings and hence a place in which nothing properly ends, but where everything is rescued from its deadened past and redestined into a future.

Cape Cod also registers what kind of epistemology this oceanic vitalism requires: neither scientific nor philosophical but, in all its fabulousness, completely ordinary. For it was related to Thoreau by the Cape Cod seamen who, like Greek poets and French scientists, not only knew the ocean to be the seat of all life but additionally believed—and acted in accordance with such a belief—that certain creatures had the special capacity to see through the oceanic workshop. As Thoreau records, Cape Cod mariners believed seafowls to be capable of discerning what was in the making in the contingent depth of the ocean and then, through their cries, of communicating to humans the change that was about to emerge on the surface, the impending transformation that, announced from the human point of view, amounts to foretelling the future. Only experienced seamen, the mariner sorcerers, who through years of watching learned the art of augury, were able to read the text left in the sky by seafowls: "It is generally supposed that they who have long been conversant with the Ocean can foretell, by certain indications, such as its roar and the notes of sea-fowl, when it will change" (CC, 98–99).

But during the years Thoreau visited the Cape and gathered his notes, it was in fact Herman Melville who truly elaborated the mariner's epistemology of seafowl augury. Seafowls fly through many of Melville's narratives to signify obscurely and ominously the future the ocean is in the process of inventing and that it is about to deliver to humans. In possession of this awesome knowledge they constitute, in "The Encantadas," the "aviary of Ocean . . . hermit birds . . . familiar with unpierced zones of air," so that to study the natural history of "strange" and "unearthly" seafowl is to come to understand the "demoniac din . . . of the eternal ocean," to pierce through the mystical origin of life.[1] In *Mardi* these unearthly birds reflect in their eyes the strange faces, "upward," as if drowned, visible in the depths of the ocean, faces that, despite their outlandishness, are suggestive of new hybrid fusions of "two souls" and offer to the observer ominous but undecipherable images of the formation of new life.[2] In "Benito Cereno" a "troubled" and "strange fowl" in an ecstatic trance that Melville's narrator calls "lethargic, somnambulistic" flight, comes to Delano to foretell the troubles awaiting him upon meeting a "strange sail" coming his way with resurrected slaves for passengers. Unlike most of Melville's experienced mariners, Delano doesn't know much about the art of augury and so fails to avoid the fatal encounter.[3] In *Moby-Dick* seafowls are the only witnesses—excepting Ishmael of course—of the *Pequod*'s catastrophe accurately predicted by so many omens and prophecies, affirming that it is to the fabulous epistemology of the seamen, which the moderns dismiss as superstition, that Melville entrusts the possession of truth. For as the *Pequod* sinks into the oceanic depths, small fowls fly "screaming over the yet yawning gulf" so that when "all collapses," and "the great shroud of the sea" continues to roll on "as it rolled five thousand years ago," only the fowls will retain the vision of the unspeakable catastrophe that has just occurred, a photographic snapshot in their eyes, turning them into living memorials existing for other marine sorcerers to interpret.[4]

But it is in *Billy Budd* that Melville—in a gesture that shows his ontologies to be as vitalistic as Thoreau's—entrusts to seafowls not just the role of witnessing the life that the ocean is busy revising (and of translating the seen into an obscure and ominous message for the seer), but also the more active role of intervening into death to rearticulate it as life. Wondrous if not miraculous things happen in *Billy Budd* after Billy is executed and about to be buried in the ocean. For the "emphasized silence" that followed Budd's execution was "gradually disturbed by a sound not easily to be verbally rendered," an "inarticulate," "muffled murmur" "dubious in significance." And even though this inarticulate human murmur is from the first moment heard fused with the "ominous low sound" of shrieking sea hawks, the narrator—or Captain Vere, it is hard to tell because the two points of view collapse—nevertheless decides to interpret it as human only, as the "capricious revulsion of thought or feeling such as mobs ashore are liable to," a sound that possesses the crew and divines rebellion. To preempt any revolt the order is issued to the "sea undertakers" to slide Budd's "canvas coffin" into the water.[5] As that is happening, the sound of the inarticulate murmur is again heard, only this time there is no mistaking that it is a compound of human and aviary voices. Now, the human murmur is caught in the moment of becoming the shrill cry of a seafowl:

> When the tilted plank let slide its freight into the sea, a second strange human murmur was heard, blended now with another inarticulate sound proceeding from certain larger seafowl, who, their attention having been attracted by the peculiar commotion in the water resulting from the heavy sloped dive of the shotted hammock into the sea, flew screaming to the spot. So near the hull did they come, that the stridor or bony creak of their gaunt double-jointed pinions was audible. As the ship under light airs passed on, leaving the burial spot astern, they still kept circling it low down with the moving shadow of their outstretched wings and the croaked requiem of their cries.[6]

Two very different epistemologies are at work in this scene. To the enlightened narrator voicing the stance of the moderns, the form of the sea fowls "suspended in air, and now foundering in the deeps" represents nothing but "mere animal greed for prey." According to his savvy rationalism the bird is a scavenger, hungry for dead flesh. He ridicules the sailors' rude epistemology, "superstitiously" regarding the crying bird diving into the ocean after the dead body as some "prodigy": "to such [superstitious] mariners the action of the seafowl . . . was big with no prosaic significance."[7] In contrast, the seamen witness how their human grief, mixed with anger, and the feeling of injustice is transformed into a cry of birds, taken up by the birds and dissipated by them. What they ascribe meaning to is how the cry of grieving nature, the inarticulated "requiem" of the birds, follows the departed into the oceanic depths. In an ontology that echoes what Thoreau develops in the years following John's death, and which he later encounters among seamen on Cape Cod, this means: when the bird flies into the water, the oceanic chaos that generates new life is confused with the materiality of the grieving cry, so that both become one animated flow. What that meant for the future only the augurs among the seamen could tell.

Part II

Cambridge, Massachusetts, circa 1837: The Science of Life

Introduction: Harvard Vitalism and the Way of the Loon

Following John's death in 1842 Thoreau resumed the translation of Aeschylus's *Prometheus Bound*—which he had started in 1839 but put aside—publishing it in *The Dial* in January 1843. In one of the most extensive engagements with the translation, Robert D. Richardson Jr. has argued that the play's importance for Thoreau's thinking lay in its promotion of bravery understood less as a heroic "rebellion against the rational, imperial authority of Zeus," than as the "moral courage [to] grasp something, harden it, and make it thereby one's own."[1] But *The Dial's* note prefacing the translation suggested that the tragedy was less about bravery than about the "mystical" representation of "human life."[2] In emphasizing that the tragedy was about the origination of human life, Thoreau therefore accurately evoked Prometheus as a figure who, in the variety of formulations from Hesiod to Plato and Aeschylus, revives an earth devitalized by the gods, bringing life back to humankind. Prometheus is a vital force that restores life.

Modernity has transformed Prometheus into a hero who, in stealing fire from the gods, brings technology and crafts to still natural humans, thus cultivating them into the structure of artificiality. It is, however, less often emphasized that the gods hid the fire from humans as only one in a series of acts meant to keep life from them and devitalize the earth, hiding "cereals" being another act toward the same end. As Vernant explains, according to the oldest version of the Prometheus myth preserved by Hesiod's *Works*, "the gods have hidden *(krupsantes)* men's life, *bion*, from them, that is, food in the form of cereals," which are "'cooked' plants as opposed to raw grasses that grow of their own accord," hence, what nourishes life, giving humans their "livelihood."[3] Fire was therefore also hidden to prevent cooking and eating. That is why Vernant can say that "'to hide [men's] livelihood,' or cereals," and "'to hide the fire'"—which are two phrases Hesiod uses in the Prometheus myth to explain what it was that the gods hid from humans—are but "two aspects of a single operation," meant to exhaust the vitality of earthly life.[4] To steal fire back from Zeus, as Prometheus does, is thus less about technologizing humans than about enlivening them by feeding them; or, in ontological terms, it is about making the powers of "force" and "life" *(Kratos and Bia)*—which together generate formed life and stand on Zeus' side against humans—come to the side of the terrestrial and so turn the earth into a force capable of engendering life.

In the years during which Greek translations offered Thoreau mythological thinking concerning life's capacity for transmutation and literal continuity, contemporary science was entertaining similar ideas. The mythological earth revitalized by Prometheus finds itself strangely echoed by contemporary geology, mineralogy, chemistry, biology, pathology, and even medicine. In *The Poetry of Science*—published in 1848 and read by Thoreau in 1851—Robert Hunt, scientist and secretary to the Royal Cornwall Polytechnic Society at Falmouth, summed up the contemporary scientific stance on the nature of matter, claiming that "transmutation" and "conversion" were understood to constitute universal "natural operation."[5] Summoning scientists as diverse as Lyell, Faraday, Lamarck, and Mohl, and corroborating his discussion with clusters of quotes from contemporary European scientific journals, Hunt argued that the science of the nineteenth century accepted as certain that "matter . . . is ever in a state of change," which is not mechanical but "vital," so that "death is but the commencement of a new state of being."[6] Animation, alteration, and transformation are inherent not just to vegetal, animal, or human life but to all phenomena typically considered as inanimate. Metal, rocks, stones, and dust particles were all considered to be vitally animated: "[this grain of dust] quickens with yet undiscovered energies; it moves with life: dust is stirred by the mysterious excitement of vital force; and blood and bone, nerve and muscle, are the results."[7] On Hunt's understanding, then, science around 1840 was of the opinion that the division of phenomena into organic and inorganic didn't coincide with the distinction between animate and inanimate, because everything is animated. The difference between organized and unorganized beings is not animation but complexity; organized beings—those with "nerves, muscles, bones and blood"—occur as a result of a more complex arrangement of simpler material and vital particles. According to Hunt's summary, then, the radicalism of new science consisted precisely in transferring "vital energies" from the realm of incorporeal "divine emanation," from which they were traditionally thought to derive, into "materialities by which they are themselves controlled." "The metaphysical hypothesis, which resolves all matter into properties, and refers all things to ideas," is abandoned by contemporary science in favor of "powers of life" coincident with the material, rendering materiality mutable and animated.[8] The materialistic vitalism that Hunt diagnoses as pervading the sciences of the 1840s has in fact been emblematically announced for nineteenth-century science by Lamarck's cancelling "dead matter" from the second edition of his *French Flora.* While the 1778 edition recognized the division between organic and inorganic as coincident with the difference between life and death ("If one observes the different beings that compose the interior structure of our globe one will first remark a large number of bodies composed of raw, dead material . . . namely earth, stones, metals, salts"),[9] the 1805 edition still held it raw but not dead:

> If one observes the different beings that compose the interior structure of our globe . . . one will first remark a large number of bodies composed of raw material. Those beings . . . include not only the productions to which the name of minerals is given, namely earth, stones, salts, metals, but also certain substances, in appearance very different from the preceding ones, such as water, air, and the elements themselves.[10]

Thus, at the beginning of the nineteenth century even earth, water, and air came alive.

As Lamarck's cancellation of death in the natural suggests, and as Hunt's overview of the theory of "vital forces" demonstrates, behind the specific questions raised by particular sciences there loomed a larger and properly philosophical interrogation into what life is and what constitutes it. In what follows I argue that this question is precisely the one guiding and connecting Thoreau's investment in a variety of sciences. This is suggested by Channing's observation that in his naturalistic investigations Thoreau desired to gain "glimpses at life"[11]—that is, at how life creates, transforms, and proceeds—and is explicitly confirmed by Thoreau himself throughout his writing. For instance, when at the end of *A Week* he argues that there is a life going on "between the lichen and the bark," of which botanists are ignorant, he is maintaining that "all that we call science, as well as all that we call poetry," being only a "particle of . . . information" of what such a life is, is in the service of a more general theory of life (W, 385).

Following Thoreau's suggestion that particular sciences have to lead toward a comprehensive theory of life, my readings of what are typically classified as Thoreau's naturalistic writings focus on the concept of life rather than that of nature. Although my substitution is not particularly obtuse, since, as Maurice Merleau-Ponty reminds us, the word "nature" comes from Latin "*nascor*," which means precisely "to be born, to live," so that "there is nature wherever there is life,"[12] it seemed to me necessary precisely because the phenomenal and conceptual coincidence of "nature" and "life" has been obscured by centuries of aestheticization of the natural, rendering it the object of beauty, generator of the sublime, concretization of what the mind graduated from long ago; or, in its pastoral version, the homely site, from which humans were expelled, foolishly entrusted to their own strategies of self-cultivation. But my substitution of life for nature is not meant as an etymological remainder alone. To turn life, rather than nature, into a central notion of Thoreau's thought allows one to engage his theory of the living without having to evaluate its (post-Darwinian) scientific correctness. When the notion of life comes to replace that of nature, the dilemma of whether Thoreau's naturalistic writings are meant as scientific treatises (achieving, as so many excellent readers had demonstrated, only modest results) or as a poetics of science can be left aside. Instead, an astonishing philosophy of life—with its regime of truth that is different from either the scientific or the literary—emerges. Thinking about what life is can take clues from science but need not be itself scientific; instead, like any question that raises the question of being it must be considered as properly philosophical. As Foucault pointed out, discussing changes in the science of biology at the end of the eighteenth and the beginning of the nineteenth century: "the notion of life is not a *scientific concept;* it has been an *epistemological indicator* of what classifying, delimiting and other functions had an effect on scientific discussions, and not on what they were talking about."[13] What Foucault's retrospective study concluded about the notion of life in the nineteenth century was exactly what Emerson articulated prospectively when he argued in 1836 that the sciences should enable us precisely to formulate a theory of life that itself wouldn't be scientific: "All science has one aim, namely to find a theory of nature. We have theories of races and of function, but scarcely yet a remote approach to an idea of creation."[14] For Emerson, then, scientific knowledge of the "functions" of nature should allow for a theory of creation, a philosophy of how life comes to be and what it is. That scientific knowledge should lead to a philosophy of life was, I argue, also Thoreau's belief, such that the question of what life is can be understood to invigorate every empirical observation he makes. The answers that he formulated in response to this question constitute a theory of life that merits the name "vitalism."

Attuned to contemporary scientific vitalistic materialism, Thoreau's vitalism, I argue, differs from the most famous form of it in the West, that developed in the philosophies of Aristotle, Leibniz, and Kant, which understands matter as energized by a network of dynamic forces. In Aristotle, for instance, vitalism is a force *(dynamos)* that, different from inert matter, is the source of all change and movement; as Heidegger sums it up, in Aristotle, "force . . . the origin of change is in something other than the change, which means in a being that is not the same as the one that changes."[15] Animating and transformative forces thus do not belong to matter even if they inhabit it and guide it into a form, thereby generating individuated beings. Immateriality is bestowed on matter as a telos toward which formless material is driven by the forces that move it.

Leibniz revived Aristotle's philosophy of animating forces, insisting, in opposition to Descartes, that "the distinguishing essence (character) of substances" is in forces of "striving, striving against, resisting." These vital forces of "doing and forming"—which Leibniz thinks of as ontological causes preceding and enabling individuation, not as a creature's capacity to be[16]—are, as in Aristotle, different from the matter they enliven; as Heidegger explains, in Leibniz "it is not the individual beings which are endowed with force, but rather the reverse: force is the being which first lets an individual being as such to be."[17] Similarly, in Leibniz life and what lives are not the same, because vitally dynamical forces are not material but are understood as an immateriality that "besouls" matter: "In Leibniz . . . all things are, so to speak, besouled: everything that is, is endowed with force."[18] In those traditions of vitalism, then, life and the forms it creates are rendered immaterial, hence ideal and substantially different from what is embodied.

In this part of the book I argue that Thoreau's notion of life is formulated in complete opposition to idealistic vitalism. It seeks precisely to overcome the divide between life and living, animation and matter, form and what is embodied. But it is also different from the holism and hylomorphism that it is sometimes taken to be. For in contrast to Thoreau's understanding of animation, holism and hylomorphism conceive of formation as taking place within a totality out of which they emerge, and in so doing these theories presuppose an already existing—somehow already formed—entirety of being (Greek "*holos*" means "all, whole, entire, total"). Logically put, the inconsistency of these approaches is in their presuming an individual that itself needs to be individuated. As Muriel Combes explains, hylomorphism thinks of "being on the model of the One and thus, at some level, assumes the existence of the individual it seeks to account for."[19] But that simply means, ontologically speaking, that these approaches can't imagine generation and life outside the model of the one and formed, which remains unaffected by the question of what Combes calls the "reality of being *before* all individuation."[20] Such approaches, then, exclude the possibility of nonindividuated realities and instead seek to establish a principle on the basis of which many individuals emerge out of an already existing one.

Thoreau's vitalism is different in several central ways: how it understands the nature of matter, causation, and animation; how it negotiates the tension between form and generation; how it understands teleology; and finally, how it articulates life's relation to death.

Animated matter, pantheism, univocity. In Thoreau life individuates but without ever organizing itself into a closed whole, and thus it never itself becomes an individual, single totality. The name that Thoreau gives to this process of individuation, which unfolds without a hylomorphic reality already being posited as the whole in which such unfolding occurs, is "crystallization." In his "Natural History of Massachusetts" and elsewhere in the *Journal*, crystallization—far from being reserved for gemstones, stalactites, snowflakes, and honey—is presented as the strategy of all individuation:

> This foliate structure is common to the coral and the plumage of birds, and to how large a part of animate and inanimate nature . . . vegetation is but a kind of crystallization . . . but as in crystals the law is more obvious, their material being more simple, and for the most part more transient and fleeting, would it not be as philosophical as convenient, to consider all growth, all filling up within the limits of nature, but a crystallization more or less rapid?[21]

As Anne Sauvagnargues explains, crystallization is the concept that, in science and philosophy equally, "combats . . . the hylomorphic schema" by accounting for individuation as a "modulation" occurring out of its own movement: "[the crystal] grows in every direction within its pre-individual milieu, each already formed layer serving as the structuring basis of the next. . . . Crystallization manifests the appearance of . . . structures in the process of becoming."[22] It is precisely that meaning of crystallization as incessant becoming out of preindividuated reality that Thoreau presumes in "Natural History of Massachusetts." Crystallization is there claimed to be so apposite to his philosophy precisely because it is a rapid individuation spreading in all directions, generating, through its own folding and unfolding, forms on the move—of corals, birds, and foliage—but in such a way that each generated form is only "transient and fleeting"; or, as Thoreau also puts it, it is "migration . . . without intermission" (J, 1, 119). That phrase signifies life's capacity to reform and mold itself rather than, as in the vitalisms of hylomorphism and dynamism, adding stable form to inert matter through the mediation of vital forces.

But if in Thoreau forms are not idealized molds clapped onto immobile matter, it is because matter itself was never lifeless. Instead it is always vitalized, life being indistinguishable from the incarnated and embodied. Thus, in Thoreau everything commonly classified as inanimate and inorganic—stones, rocks, water—is rendered sensitive. Everything becomes alive even if not all life is organic. Hence his question: "Is it not a satire-to say that life is organic?" (J, 1, 392). For him, the inorganic lives and is afforded the capacity for organization, as Thoreau witnesses when, passing through a deep cut on the railroad, he registers "material sand" becoming a "hybrid product," turning itself into a grotesque vegetation that goes on to become a plant. The identity of animation and incarnation described in the passage from *Walden* where clay grows into a plant, to which I will return, is not, of course, an affirmation of animism, for animism precisely presupposes a substantial difference between spirited life and inert matter; in animism, if matter moves, it is because it is moved by what it is not—the powers of spirits merely hosted by matter. Instead, as I argue in this part of the book, Thoreau's identity of animation and incarnation is an affirmation of a different kind of materialism that, as Jane Bennett puts it, "eschews the life-matter binary . . . doesn't believe in . . . spiritual forces [but] nevertheless also acknowledges the presence of an indeterminate vitality."[23]

The more traditional name for this vital materialism is "pantheism," which Thoreau wants to revive in the context of new scientific developments. Playing on the name for the god of fauna, "Pan," he seeks to universalize the natural across theistic activity: "Pan is not dead, as was rumored" (W, 65), and with which he identifies his philosophy ("In my Pantheon, Pan still reigns in his pristine glory" [W, 65]). Thoreau's pantheism, indeed his whole project, is sometimes read as the deification in the sense of spiritualization or idealization of the natural. But the opposite is true, because pantheism declares the identity of God and nature to be the identity of substance (*Deus sive natura* in Spinoza's famous formulation), which makes the natural or material the very substance of God. The substantial identity of divine and material—what medieval proto-pantheists called "univocity"—means that there is no real, hence no categorical, difference between the transcendent being of God and corporeal nature, just as there is no categorical difference between "being" and beings. As Gilles Deleuze puts it in reference to the medieval philosophies that preceded Spinoza's pantheism: "being is univocal, this means: there is no categorical difference between the assumed senses of the word 'being' and being is said in one and the same sense of everything which is. In a certain manner this means that the tick is God; there is no difference of category, there is no difference of substance, there is no difference of form. It becomes a mad thought."[24]

Thoreau expresses explicitly this substantial coincidence of the divine and the material when in *A Week* he insists that we must "speak . . . in the same sense . . . of 'the Earths' and [the] Father who is in them" (W, 382). In the manner of medieval philosophies of univocity and Spinoza's pantheism, in Thoreau also the earth and God are said in one and the same sense to signify what is material and embodied and to claim an immanently material creation. For—and this is what joins vitalism to pantheism—holding that nature is substantially God means affording the power of self-generation and organization to the material. In pantheism the supernatural does not cause the natural while being transcendent to it, but rather the natural becomes its own cause. As Eugene Thacker explains, "this kind of pantheism not only entails the negation of the divine, but it also entails a radical distribution of the divine, such that it cannot be separated from the earthly, or even the material (thus the horizon of pantheism is materialism)."[25]

Yet, if this vitalism cancels any difference in substance between ideal and incarnated, it doesn't abandon differences among individual phenomena, turning them into pure flow and turning life into the amorphous Dionysian drift of which Nietzsche spoke, in which transformations finally overcome all individuations.[26] In a passage from *A Week* that I have already discussed, Thoreau explains our senses as a type of material germ, "rudiments" of what they will develop into, imagining that future development as more refined matter capable of the extraordinarily sophisticated perception of minute and currently imperceptible movements. All life is germinal matter that—without hylomorphically and idealistically preceding individuations—articulates itself into delicate phenomena that are numerically and qualitatively different yet substantially the same. In understanding life as germinal matter, Thoreau is relying on the tradition already established for American thought by Emerson, who also maintained that everything existing was created of the same matter that endlessly refined itself: "the addition of matter from year to year, arrives at last at the most complex forms; and yet so poor is nature with all her craft, that from the beginning to the end of the universe, she has but one stuff,—but one stuff with its two ends, to serve up all her dream-like variety."[27] For Emerson, differences between dreams and bodies, elements and thoughts are qualitatively irreducible while nevertheless being two manners of the same "stuff," so that nature "still goes back for materials, and begins again with the first elements."[28] There is, then, something earthly even in dreams. This tradition of thought will not end with Thoreau but, gaining in complexity, continues with William James, who, using Emerson's exact words, formulates what he calls his "*Weltanschauung*" to mean that "sensible natures"[29]—thoughts, feelings and things, for instance—remain irreducibly different, and accordingly the universe is pluralistic, with no "whole preceding its parts,"[30] while simultaneously being made of the same stuff: "thoughts in the concrete are made of the same stuff as things are."[31] The "stuff," which, once again as in Thoreau and Emerson, empirically appears as the plurality of what is generated, defies the "dualism of being represented and representing," thus also defying the "self-splitting . . . into consciousness and what the consciousness is 'of.' "[32]

Incomplete determination. What holds for life in general holds, in Thoreau, for individuals. Just as life is open ended, never enclosing itself into a holistic totality, so each individual, while being distinctly different from all others, is nevertheless never a closed system or figure. That is possible thanks to another central premise of Thoreau's understanding of life, according to which forms are no more bestowed on matter—as they are for Aristotle—than are vital forces, but instead matter generates them through its own animation. Thanks to this fundamental coincidence of formation and generation, forms are rendered living too, which is why Thoreau imagines them more as outlines than as forms proper. As something living, and as the non-individuated becoming that animates it, each individual form is mobile, ceaselessly differentiating itself into something else. Individuals are never simply created in the stability of their existence but are always in the process of being created; in that way their becoming so often not only brings them to the edge of what they are not but also allows them to experience radical reformation of what they were. As I detail in what follows, in Thoreau very different beings can be fused and a variety of transformations enacted, precisely because individual differences are not conceived of as complete formations. Each being in Thoreau is incomplete and relates to other beings by relations that are themselves objective and vital, and each moves toward becoming something else. Existences are always *in hubris*, always approaching the extreme and in excess of themselves, passing into something else. And it is this "passing into"—the real metamorphosis and ecstasy of beings—that in Thoreau's world makes life beautiful to give and disastrous to leave. Of Thoreau's understanding of life one can thus say what Deleuze says of Nietzsche's eternal return: "when Nietzsche says that hubris is the real problem . . . he means one—and only one—thing: that it is in hubris that everyone finds the being which makes him return."[33]

There is, however, no contradiction in saying that beings can exist as individuated but not completely determined, and so are capable of becoming. As Deleuze explains, complete determination—the idea that a being assumes its identity only if completely distinguished "from everything that it is not"—is a relatively novel logical and philosophical proposition, being "formulated for the first time by Kant."[34] Before Kant, beings were considered to be incompletely determined, that is, it was possible for a being to exist as an individual, as numerically one and qualitatively specific, yet in commonality with other beings. Isolation generated by complete determination was thus something that came later in the Western understanding of what being individuated means. It cancelled the individual's instability and a continuity with other beings that had allowed for their becoming something else. In that sense, much of pre-Darwinian science was pre-Kantian, entertaining ideas of literal metamorphoses. As Gillian Beer explains, "metamorphosis turns out to be a concept as crucial to physiology, geology, or botany, as to myth. Linnaeus published the *Metamorphosis Plantarum* in 1755 and Goethe's *Versuch die Metamorphose der Pflanzen zu Erklären* was published in 1790."[35] Darwin too will turn metamorphosis into a crucial concept for understanding life, but he will conceive of it developmentally, and as irreversible. However, in much of the pre-Darwinian scientific imagination, reversibility is still preserved, so that, for instance in Lyell, extinct species return to earth,[36] while in Jules Michelet all kinds of cross-species mixtures are possible. For Michelet, according to Barthes's summary, there existed "ambiguous zones of process in which flint gives way to grain, then to the Frenchman whom it nourishes; in which plant extend[s] to animal, fish to mammal, swan to woman . . . ; in which the baby's brain is nothing but the milky flower of the camellia; in which man himself can be substituted for woman in the transhumance of marriage."[37]

Thoreau opens *A Week* with an epigraph from Ovid's *Metamorphoses*, thereby announcing an engagement with not only Greek mythology but also with contemporary science. Similar to Lyell and Michelet, Thoreau's scientific imagination is invested in all sorts of transformations that are most famously referenced throughout *Walden* by the loon, whose cunning molting came to emblematize the capacity of all life to mutate. Metamorphosis thus assumed the status of a task that humankind also was supposed to be capable of performing, following the example of the loon (Wa, 24). When there and elsewhere he tasks humans with the obligation to transform, he often means it less materially than as a capacity to enact a change of mind or heart. But no less often he means it disconcertingly literally, as when, for instance, after drinking water from Nut Meadow brook, he wonders whether he has swallowed eggs that will generate a new crossbreed in his stomach: "How many ova have I swallowed—who knows what will be hatched within me? . . . The man must not drink of the running streams the living waters-who is not prepared to have all nature reborn in him-to suckle monsters" (J, 3, 369–370).

Teleology. When Thoreau imagines humans to be capable of hatching fish he annuls the ideology of human exceptionalism and signals one of the major premises of his understanding of life: namely, no living form is more accomplished than another, and life doesn't therefore unfold hierarchically and progressively but, more democratically, moves simultaneously in a variety of directions. Similarly, each of the directions life takes is equally valuable, none leading to a superior or final end, but conversely, each "goal" or condition transcending itself into new orientations. This radically anti-teleological understanding of life was shared by both Thoreau and Emerson, who had himself—in rarely registered opposition to both Kant and Goethe—argued that one can talk about life's "goals" only on condition one understands that life moves "simultaneously . . . throughout the whole body, the equal serving of innumerable ends without the least emphasis or preference to any."[38] The movement of life doesn't describe a hierarchical line but charts an egalitarian network along which beings transform. That there is no "universal end" but a "universe of ends" in the way life proceeds, meant to Emerson, as it did to Thoreau, that no form or goal can ever be imposed as a norm that life should reach. Life in Thoreau therefore remains fundamentally nonhierarchical and non-normative.

Thoreau's politics are typically discussed in light of his civil disobedience and the John Brown essays. And although such an approach is, of course, valid, in what follows I suggest that the centrally important site of Thoreau's politics is in fact his philosophy of life. The question one should ask about Thoreau's understanding of life is not only whether it is scientifically grounded or even rationally plausible, but also what idea of justice and democracy it outlines. In that respect, Thoreau is not unusual, for it is in and through philosophies of life that so many thinkers of the nineteenth century—from Schelling to Coleridge—formulated their political views, and it is in response to such formulations that Thoreau articulates his own ideas.

In 1848 Thoreau read Coleridge's *Hints towards the Formation of a More Comprehensive Theory of Life*, predicated on a rigorous hierarchical ordering of the living, "rising in ascent"[39] from inanimate metals and "various forms of crystals," to geological formations in which the "simplest forms of totality evolved," to the appearance of life in "vegetation or animalization."[40] But against Aristotle—who rendered plants living on the basis of their capacity to nurture and grow, thus turning vegetal life into something like a basic form of life shared by all other living beings—Coleridge introduces more specific criteria. In his theory of life, the realm of the "vegetable and animal Life" are rigorously differentiated, according to the capacity of beings not to grow but to sense. Thus, even though "fungus or a Mollusca . . . the very lowest species" are capable of assimilation and growth, they are judged inappropriate for studying life, for they do not have the capacity that defines it, namely, irritability or sensitivity.[41] The new proposition Coleridge introduced was that growth and nutrition are not criteria of what is alive, for "the mere act of growth does not constitute the idea of Life."[42] Beings might be eating and growing yet not be alive; not all growth is alive, because not all growth is sensitive. The ladder of Coleridge's hierarchies ends with animals, which are both irritable and sensitive, but it does not include humankind. For in his polygenetic imagination, humans do not belong to the life that proceeds from fungi to animals. On the contrary, with humans "a completely new series" of individuation begins, "beyond the appropriate limits of physiology."[43] In Coleridge's strictly teleological understanding humankind does not share the same life with other living beings, not being an animal at all. Humankind belongs to a form of life that is not rooted in matter but is purely spiritual and therefore passes the limits of physiology. With humans, life is transformed from what is corporeal into something ideal, a "true idea of life."[44]

The ideated status conferred on the human—formulated as Amigoni argues, precisely in response to the "materialism of vitalist thought"[45]—doesn't, however, exempt it from teleological ordering. Whereas in vegetal and animal life sensibility marks the highest beings, in the realm of the human, reliance on the senses will be graded as the lowest form of life, signaling a proximity to the physiological, corporeal, and animal. Thus, although all humans are said to be exempt from the animal life, some will be more advanced than others. Advanced humans are removed from the corporeal and consequently "connect with the objects of our senses . . . notions . . . which are due only to Ideas of immortal and permanent Things";[46] in contrast, less advanced humans obtain their notions from the senses and unstable material reality, thus remaining close to the animal, vegetal, and material, existing as "numerous Tribes of Fetish-Worshippers in the streets of London or Paris, as we hear of on the Coasts of Africa."[47] The biological discourse that ranked all humans as exceptional, as beings having realized the goal of the teleological unfolding of life, thus gives rise to an ethnographic imagination that ranks peoples according to their capacities and relegates certain groups to zones of life where proximity to the body, perception, and concreteness renders their human status uncertain.

Far from fashioning his own philosophy under the "significant" influence of Coleridge's "Theory of Life," as Sattelmeyer and Hocks argue,[48] Thoreau in fact completely reverses it. Already after John's death, struggling to understand what life is, Thoreau concludes that the best way to understand it in its human and animal formations is to address vegetal manner: "We can understand the phenomenon of death in the animal better if we first consider it in the order next below us—the vegetable" (J, 1, 368). And only a few days later he will more explicitly claim that the secret of all life is contained in fungi, an insight that will propel his decision to dedicate himself to what he calls the "botany of fungi" (J, 1, 381). Thus, in complete reordering of Coleridge's teleology, rather than considering humans to be "degraded" by the insight that they possess something "in common with oyster and the mushroom,"[49] Thoreau will precisely insist on such a "commonality."

In his theory, the "lowest" and most "degraded" are never overcome but always remain essential to how life operates. For that reason, and even though "at first we are not disposed to believe that man & plants are so intimately related" (J, 3, 234), the "lowest" forms of vegetal life will, in the "Natural History of Massachusetts," be identified as embodying the logic of all life: "Vegetation has been made the type of all growth."[50] And human life will be declared to be essentially one with the vegetal, the "manifest" relation of plants and humans "suggest[ing] a history to Nature-a Natural *history* in a new sense" (J, 4, 6). The new natural history Thoreau wants to write is not the story of hierarchies and the ordering of forms of domination; it is a history of radical, anarchic equalities. It is not a story that degrades the human into the "lowest" as Coleridge understood it, but, rather one that elevates everything to the status of something precious: "The mystery of the life of plants is kindred with that of our own lives, and the physiologist must not presume to explain their growth according to mechanical laws, or as he might explain some machinery of his own making. We must not expect to probe with our fingers the sanctuary of any life, whether animal or vegetable" (Jn, 12, 23).

Death. The statement that life is incessant individuation is twofold. In suggesting individuations to be unstable and always in the process of becoming, it claims, positively, that life constitutes a generous and abundant creativity. But because to become is to undo the being that one was, the statement also understands life negatively as a force of decreation. In my view Thoreau's vitalism is acutely aware of this duality. On his account, life is never a flowering abundance without also being illness, suffering, and decay. Life is never imagined as a force of power, and health is never elevated to the status of the norm; instead, life is always considered to be immanently traversed by disease and always rendered fragile. His vitalism, then, is never a naively optimistic celebration of life that remains unaware of the forces of death and dissolution. When, he talks about life's continuity and opposes it to the transience of death, his statement should be understood as ontological rather than existential. Ontologically speaking, it proposes that life as such is never interrupted, whether by the general catastrophe that Agassiz imagined or by individual deaths. All dramatic changes and all individual deaths are assimilated into it. Yet that doesn't relativize the value of individual lives. Quite the contrary, because individuations are unstable—because everything is an event rather than a fact—every phenomenon becomes in Thoreau unique and unrepeatable; each assumes absolute value. That is why vitalists of Thoreau's sort mourn with a crazed intensity, as I detail in the first part of this book. But it is also why Thoreau's work is so obsessively devoted to the perception of nuances. For everything he sees—thanks to its absolute uniqueness—can be seen only once and always for the last time. Infused with life as they are, all his perceptions are thus facts of mourning, since each perception is not only a joyous encounter with another but also a parting from it.

Yet, even if it doesn't attenuate the disaster that is each death of an individual, Thoreau's claim that life is never interrupted by death does not reduce to a vacant ontological proposition. To the contrary, it changes the way life is affiliated with death and so reshapes the way we relate to death. As Deleuze put it in another context, vitalism is not about the denial of an individual death but about a new dynamic between life and death in which "life screams at death, but death is no longer the all too-visible thing that makes us faint; . . . death is judged from the point of view of life, and not the reverse."[51]

Here I pursue a reading of the archive that corroborates those general claims about Thoreau's vitalism. In so doing I situate Thoreau in the context of contemporary scientific theories of life. In my account of Thoreau's philosophy of life some figures judged by others to be important for his development play a lesser role. For instance, Humboldt resides on the margins of my account, not however because I wish to minimize his importance but because his influence has already been well documented by Laura Dassow Walls's excellent work. Similarly, I rarely and only obliquely gesture toward Darwin, believing the value of Thoreau's writings on life shouldn't be judged on the basis of how much of evolutionary theory they were able to predict. As Sattelmeyer and Hocks correctly observed, much of what has been written on Thoreau as a naturalist "tends to suffer from a sort of post-Darwinian tunnel vision that neglects the lively and vital climate of competing paradigms of natural science that prevailed during the first half of the nineteenth century."[52] In accounting for Thoreau's understanding of life, I thus take into consideration scientific developments in Europe, specifically in Paris and London of the 1830s and 1840s. But above all I concentrate my attention on how those scientific developments were accepted and revised by scientists—geologists, physiologists, biologists, and physicians—closely related to or teaching at Harvard University.

When commentators relate Thoreau's naturalist writings to developments at Harvard it is typically to reference his relation to Louis Agassiz. Agassiz has traditionally occupied the central place in our understanding of natural science at Harvard in the 1840s. This assumption leads to the erroneous conclusion that most geologists and paleontologists were catastrophists like him, and—in radical opposition to the vitalist belief in life's continuity—advocated the theory that life was completely interrupted on many occasions in the history of the Earth. In catastrophist naturalism there was also embedded a theory of polygenesis, with all the disastrous political effects of such thinking. In contrast to that understanding, I maintain that Harvard was a center of vitalistic progressive philosophies at least as of 1824, when John Ware—first editor of the *Boston Medical and Surgical Journal* and the *Journal of Philosophy of Arts*, both of which were publishing reports on contemporary developments in medicine—published William Smellie's *Natural History*, with significant changes that Ware himself inserted. Ware not only renamed Smellie's book *The Philosophy of Natural History* but also omitted Smellie's first chapters on the general theory of life, animals, and plants, substituting for them a long introduction to bring the book in tune with the "progress more recently made in Comparative Anatomy."[53] For at least two decades after its first publication Ware's edition remained the centrally important handbook in courses on the natural sciences at Harvard. It was therefore also the major text Thoreau read when he studied natural history with Thaddeus William Harris in 1837, and a copy of it remained in Thoreau's own library.[54]

In the new introduction that Ware wrote, he formulated a vitalistic philosophy through analysis of vegetal life. Anticipating and influencing Thoreau's philosophy to come, Ware argued that what makes vegetal life so crucially important for understanding life in general is both its nonhierarchical logic ("vegetables vary . . . but you cannot say of one, that it is of a higher order in the scale of creation than another")[55] and its continuity. Ware found proof of the continuity of the vegetal life in trees that annually add "new matter" or "growth" to their bodies, which is then deposited between bark and alburnum, operating there as a "seat of the growth and nutrition" for years to come. Thus, "the growth of one year is only subservient to the circulation of the next, and is ever afterwards of use . . . in giving strength and stability to the trunk, in order to support the increasing size and weight of the branches," which can go on growing endlessly, because plants "do not . . . arrive at a definite size, and there cease, but go on growing to an indefinite extent."[56] What Ware called "the wisdom" of trees is such that their past life is never gone but is included in the generation of new life, becoming its very seed and enabling it to spread and defy the boundaries of form. Trees teach us that life never discards anything as dead but renews itself through its past, as if present and commemoration, regeneration and memory had become one. Ware eventually generalized his theory of the circulation of vegetal life, whereby "past" matter is included in current growth, to apply it to humans, developing a theory of homeopathic healing predicated on the faith that human life, like the life of trees, can revitalize its numbed matter, recirculating it as new life. As Jacob Bigelow testified, Ware eventually developed an "instinctive aversion" to over-medication, claiming that "hyper-practice"—artificial intervention into self-regenerative life—was among one of the most important causes of mortality.[57]

Rumford Professor of Medicine and Botany at Harvard, Jacob Bigelow is himself a crucial figure in my reading of the story of vitalism in Cambridge, Massachusetts. As I see it, Bigelow doesn't appear as a genteel figure who advocated homeopathy in resistance to modern pathology and experimental medicine. To the contrary, in this part I show how much of Bigelow's homeopathic and vitalistic approach to life is predicated on the particular way that he understands the new science of pathology, which he first encountered while studying at University of Pennsylvania, Philadelphia, and later more directly, when he traveled to London and Paris. It was specifically Xavier Bichat's claim concerning the existence of two forms of life, one common to plants, animals, and humans, the other specific to animals and humans only, that influenced Bigelow.

Vegetal life in Bichat is not capable of relating to forms of life external to itself, as animals and humans are able to do. Yet, even though vegetal life is nonrelational and unconscious in Bichat, it is nevertheless understood as organic, that is, capable of self structuring and generating, and endowed with a series of complex functions and qualities, such as digestion, circulation, respiration, nutrition, exhalation, and secretion.[58] However, one feature of vegetal life, specific to it alone, astounds Bichat for what it possibly suggests about the nature of organic life in general, namely, the capacity for continual becoming. What becomes evident in the animal mode of existence "that what it was at one period he ceases to be at another; his organization remains the same, but his elements are continually changing,"[59] is the basic, "vegetal" principle of all life. Vegetal life manifests the capacity of certain beings to change completely while retaining the same form or to change form while remaining organized. Thus, in this type of life, forms change into other forms without collapsing into the pure flow of formlessness that cancels individuated existences, which is precisely why vegetal (organic) life is on Bichat's understanding so "well suited to this continual circulation of matter."[60] Vegetal life thus circulates through matter, vitalizing and transforming it, and so proving life's literal capacity for becoming.

It is this capacity of vegetal life to enact metamorphoses of matter that Bigelow will reformulate and generalize. His theory of the natural will come to understand both vegetal and animal life as being in constant circulation, capable of self-recovery, while his medicine will apply vegetal's life capacity for transformation to his understanding of human life, researching its ability to heal and restore itself. That medicine, which so centrally influenced Thoreau, will thus be formulated at a complex intersection of vitalist ontology—which argues that all matter is in circulation—botany, pathology, traditional healing practices, and contemporary developments in medicine.

In the late 1840s, other Harvard professors, or scientists otherwise connected to Harvard, afford inanimate matter the features of vegetal life. For instance, in 1849 Arnold Guyot, a former professor of physical geography at the College of Neuchâtel, Switzerland, living at that point in Cambridge, Massachusetts, gave a series of talks at the Lowell institute, which Cornelius Conway Felton, professor of classics at Harvard and later its president, immediately translated into English and published as a book the same year. In that book, which

Thoreau read in 1851, Guyot argued, in opposition to Humboldt and Agassiz (who was otherwise his close friend) that the earth is living and that its life is continuous rather than interrupted. His account of the vital relations that traverse inorganic nature as well as artificially made objects—Guyot's examples are gas lamps and desks, metals, and stones—was a radical formulation of materialistic vitalism. Guyot argued that any encounter between two "simple bodies," such as hydrogen and oxygen, enacts a vital movement generative of new bodies, for "everywhere a simple difference . . . be it of matter, be it of condition, be it of position, excites a manifestation of vital forces."[61] Excitability—which, together with sensibility, Coleridge ascribed to the "highest" forms of natural life—is thus attributed by Guyot to everything, from earthly elements to plants and animals. Such a general "excitement [of] vital movement" results for Guyot in a world in constant animation, full of differently speeded "rhythms" that fill everything with "life . . . [and] mutual exchange of relations."[62] Relations are incessant and can be exchanged between minute bodies as well as gigantic ones. Thus entire continents can enter new relations that affect their level of excitement, moving them into new forms of existence. In Guyot's account—which Thoreau found relevant enough to refer to it in the *Journal*—America is a continent monotonously encountering only oceanic water, which, even though it provides encounters that were always different and enlivening, doesn't bring about enough terrestrial trepidation to excite it intensely. Overwhelmed with boredom, life on the American continent appeared to Guyot's vitalistic imagination as less excited, hence closer in its nature to the oceanic and vegetal.[63] In other words, in Guyot's understanding, which attracted Thoreau, all forms of life on the American continent—mineral, vegetal, animal, and human—were oceanic and vegetal in that the level of their agitation was low; as a result, because agitation is what gathers matter into form, their forms were less focused and stable.

There were yet others in Cambridge, Massachusetts, and Boston in the 1830s and 1840s who argued that everything is alive, vitalized by either vegetal or some other form of life. For instance, that was the position of someone else who influenced Thoreau, namely, Edward Tuckerman, a Harvard alumnus, affiliate of the Boston

Society of Natural History, and good friend of Thaddeus Harris, with whom Thoreau studied natural history. As a botanist, Tuckerman didn't say much about oceans, rocks, and metal; yet, his main interest was in lichens, a form of life that, together with moss and fungi, was considered to populate the border delimiting animate from inanimate. And it is on the basis of the life of lichens that Tuckerman was to take a radical position regarding the natural in general, arguing that all forms of life—across family and species boundaries—are connected by more or less intense but always vital affinities, irrespective of taxonomical ordering. Similarly, Thomas Nuttall, curator of the botanical gardens at Harvard University from 1825 to 1834 (and who, as I show in detail in this part of the book, profoundly influenced not only Thoreau but also his brother John), was also to formulate a vitalistic theory on the basis not of geology or botany but of ornithology. To him, it was bird life that proved that everything is alive and that life is continuous, transcending taxonomical boundaries and enabling anomalous becomings.

None of the vitalists I discuss or mention is unknown, even though some, such as Ware and Guyot, are understudied. But they are always discussed separately, in the context of their respective disciplines. Nuttall is thus typically discussed in relation to the ornithology of his friend Audubon, with whom he also collaborated; Guyot is understood to be a much lesser Agassiz, mainly repeating Agassiz's catastrophic chant and not advancing beyond Humboldt. Tuckerman is known as a great lichenist, a lover of botany, and supporter of science, but he is rarely if ever recognized as proposing a novel philosophy of life; Ware was recognized by Bigelow but is now buried in the medical archives as a minor homeopath. Bigelow is still registered as an important cultural figure, and histories of American medicine always mention his important contribution to the profession. But attention is still not paid to his work on pathology, self-recovery, and decay in the context of a more comprehensive theory of life.

My reading emancipates those scientists from the boundaries of their disciplines and recontextualizes them in an interdisciplinarian discourse on life. In so connecting them across disciplines, and on the basis of the similarity of ideas they shared regarding life, I chart a different map of the sciences of life at Harvard in the 1830s and

1840s. On that map the Agassiz-territory shrinks to a more modest size, his polygenetic catastrophism being subverted and in fact regarded as dated by a group of scientists who were mobilized by theories of vitalistic materialism and resistant to ideologies of scientific taxonomy. In the picture I outline, the Boston and Cambridge of the 1830s and 1840s appear both philosophically and scientifically less as parochial towns awaiting news from Switzerland than as important centers of Western modernity, in which radical, albeit often strange, ideas were being probed. My picture must, of course, remain one in outline only, for it is always in the service of understanding Thoreau. I have thus excluded many important scientists who deserve to be included in the picture only because I couldn't determine whether Thoreau had read them. Those I have included are discussed only insofar as they help me unfold Thoreau's philosophy.

Because of the special investment both John and Thoreau had in birds, I begin by discussing Thoreau's ornithology, moving in my second section to his ichthyology. In interpreting his famous discourse on fish from *A Week*, I reveal the tacit polemic against Agassiz that is embedded in it, as well as how Thoreau's understanding of fossils signals his more general understanding of life. I move in the third section to geology. But there again, my interest is less in a scientific assessment of it—for it is hard to imagine that anything could be added to Robert M. Thorson's definitive study[64]—than in unfolding Thoreau's understanding of the inanimate and positioning it in his more general philosophy of life. I thus discuss Thoreau's vitalistic geology as a function of his understanding of death and even burial rites, which leads me to relate it to his archaeology and pioneering work on Native American middens. In the fourth section I move from Thoreau's lively stones to his understanding of photography, in which stones and monuments again play a central role. When Thoreau read Nicéphore Niépce's letters to Robert Hunt, reprinted in Hunt's books on photography, he understood there to be a profound difference between the operation of heliography and that of photography. Because photography was understood to register transformation by recording changes in the shadows cast by objects during the long time of exposure, Thoreau concluded that photography—unlike heliography, which registers only sunlight—

is committed not just to light but also to shade and darkness in which the forms of objects are destabilized. He thus concludes that photography's alliance with night and shadows is restorative. In reconstructing Thoreau's understanding of what a photographic image does, I thus seek to answer central question he poses: what is meant by the photograph's recovery of the body? And finally, my last section entertains the question of recovery and life through processes of dissolution, illness, and decay. I reconstruct Thoreau's obsessive thinking about pathological processes in all—but especially in vegetal—life, to complicate our understanding of how he related to the natural, and to demonstrate that far from believing in a happy pastoralist aesthetisizing nature, Thoreau understood it to be a dark laboratory in which flowers are mixed with offensively smelling decaying fungi, in which lilies grow galls, in which all health is shot through with disease. It is precisely Thoreau's theory of disease and dissolution that renders his vitalism so complex and, as I also argue in this part of the book, politically progressive.

A whole genealogy of vitalistic philosophies similar to Thoreau's developed later in the nineteenth and throughout the twentieth century, both in American and Continental traditions. Most notably this genealogy includes the philosophies of William James, Charles Sanders Peirce, Henri Bergson, Maurice Merleau-Ponty, Gilbert Simondon, and Gilles Deleuze. None of these individuals was influenced by Thoreau, even if James and Peirce of course read him. But it is not my concern here to offer a comparative analysis of Thoreau's work and other vitalist philosophies appearing in his wake, because it seems to me that such a comparatist approach becomes possible only once we have patiently attended to what it is that Thoreau's vitalism in fact says. However, even though I avoid a comparative analysis, I do emphasize that the vitalisms in the genealogy just charted had nothing to do with what, in the wake of Foucault's analysis, became known as the biopolitical organization of power, and that for several reasons. First, because biopolitics, as Roberto Esposito convincingly demonstrates, far from being established by the propositions of nineteenth-century vitalists, was in fact formulated by modern and nonvitalist political philosophers, such as Locke.

These philosophers imported it into modern liberal democracies when they argued that the political body functions according to the logic of immunization, that is, the way life operates to protect itself from what is foreign and thus unsafe.[65] Second, the vitalist genealogy I have in mind cannot be related to nineteenth-century biopolitics, which Foucault defines as "state control of the biological,"[66] because their major orientations were radically opposed. Nineteenth-century biopolitics, again on Esposito's account, was enabled less by the sciences of life than by a cluster of humanistic disciplines that interpreted certain scientific propositions from the viewpoint of racialized and racist politics. Such politics were predicated on the hierarchization of all living beings, hence also human beings. It is those very humanist discourses, which ranked humans according to how they fit in the politically formulated norm of the able, healthy, and racially pure, that provided the ideology for what Esposito calls the "thanato-political" drift of the Nazi regime. At the core of its lethal political imaginings was the idea that life that is normed as able and healthy can be maintained only if the lives of those who do not fit the norm are destroyed. Nazi biopolitics was in this way predicated on an antinomy summarized, as Esposito puts it, "in the principle that life defends itself and develops only through the progressive enlargement of the circle of death."[67] At the heart of the biopolitics that culminated in Nazism one finds death, not life, in complete opposition to the vitalisms to which I refer, which were formulated in radical resistance to any hierarchization and normativization of life, working toward a philosophy of life that is immanently constituted by the anomalous and the strange.

But regardless of how one details the relation between the state and sciences of life in the second half of the nineteenth century and throughout the twentieth century, biopolitical theories that served an ideology supporting the state's catastrophic appropriation of bodies—which Esposito terms "negative biopolitics"—were always only one aspect of the way philosophers thought about life. There were thus theories of life formulated otherwise and identified by Esposito as "affirmative biopolitics." The third reason, then, why the vitalist genealogy I have in mind must not be understood as girding the destructive organization of modern biopolitics, is that its proponents belong to that different tradition, formulating precisely an affirmative thinking about life, that is, a philosophy of *bios* that doesn't think about the power over life but about power *of* life: "biopolitcs that is finally affirmative. . . . No longer over life but of life, one that doesn't superimpose already constituted (and by now destitute) categories of modern politics on life, but rather inscribes the innovative power of a life rethought in all its complexity."[68] Some of the features of those affirmative vitalisms are: cancellation of the "organic model that joins every member of the body to its assumed unification" in favor of the inorganic or what is without organs;[69] abandonment of the "philosophy of history" in favor of an effort to "grasp the nexus—neither 'historical' nor 'geographic'—of history and transcendental geology";[70] abandonment of the human exceptionalism promulgated by anthropology in favor of the idea that the human is always a "living human," and that "between the psychic and biological, just as between the biological and the physical, a difference passes through not of substance or nature but of level and function. This means that between man and animal—but also, in a sense, between the animal and the vegetal and between the vegetal and the natural object—the transition is rather more fluid than was imagined";[71] and finally, as is evident in Deleuze's vitalism, presumption that there is "a modality of *bios* that cannot be inscribed within the borders of the conscious subject, and therefore is not attributable to the form of the individual or of the person."[72] A life not confined to the boundaries of personhood is an impersonal life, traversing "men as well as plants and animals independently of the matter of their individuation and the forms of their personality."[73] This vital, transversal animation of everything is of crucial political importance, as it allows us to understand "that any thing that lives needs to be thought in the unity of life— . . . that no part of it can be destroyed in favor of another."[74] The discussion in this part of the book positions Thoreau as one of the most important thinkers of the power of life—as opposed to power over life—that the nineteenth century had to offer.

Birds

On January 16, 1842, Barzillai Frost preached a funeral sermon for John Thoreau. The peculiarity of John's death convinced him that "God's hand" was involved in that death in some "special manner," producing "a deep impression on all, and especially the young."[1] Hence the need for typological interpretation of John's life. But Frost's wish to turn John into an emblem of the exemplary life is subverted by the oddities of the very life he wants to commemorate. For while some features of John's life are fit to impart "salutary lessons" (John's love of children and the elderly, his feeling for "all classes of suffering humanity," his speaking for "the cause of the poor inebriate, the slave, the ignorant, and depraved"),[2] others are disturbing enough to turn Frost's celebratory tone into a public critique of John's values and even a polemic about transcendentalism, such as when he voices his dissatisfaction with a lack of religiosity that led John to a radical, revolutionary politics ("Of his religious opinion I must speak with less confidence. He has been affected no doubt by the revolutionary opinions abroad . . . he adopted transcendental views").[3] Still other features of John's life—indeed those interests and tastes that Frost recognizes as "most prominent"—are so eccentric that Frost has difficulties fitting them into any type.

Such was the case, for instance, concerning John's love of graves and cemeteries.[4] But as Frost specifies, John's frequenting cemeteries wasn't prompted by morbid melancholy, for John was, to the contrary, "habitually cheerful." Moreover, love for cemeteries didn't signal a preoccupation with death at all; instead it was generated by John's belief that there was something "undecaying" in nature; that, strictly speaking, nature doesn't know death: "Mr. Thoreau had also a taste, for scenes which to many, seem gloomy and melancholy. He was fond of frequenting the graveyards. He said it was devout; that the emblems of decay, did not take hold of his mind: he thought only of that which is undecaying. He felt the influence which came down from first generations, and thought of the dead as still living."[5] Graves, as "emblems of decay" didn't take hold in John's mind, because in the cemetery he clearly saw how inaccurately they signify the logic of natural life, by marking its end. For John, it seems, natural life was never interrupted by death. It was his belief in the continuity of natural life that allowed him to think of the dead as "still living," if not in their original form, then as intense forces acutely influencing him. Cemeteries thus became less the sites of death than of the self-transformation of life.

Fascination with transformation explains John's other life-long preoccupation, his "enthusiasm" for birds. As Frost testifies, "any new bird . . . [John] discovered gave him the most unfeigned delight, and he would dwell up [*sic*] it and seem to commune with it for hours."[6] Frost speaks here about a perception so intense that it eventually produces an identification, enabling John to "commune" with the bird, speaking its language so that he can "dwell up" with it, as if in a community. Generating the transformation of a human person into a human-looking bird, John's experiments were a veritable exercise in becoming.

In 1836 John began a notebook on the birds of Massachusetts, now known to editors as "Nature and Bird Notes."[7] However, far from being a scientific success, the notebook testifies to John's increasing frustration with scientific methodologies of classification, registering his shift from ornithological to epistemological questions regarding the legitimacy of taxonomies and offering a glimpse into how his understanding of life changed.

John's first idea was to classify birds according to models proposed by Audubon, Nuttall, and Peabody, accounting specifically for their habitats and migratory patterns ("Gaul's Owl was found here by Audubon"; "The Golden Eagle, here: Nuttall"; "Goshawk sometimes follows the ducks into Massachusetts: Peabody").[8] But the topographical classification is soon replaced by a temporal one that, in minimizing Audubon's presence while intensifying Nuttall's, ordered birds according to the season in which they were most likely to be observed. As John moves from January to December, his lists of birds, however, become shorter (July lists only the Little Sandpiper and Red Breasted Snipe, said to in fact be characteristic of April), returning ad hoc and briefly to April and May birds, before ending abruptly. Starting from the opposite side of the notebook, John will try again, this time classifying according to alphabetical order. But it seems that strategy pleased him the least, for in allowing different families and genera to be mixed in the same linguistic set (the list for B thus includes Bobolink, Black bird, Black bird crow, Blue bird, and Butcher Bird), his alphabetical ordering in fact subverted the very logic of scientific differentiations; thus, he abandons it too. There will be one last attempt after that. John's birds will migrate from topological to calendric to alphabetical classification, to settle finally into thirteen ornithological orders proposed by Nuttall. What, then, was the reason for John's taxonomic frustration, and why is it that Audubon progressively fades in favor of Nuttall?

John's realization that Audubon's and Nuttall's ornithological taxonomies couldn't be referenced alternatively, as if relying on the same scientific regimes of knowledge, testifies to his accurate sense of their differences. Even though Audubon and Nuttall collaborated for many years and on many projects, their approaches to ornithology are different, indeed so much so that one can claim their methodologies belong to different centuries. Audubon's is from the eighteenth century. Following eighteenth-century models of natural history—the classicist's belief in allotting to each creature its proper category, trusting that the world can be exhausted through extensive descriptions—Audubon seeks to distribute life into fixed tiers or, as he puts it, to organize specimens "according to [his] notions." Since the specimens are most likely to fit notions when dead, Audubon contrives an ideated method of representing birds, rendering them as programmable automatons and, moreover, in Christoph Irmscher's phrasing, turning the whole of nature into a mere "wire construction."[9] How this transsubstantiation of animated being into mechanism was reasoned can be sensed through Audubon's description of mounting a Kingfisher. After killing the bird, he

> pierced his body . . . and fixed it on the board; another wire passed above his upper mandible held the head in a pretty fair attitude, smaller ones fixed the feet according to my notions. . . . The last wire proved a delightful elevator to the bird's tail and at last—there stood before me the real Kingfisher. . . . This was what I shall call my first drawing actually from nature, for even the eye of the Kingfisher was as if full of life whenever I pressed the lids aside with my finger.[10]

For Audubon, as if parroting Descartes, a living bird is replaced by a lifelike puppet, but one whose "full" life derives from being skinned, stuffed, and wired. Nature becomes a technological device and life a mechanism, appearing at its most vital when dead. What is dead becomes truly real for Audubon ("there stood before me the real Kingfisher") and that reality is identified as "actual nature" (Audubon draws the Kingfisher's carcass and calls it "my first drawing actually from nature").[11] That is because for him, as for any eighteenth-century natural historian, "life itself didn't exist," as Foucault put it. This is not to say that a natural historian treated individual creatures as dead but rather that he lacked an ontology that would see beings connected by a single life that transcends their individual forms by means of its continuity. Hence living beings appeared isolated through a grid of representations and descriptions that included not just what was observed but also what tradition and language deposited in them. As Foucault explains, "the descriptive order proposed for natural history by Linnaeus . . . is very characteristic. According to his order, every chapter dealing with a given animal should follow the following plan: name, theory, kind, species, attributes, use, and, to conclude, *Litteraria*," the last being "a sort of supplement in which discourse is allowed to . . . record discoveries, traditions, beliefs, and poetical figures."[12] A classical natural historian moves like Audubon, from experiencing to describing, from perceiving to narrating, from seeing to drawing.[13]

Although Audubon and Nuttall collaborated and even "jointly publish[ed] a description of 12 new species," the volume by Nuttall that so influenced John, namely, *The Genera of North American Plants*, registers his distance from Audubon.[14] Unlike a natural historian for whom there are living beings but not life, for Nuttall, who became the curator of botanical gardens at Harvard University in 1825, there is one life preceding and animating all living beings, rendering the differences among them less stable and the taxonomies that order them prone to shifting. In Nuttall's world creatures never "forget," as it were, the oneness of life out of which they emerge, and regardless of the differences separating them, they remain able to recognize something of themselves—precisely the same life they share—in all others, thus remaining mutually related through such affinities. As Nuttall formulates it against the "artificial" taxonomical systems "still exclusively taught throughout the United States," his understanding of life as one for all beings that manufactures inter-species activities generates volatile taxonomies:

> We are at length inclined to believe that the last and the most perfect of systems, perfect because the uncontaminated gift of Nature, is about to be conferred upon and confirmed by the Botanical world. The great plan of natural affinities, sublime and extensive, eludes the arrogance of solitary individuals. . . . Can we deny the perception of a prevailing affinity through the vegetable kingdom, and carp at the anomalous character of a few individuals? But even here the science begins to triumph, when we perceive that the anomalies diminish by the accession of objects.[15]

We carp at the anomalous—at the unclassifiable—only so long as the taxonomical fiction of fixed and separated identities, which standardizes beings, is understood to mirror their truth. But everything changes once we presume that, far from being rigorously separated, beings are rather changeable sites traversed by life's mobility; they then become processual, moving toward other beings so that "all the arrogance of solitary individuals," as Nuttall puts it, is eluded. Nothing remains isolated in the solitariness of its identity, but all individuals are turned into humble—because incomplete—processes of individuation. Far from fearing "monstrous" combinations of beings and hoping that they would be limited by the environmental control of the "arrangement of material forces"—as Monique

Allewaert suggests was the case with eighteenth-century proponents of Goethe's theory of elective affinities—Nuttall develops a sense of anomaly as nothing other than an existence en route to becoming something else.[16] For that reason he announces that his new science of affinity will produce a "concussion of Botanical nomenclature" comparable to any "revolution whether in science or politics,"[17] reducing the "limits of genera" to permeable outlines: "The limits of genera . . . have now been greatly reduced."[18] When limits become porous, what has been harmoniously separated by them will contagiously mix. Accordingly, Nuttall's birds migrate too, not just geographically, but from species to species, from one genus to another. His *Manual of the Ornithology of the United States and of Canada*, which also influenced John, thus starts by distinguishing two large classes of birds ("Birds . . . may be generally distinguished into two great classes from the food on which they . . . subsist")[19] but recognizes that the distinction between them is blurred by some birds who partake in both ("Some also hold a middle nature, or partake of both").[20] It also admits that other possible classifications are similarly blurred by the "middle nature" of various birds (for instance, the great divide between aquatic and airy classes is confused by so many amphibious birds); moreover, the book recognizes that even "birds of the same genus differ much in their modes of nidification,"[21] in their migratory habits (birds of a same tribe can be either migratory or sedentary), and in the ways that they communicate. Above all, birds manifest the capacity for astounding transformations and habit alteration, so that when exposed to human life, some become human. Nuttall's list of human birds is long and includes William Bartram's Brown Thrush, who developed aesthetic preferences regarding decorated glass;[22] Colonel O'Kelly's ash-colored parrot, an "individual" (a word that in Nuttall signifies a self) who "answered many questions," engaging in conversation and "[giving] orders in a manner approaching to rationality"; and "twenty-four Canary birds who acted in an 1820 London play directed by a Frenchman named Dijon."[23] When he concludes by arguing that "we cannot deny" that birds share "rational intelligence"[24] and are capable of causal knowing, he is not suggesting, anthropocentrically, that all creatures have potential to become humans. Instead, his conclusion is more general, and that is what will influence the Thoreau brothers: "Nature probably delights less in producing such animated machines than we are apt to suppose; and amidst the mutability of circumstances by which almost every animated being is surrounded there seems to be a frequent demand for that relieving invention."[25] Nuttall's understanding of life is here explicitly distanced from Cartesian mechanics and its understanding of beings as automatons, such as will turn Audubon's Kingfisher into an apparatus. Instead, life appears as sheer "mutability," animating certain beings in a way that allows them to reinvent themselves time and again. The reinvention is sometimes so radical that it "demands" that some creatures become something else entirely, engaging in inter-species crisscrossing, migrating from their own instincts into those of other species. Far from being creatures of destiny—of instinct and habit—birds are thus creatures of freedom and passage. Hence, when Nuttall finally comes to propose his taxonomy of thirteen ornithological orders, he will caution us to remember that birds are creatures characterized by "aberration from [their] usual course," some of them not even being able to fly, that is, not even having the capacity considered essential to their "nature." There is, then, nothing essential about birds, nothing save the capacity for mutation.

As John's "bird notebook" moves away from Audubon to Nuttall—for, as I mentioned, Nuttall's thirteen orders will be John's last effort in the notebook to classify birds—he signals not a simple methodological preference but rather his affinity with Nuttall's transformationist understanding of life. Moreover, John saw boundaries separating classes and families as more porous still than did Nuttall, as if he wanted to take Nuttall's unstable taxonomies to their logical, albeit extreme, consequences; that is, as if he wanted to break up what Foucault describes as "all the ordered surfaces . . . with which we are accustomed to tame the wild profusion of existing things."[26] Whereas Nuttall does offer exhaustive descriptions of families of each order, John's orders are hesitant, short of descriptions, sometimes listing just one or two families in an order, at other times leaving the order empty.[27] Why John leaves so many pages blank, with no general description of bird families, can be understood on the basis of the descriptions that he does offer. For instance, in the second order—"Omnivorous birds"—John lists in 1836 only the Baltimore Oriole, Golden Robin, and Fiery Hangbird and leave it at that, without providing comments or descriptions of the birds. A year later, however, in May 17, 1837, he inserts below the Baltimore Oriole a note on a particular Oriole he had observed: "May 17 '37 An adult male; upon "Chesuncook" in company with the males of last season whose lighter plumage contrasted pleasingly with the brilliant dress of this more advanced specimen."[28] As if to suggest that the wild profusion of existing things can't be tamed by classifications, John moves from a family to an individual specimen, from a category to a body. What interests him is what no nomenclature, however invested in nuance, can offer: the particular and the relational (how a particular Oriole contrasts with others in the observed group), the *that*. Because all life is one, it is known simply by its particulars. There is no one total history of a class or a species. Instead, history is made of charts of migratory individual beings.

For more than a month after John's death, from January 11 to February 20, 1842, Thoreau didn't write in his *Journal*. When he finally did, it was, as discussed in the first part of this book, to announce that the highest ethical injunction one can put to a person is "consent to have no character," because only de-individuation enables one to migrate through other beings ("he who runs through the whole circle of attributes can not afford to be an individual" [J, 1, 367]). As if deciding to act on that ethical injunction, a mere three weeks after pledging himself to flight through "attributes," Thoreau wrote his first inscription in John's bird notebook. Excepting different handwriting, nothing—no clear announcement, no signature, not even a vague reference—signals that the author has changed. It is as if Thoreau were continuing John's entries *as* John. The only commentary on that decision is encrypted in Thoreau's own *Journal*. On March 15, 1842, as he is getting ready for a walk, he makes a remark about taking with him the notebook of another: "As I am going to the woods I think to take some small book in my pocket whose author has been there already.- whose pages will be as good as my thoughts" (J, 1, 375). As the insertions in John's bird notebook

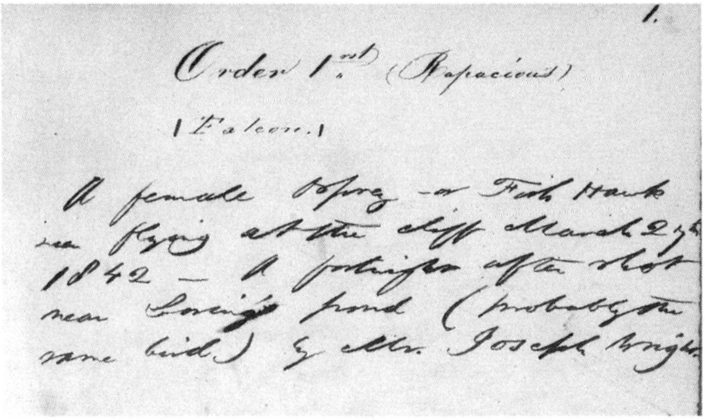
1.
Order 1st (Rapacious)
(Falcon.)
A female Osprey — or Fish Hawk seen flying at the cliff March 27th 1842 — A fortnight after shot near Loring's pond (probably the same bird) by Mr. Joseph Wright

Henry D. Thoreau, “Order 1st,” in “Nature and Bird Notes: Journal and List of Birds, kept by Sophia, John and H. D. Thoreau,” manuscript, Henry W. and Albert A. Berg Collection of English and American Literature, The New York Public Library, Astor, Lenox and Tilden Foundations.

testify, Thoreau doesn’t go just any where with the notebook in his pocket but visits sites where its “author has been already.” He travels the paths John walked to perceive what John saw. To make things even more literal, while walking John’s walks, with John’s notebook, Thoreau doesn’t search for just any bird from among John’s classifications. Instead, he is on the lookout for the very bird that John’s notes suggest he saw, as if Thoreau were hoping to be able to look into the eyes of the bird that looked at John’s eyes. Thoreau is enacting a migration: inhabiting John’s perceptions, he is en route to becoming the bird that, according to Frost’s testimony, John had a habit of becoming.

The first entry is from March 27, 1842. Below John’s entry “Falcon,” Thoreau inscribes: “A female Osprey—or Fish Hawk seen flying at the cliff March 27th 1842—A fortnight after shot near Loring’s pond (probably the same bird) by Mr. Joseph Wright.”

The second entry is from May 1842 (undated). Below John’s entry describing a male Oriole he observed on May 17, 1837, in the company of the males of “last season,” Thoreau writes an entry about a “female Oriole” now nesting near his house, “constantly [coming] to the fir-trees by the window.”[29]

On May 31, below John’s entry “Owls,” Thoreau writes: “Great Horned or Cat Owl—apparently a young female—22 inches by 4 feet—shot in Lincoln woods by Mr. Geo. Farrar—sitting upon a tree at 3 o’clock P.M. May 31st 1842.”[30]

On June 2, Thoreau spots a St. Domingo Cuckoo, a specimen of which John had observed in Sleepy Hollow woods, as records in the

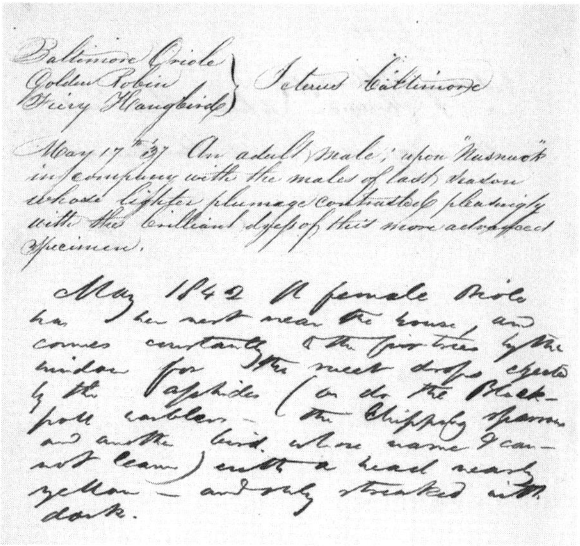

Baltimore Oriole
Golden Robin
Fiery Hangbird } Icterus Baltimore

May 17th '37 An adult male; upon "Nasnuck" in company with the males of last season whose lighter plumage contrasted pleasingly with the brilliant dyes of this more advanced specimen.

May 1842 A female Oriole has her nest near the house, and comes constantly to the fir trees by the window for the sweet drops ejected by the aphides (as do the Black-poll warblers—the Chipping sparrow and another bird whose name I cannot learn) with a head nearly yellow—and only streaked with dark.

Henry D. Thoreau, "Baltimore Oriole," in "Nature and Bird Notes: Journal and List of Birds, kept by Sophia, John and H. D. Thoreau," manuscript, Henry W. and Albert A. Berg Collection of English and American Literature, The New York Public Library, Astor, Lenox and Tilden Foundations.

notebook suggest, on June 5, 1837. He writes: "June 2nd 1842 found nest of this bird—two feet from the ground in a small white pine—very rude and hollow—with one egg. . . . The bird flew out on my approach, and I heard only a faint and distant—cocu-cocu -cocu-cocu-."[31]

On June 1, 1842, Thoreau finds a Night Hawk—a specimen of which John observed on June 1840—and notes that the bird was "formerly wounded [but is] now recovering." And, finally, under John's heading "Partridges," Thoreau writes a long description of the encounter with several specimens of the bird. He later includes a version of that encounter in *Walden* (Wa, 226–227), as if to relocate John's bird notebook there and so make John speak in his own major achievement:

> June 2nd 1842 found three of the brood of a partridge that flew up close to me—not more than 2½ inches long—squatting close to the ground upon the leaves which they so exactly resembled that after searching for quarter of an hour I found no more, though experience taught me that my eyes might have rested on them repeatedly. They squatted in my hand and on the leaves without the slightest motion, or betraying the least agitation. So perfect was this instinct that when I laid them upon the leaves again and one accidentally fell on its side—it was found

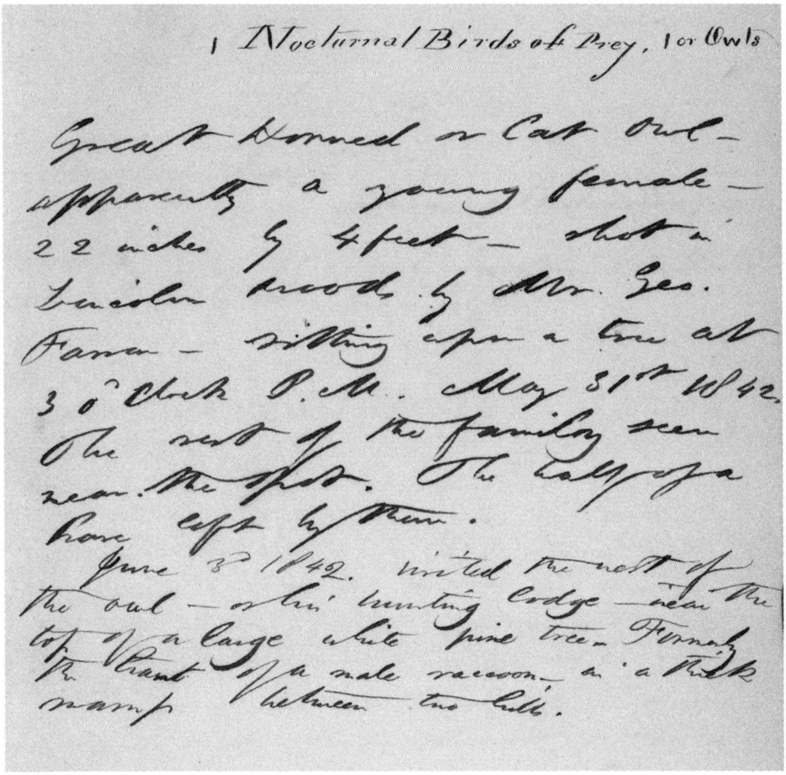
1 Nocturnal Birds of Prey, 1 or Owls

Great Horned or Cat Owl—
apparently a young female—
22 inches by 4 feet— shot in
Lincoln woods by Mr. Geo.
Farrar— sitting upon a tree at
3 o'Clock P.M. May 31st 1842.
The rest of the family seen
near the spot. The half of a
hare left by them.
June 3d 1842. visited the nest of
the owl—or their hunting lodge—near the
top of a large white pine tree. Formerly
the tenant of a male raccoon—in a thick
swamp between two hills.

Henry D. Thoreau, "Nocturnal Birds," in "Nature and Bird Notes: Journal and List of Birds, kept by Sophia, John and H. D. Thoreau," manuscript, Henry W. and Albert A. Berg Collection of English and American Literature, The New York Public Library, Astor, Lenox and Tilden Foundations.

> with the rest in exactly the same position ten minutes afterward. The parent made a loud noise like a cat in the meanwhile and then kept silent. . . . June 3d 1842. Found another brood to-day of the same age. near the owl's nest.

There is a pattern in those entries. Birds fly to life from death—the Fish Hawk comes back after being shot, and the Cat Owl survives its own death—as if blurring the boundary between existence and nonexistence; like magical healers, they recover after being fatally wounded; like existential sorcerers, they speak the language of other

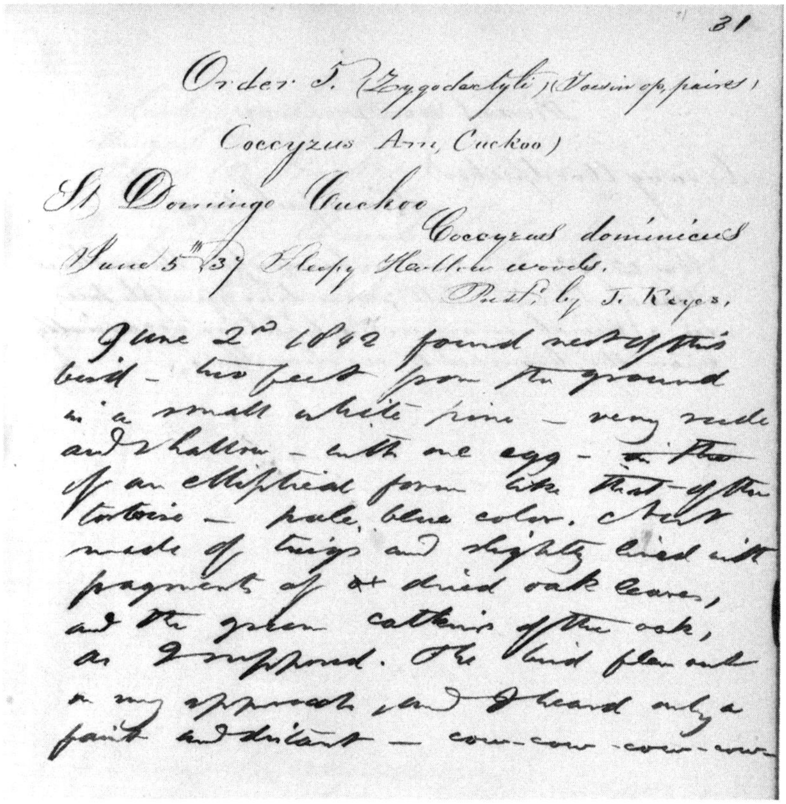

31

Order 5. (Zygodactyli) (Toes in pairs)

Coccyzus Am, Cuckoo)

St Domingo Cuckoo

Coccyzus dominicus

June 5th 37 Sleepy Hollow woods.

Presd by J. Keyes,

June 2d 1842 found nest of this bird – two feet from the ground in a small white pine – very rude and shallow – with one egg – of an elliptical form like that of the tortoise – pale, blue color. Nest made of twigs and slightly lined with fragments of dried oak leaves, and the green catkins of the oak, as I supposed. The bird flew off on my approach, and I heard only a faint and distant – cow-cow-cow-cow

Henry D. Thoreau, "Order 5," in "Nature and Bird Notes: Journal and List of Birds, kept by Sophia, John and H. D. Thoreau," manuscript, Henry W. and Albert A. Berg Collection of English and American Literature, The New York Public Library, Astor, Lenox and Tilden Foundations.

species, so that a parent-partridge becomes a cat; like analogists of being, they move between two irreconcilable ontological states, such as when partridges suddenly turn invisible while remaining present, suspending their mobility in absolute stillness, or such as when, in *Walden's* contrary variation of the same scene, partridges become capable of shape-shifting, "roll[ing] and spin[ning] round before you in such a dishabille, that you cannot, for a few moments, detect what kind of creature it is" (Wa, 226). They become pure mobility, a mobility whose rapidity doesn't allow any shape to consolidate, a non-classifiable creaturely life. What John, following Nuttall, registered

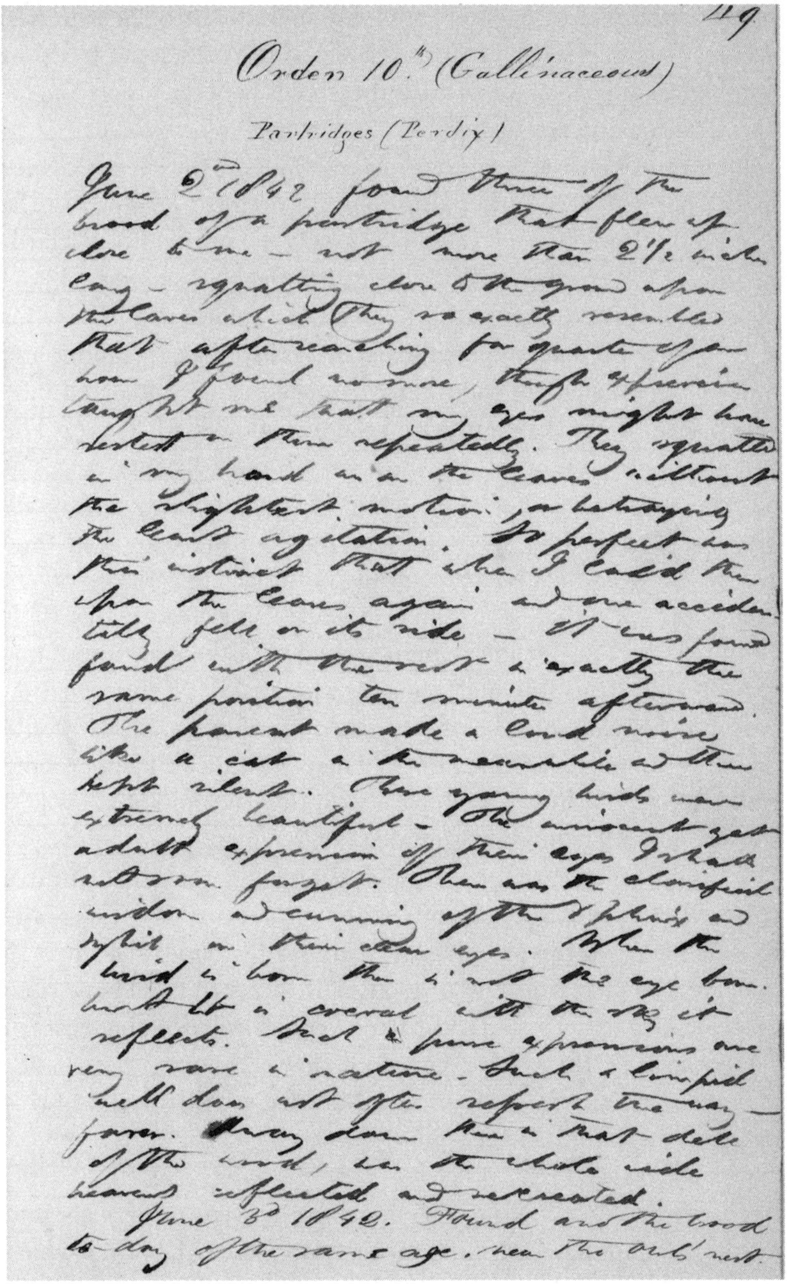
49

Order 10th (Gallinaceous)

Partridges (Perdix)

June 2nd 1842 found three of the brood of a partridge that flew up close to me — not more than 2 1/2 inches long — squatting close to the ground upon the leaves which they so exactly resembled that after searching for quarter of an hour I found no more, though experience taught me that my eyes might have rested on them repeatedly. They squatted in my hand or on the leaves without the slightest motion, or betraying the least agitation. So perfect was this instinct that when I laid them upon the leaves again and one accidentally fell on its side — it was found found with the rest in exactly the same position ten minutes afterward. The parent made a loud noise like a cat in the meanwhile and then kept silent. These young birds were extremely beautiful — The innocent yet adult expression of their eyes I shall not soon forget. There was the clarified wisdom and cunning of the Sphinx and sybil in their clear eyes. When the bird is born then is not the eye born but it is coeval with the sky it reflects. Such a pure expression is very rare in nature. Such a limpid well does not often reflect the wayfarer. Away down there in that dell of the wood, was the whole wide heavens reflected and recreated.

June 3d 1842. Found another brood to-day of the same age, near the owls' nest.

Henry D. Thoreau, "Order 10," in "Nature and Bird Notes: Journal and List of Birds, kept by Sophia, John and H. D. Thoreau," manuscript, Henry W. and Albert A. Berg Collection of English and American Literature, The New York Public Library, Astor, Lenox and Tilden Foundations.

as the capacity of beings to relate inter-specially through affinities, Thoreau's empirical observations detect as the capacity of birds for ontological migration: birds cross thresholds between visible and invisible, being and non-being.

John's and Thoreau's empirical observations are thus in accordance with Thoreau's very early philosophical statement recorded in the *Journal* on September 25, 1840. Discussing the status afforded birds in ancient Greece, he explicitly identifies them as ontological borderers: "Birds were very naturally made the subject of augury—for they are but borderers upon the earth—creatures of a subtler and more ethereal element than our existence can be supported in—which seem to flit between us and the unexplored" (J, 1, 179). The idea that all birds (not just the most volatile ones, like partridges or falcons) are subjects of augury by virtue of their ability to migrate into the "unexplored" was suggested by Thoreau in another 1840 *Journal* entry—which is later centrally positioned in *A Week*—where even a "patient," "dull," and "sluggish," bittern is a creature that "sails away" (J, 1, 126). Obviously preoccupied with the question of migration and shape-shifting, Thoreau wonders, in an entry written only a couple of months before the one on the bittern, what it would take for a human to become a bird: "Thank fortune we are not rooted to the soil. . . . Why not keep pace with the day and not allow of a sunset—nor fall behind the summer and the migrations of birds. . . . Our limbs indeed have room enough but it is our souls that rust in a corner. Let us migrate interiorly without intermission—The really fertile soils lie on this side the Aleghanies" (J, 1, 119–120). Thoreau imagines the becoming-bird of a soul as an incessant flight away from fixed identity (from a soul incapable of transformation that rusts in the corner); he imagines a bird-soul that is a nomadic existence moving ever closer to "fertile soils." But what are we to make of this strange soul-flight into the soil?

Between the entry inviting the migratory movement of souls into fertile soils and that describing the migratory bittern, both of which signal Thoreau's preoccupation with soul-metamorphoses, one finds an entry on Thales's doctrine of souls: "April 22nd '40 Thales was the first of the Greeks who thought that souls are immortal—and it takes equal wisdom to discern this old fact to-day. What the first philosopher taught the last will have to repeat" (J, 1, 123). But while indeed immortal, souls in Thales are neither humanized selves with an individualized psyche nor spiritual Platonic essences. Instead, they were for Thales completely material. As Aristotle explains in his *Metaphysics*—a text well known to Thoreau, in which the longest surviving analysis of Thales's soul-doctrine is to be found—Thales, and most of the first philosophers following him,

> thought the principles which were of the nature of matter were the only principles of all things. That of which all things that are consist, the first from which they come to be, the last into which they are resolved (the substance remaining, but changing in its modifications), this they say is the element and this the principle of things, and therefore they think nothing is either generated or destroyed, since this sort of entity is always conserved, as we say Socrates neither comes to be absolutely when he comes to be beautiful or musical, nor ceases to be when he loses these characteristics, because the substratum, Socrates himself, remains. Just so they say nothing else comes to be or ceases to be. . . . Thales, the founder of this type of philosophy, says the principle is water . . . , getting the notion perhaps from seeing that the nutriment of all things is moist, and that heat itself is generated from the moist and kept alive by it. . . . He got his notion from this fact, and from the fact that the seeds of all things have a moist nature, and that water is the origin of the nature of moist things.[32]

There is nothing but matter (all things are of the "nature of matter"), and thus all life is corporeal and undying (for the "substance" always remains animated). Individual beings are thus only more or less visible vestiges on the surface of this material life, and what is called their life and their death is simply a modification of life, a change in its manner, as it were. Individual beings thus never completely cease to be (for "nothing is destroyed") but—like Socrates's characteristics—fade a little, becoming less clear while simultaneously being reformed into other beings. As Heidegger puts it when he interprets

Aristotle on Thales, in Thales there is only "constant change and transformation, but indeed no coming to be or passing away. On the contrary, everything remains."[33] Hence, Heidegger's summary: because it is always material, life is in Thales "coincident with nature (*physis*, "to engender," "to grow")," which is "1. The ever enduring, 2. The becoming. Both."[34]

As Aristotle suggests, if for Thales the principle of this unceasing life is water, it is because he noticed that all beings are nourished through sites where moisture is condensed. Aristotle identifies seeds as these watery sites out of which all things come to be ("seeds of all things have a moist nature"). And it is precisely because these seeds keep everything in motion through nourishment ("the nutriment [what keeps a being alive] of all things is moist") that Thales alternatively identifies them as souls, for souls *(psyche)* are nothing other than what animates through nutrition. As Aristotle makes explicit in "De Anima"—another text that preoccupied Thoreau throughout his life—"Thales . . . seems to have held soul to be a motive force [something that produces motion], since he said that the magnet has a soul in it because it moves the iron."[35] Because everything, even things that we would commonsensically classify as inanimate, such as magnets and iron, grow out of seeds and are thus alive, seeds must be everywhere. Alternatively, because everything is animated by something else, souls must be everywhere. That is how Aristotle puts it in "De Anima": "Certain thinkers say that soul is intermingled in the whole universe, and it is perhaps for that reason that Thales came to the opinion that all things are full of gods."[36] However, that the universe is infused with souls or is full of gods doesn't mean that a life inhabited by soul somehow comes to seed inert matter while remaining substantially different from it. For, as Aristotle insists, for Thales everything is of the "nature of matter," or as Heidegger puts it, for Thales "the two principles [of matter and spirit]" are still "in unity";[37] gods are material and seeds are souls. Thus, when Thoreau—in discussing Thales's transformative materialism—insists in his *Journal* on our "inner" or souled migration toward fertile soils, he reinforces the extraordinary rigor of his thought by showing that the movement he has in mind is much like Thales's: it is both "enduring" and "becoming" and, above all, always material.

A passage from *A Week* connects the entries from 1840 that I've discussed so far—on birds as ontological borderers, on Thales, and on the bittern—into Thoreau's own version of Thales's philosophy. It explicitly references Thales in proposing that life is material, everlasting, and related to watery conditions; but it is specific to Thoreau in that it locates the principle of this everlasting life in birds. The passage describes how, while sailing down the Merrimack, the two brothers looked at the eyes of a smaller bittern:

> As we shoved away from this rocky coast . . . the smaller bittern, the genius of the shore, was moping along its edge. . . . It is a bird of the oldest Thalesian school, and no doubt believes in the priority of water to the other elements; the relic of a twilight ante-diluvian age which yet inhabits these bright American rivers with us Yankees. There is something venerable in this melancholy and contemplative race of birds, which may have trodden the earth while it was yet in a slimy and imperfect state. Perchance their tracks too are still visible on the stones. It still lingers into our glaring summers. . . . One wonders if, by its patient study by rocks and sandy capes, it has wrested the whole of her secret from Nature yet. What a rich experience it must have gained, standing on one leg and looking out from its dull eye so long on sunshine and rain, moon and stars! . . . It would be worth the while to look closely into the eye which has been open and seeing at such hours, and in such solitudes, its dull, yellowish, greenish eye. Methinks my own soul must be a bright invisible green. (W, 236–237)

Resting on "wet stones and sand," Thoreau's bittern is nourished by the Thalesian everlasting principle—of a moist soul/seed that is generative of beings; but by flying and migrating from one to another "glaring summer," the bittern also embodies becoming as the operative mode of that everlasting life, for it is a life that constantly modifies itself without changing its substance. The bittern is thus both what is generated by eternal life and that very life; in it the eternal cause and its particular effect, substance and mode, come together.

While the bittern thus fuses Thoreau's belief in birds as augury beings capable of existential metamorphosis with his belief in the truthfulness of Thales's philosophy, the reference to the bittern as "ante-diluvian" bird provides yet another reason Thoreau chose that bird to embody eternal life. For in the Old Testament the bittern famously becomes the bird of survival. As Isaiah prophesized (in King James's version) God will allow only the bittern to survive the complete destruction of Babylon and in so witnessing the deliverance of Israel to the Jews, to connect the moment of exile to the moment of return. Migrating from the moment of destruction of Israel, through the destruction of Babylon, the bittern flies into the time of Israel's restoration:

> For I will rise up against them, saith the Lord of hosts; and cut of from Babylon the name, and remnant. . . . I will also make it a possession for the bittern, and pools of water. . . . And the streams thereof shall be turned into pitch, and the dust thereof into brimstone. . . . It shall not be quenched night nor day; the smoke thereof shall go up for ever: from generation to generation it shall lie waste; none shall pass through it for ever and ever. But the cormorant and the bittern shall possess it . . . and he shall stretch out upon it the line of confusion, and the stones of emptiness. (Isaiah 14:22–23; 34:9–11)

In Isaiah the bittern becomes something like a total witness of history arranged as series of destructions and new beginnings; yet the bird is a cruel witness, forgetting some of the destruction it sees, turning sites of memory into stones of emptiness. In contrast, in Thoreau, the bittern doesn't turn monuments into stones of emptiness but keeps them as repositories of history by patiently studying them, contemplating their secrets, even inscribing its own tracks into them ("perchance their tracks are still visible on the stones"), gaining "rich experience" from this eternal contemplation, never forgetting anything, ready to tell it all ("what a rich experience it must have gained . . . what could it tell"). In Isaiah, Babylonian history is wiped out by geology (stones of emptiness), and the bittern becomes the relic signifying only the complete punishment and utter oblivion of the Babylonians; it becomes a relic signifying the awesome possibility that nothingness can come to be. In Thoreau, the bittern is a relic too, as he explicitly calls it, but that relic is alive: it is a life itself that remembers, an animated monument. And it is precisely because the life is an all-remembering memorial that Thoreau can say how it "would be worth the while to look closely into the [bittern's] eye which has been open and seeing in such hours." To look into bittern's eyes is to see the whole of history, to see through total memory; it is to redraw the lines of history onto the stones of emptiness.

That Thoreau relies on first philosophies to claim that birds embody enduring yet self-modifying life doesn't mean that birds merely play the role of philosophical allegories. To the contrary, birds in Thoreau are real forces. In the same year that, relying on Thales, he formulates his bird-ontology, Thoreau also makes the first gestures toward his own "scientific" answer to the question of what life is, and what kind of forces move it. In that taxonomically bizarre explanation, birds, together with fish, play a central role:

> The motion of quadrupeds is the most constrained and unnatural; it is angular and abrupt, except in those of the cat tribe, where undulation begins. That of birds and fishes is more graceful and independent—they move on a more inward pivot. The former move by their weight or opposition to nature, the latter by their buoyancy, or yielding to nature. Awkwardness is a resisting motion, gracefulness is a yielding motion. . . . Rippling is a more graceful flight. If you consider it from the hill top you will detect in it the wings of birds endlessly repeated. The two waving lines which express their flight seem copied from the ripple. (J, I, 206)

In contrast to contemporary sciences of life, Thoreau's criterion for ranking beings is predicated neither on function nor on the complexity of organisms; instead, beings are ranked in accordance with the type of locomotion or animation they engender, quadrupedism being ranked lower than gliding, flapping, or swimming, and, consequently, mammals are ranked lower than fish. However, this overturning of scientific taxonomies is generated by a meta-hypothesis about how life proceeds: in complete opposition to what was going to become the central premise of evolutionary biology, that life is strengthened through combat and opposition, Thoreau proposes the essence of animation to reside in acceding, deferring, and yielding. It is because they "move by opposition"—turn against or run counter—that quadrupeds are "awkward" and "most constrained," or even most "unnatural." In contrast, the most natural, or nonresisting life is represented by a variety of motions Thoreau understands as "undulation": floating; swimming; gliding; flying; and, above all, rippling. Unlike the mammalian motion that interrupts itself by turning around—which is why Thoreau calls it "abrupt"—rippling motion that most accurately embodies the way life moves is a continuous waving line.[38] As Thoreau puts it, the "graceful flight" that is rippling would be an "endless repetition" of the "two waving lines which express the flight of a bird":  and would therefore look something like this: . Just as in his Thalesian ontology, then, where life is embodied by birds and endures yet constantly becomes, so in his science, life, to which birds are closest, is imagined as a line that endlessly moves without ever forming a fixed shape; a constant flight that doesn't stabilize itself into a formal identity.

Fossils

In Thoreau the nonoppositional logic of life is embodied not just by the movement of birds but of fish too, because the effortless way in which their body follows the curves of water also obeys the logic of yielding: "fishes . . . move by their buoyancy or yielding to nature" (J, 1, 206). Together with birds, the fish thus become in Thoreau the emblem of all life, as the famous discussion in the "Saturday" chapter of *A Week* puts it: "it concerns us to attend to the nature of fishes, since they are . . . forms and phases of the life in nature universally dispersed" (W, 25).

Following Buell's suggestion, which represents an atypical reading of "Saturday," that "Thoreau's 'scientific' side comes to the fore" precisely in his ichthyology, I suggest that his ichthyology intentionally opposes contemporary scientific epistemologies, because it starts from a different understanding of what life is.[1] I argue that the fish discourse from *A Week* formulates Thoreau's radically materialistic brand of vitalism against the backdrop of Cuvier's and Agassiz's theories of life—Agassiz's theories in fact being explicitly referenced on two occasions (W, 28, 33). It does so to reaffirm the absolute value of living beings in their incomparable singularity. I suggest that even if, despite its scientific orientation, it doesn't end up becoming "scientific" (often, in fact, verging on the bizarre and marvelous), Thoreau's ichthyology nevertheless propels an understanding, and by extension, an ethics of life so radically democratic that in its basic orientation it *disables* the discrimination of beings—not only in families or species but also from species to species—on the basis of their difference.

Thoreau was systematically exposed to Cuvier at least starting from his time at Harvard, as the major naturalist and Thoreau's professor, Thaddeus Harris, organized his courses for a number of years around Cuvier's *Recherches sur les ossemens fossils de quadrupeds* (1812)—whose "Preliminary Discourse," summarizing the argument of the whole book, appeared in English in 1813.[2] Thus, in the spring semester of 1837[3] Thoreau studied John Ware's *Philosophy of Natural History*, which claimed that Cuvier's "universally admitted" theories were the "most accurate . . . of any yet proposed."[4]

The "general outline" of Cuvier's science concerned the nature of life. What life is, Cuvier claimed, will be revealed only once we learn how to read the structure of fossils. The centrality of fossils for disclosing the secret of life lay in the fact that they never include contemporaneous living forms (a species of trout, for instance, can't be both living and fossilized) but are instead relics of a life form now completely extinct. Because they are inanimate signs of what is not, Cuvier understood them to be monuments; and because monuments are the naturalist's central concern, naturalists, on his account, became in fact "antiquarians": "As a new species of antiquarian, I have had to learn to decipher and restore these monuments, and to recognize and reassemble in their original order the scattered and mutilated fragments of which they are composed; to reconstruct the ancient beings to which these fragments belonged . . . to compare them to those that live today at the earth's surface. This is an almost unknown art."[5]

A new antiquarian is concerned only with death, because he gathers ravaged bones signifying a life different from the one that animates us "today." Derived from a different vitality that we don't experience, the naturalist's bones are a silent but potentially decipherable archive in nature. Provided he is sufficiently skilled, Cuvier's hermeneut of osseous textuality can learn how to read the bones, becoming a fortune teller in reverse: sometimes from a fragment of one bone only he is able to read its genus and species, thus letting appear a whole world that is now terminated: "I shall expound the principles . . . of recognizing a genus and distinguishing a species from a single fragment of bone."[6] That we live substantially different lives from those documented by fossils[7] suggested to Cuvier that the forces of life proceed by interruption, that there is something like a revolutionary period during which life is on hold: "the surface of the globe has been upset by successive revolutions and various catastrophes. . . . Life on earth has often been disturbed by terrible events."[8] And since such revolutionary leaps enact an absolute separation of life forms, there was with Cuvier, as François Jacob put it, "a breach . . . opened, not only between beings and things, but also between groups of beings. . . . To display beings and fit their organization into the conditions of existence, nature proceeded by leaps and bounds." That is to say, as Edward Lurie explains, that "since it was impossible to discover transitional forms between ancient and living individuals," in Cuvier's view "species . . . were fixed and immutable."[9] Cuvier's antitransformism[10] thus presumed everything that Thoreau would come to resist: the radical ban on metamorphoses; the rejection of the crisscrossing of species; the promotion of "branches" in their unchangeable forms; and the treatment of anomalies not as thresholds of new beginnings but as autonomous forms, destined to be fossilized.[11]

In 1833 Agassiz published *Recherches sur les poissons fossiles*, which had an immediate influence on American naturalists.[12] Specifically focusing on the status of marine fossils as monuments, the book revealed Agassiz's unreserved acceptance of Cuvier's philosophy of life ("I can't repeat enough . . . how much I owe to Baron Cuvier"), confirming polygenetic hierarchies among unchangeable species, reasserting the right of fixity and identity: "Thus [the fish's] body has a single form; their head is not detached at all from the trunk . . . ; their sense organs are obtuse . . . ; their dual members are not yet principal organs of movement. . . . The reptiles that succeed fish in temporal order already offer a more perfect organization. . . . In the case of birds which come next we observe very remarkable development."[13] Life, then, proceeds by intensifying forms, from least to most articulated, so that the human—the most formed, hence most isolated creature—comes last to dignify life with its "noble posture" allowing it to finally look at the face of "its creator."

However, Agassiz's *Recherches* did not simply side with Cuvier but proved what Cuvier had left uncertain. For even though Cuvier claimed the importance of all fossils in reconstructing past life forms, which meant—in contrast to eighteenth-century geology, which denied that fish had ever been alive and organic[14]—that he admitted marine fossils into the realm of life's monuments, he also suggested that fish fossils were the least reliable of antiquarian reconstructions. Because fish live in water (the most mobile geological element), he thought it possible that some fossils were not "properly buried," that is, were not confined in their own stratum but instead were moved by the water to a lower or higher one. An antiquarian could thus find a marine fossil in a stratum pre- or postdating the life form it monumentalized.[15]

In contrast to Cuvier's vacillations, Agassiz's *Recherches* claimed that marine fossils in fact afford a more reliable deciphering of life's history than any other fossil structure, thanks precisely to water. Far from producing a mobility of marine monuments—as Cuvier's intuition suggested—water, Agassiz claimed, was a "more dense . . . milieu," hence more likely than air for instance, to confine fossils to their stratum:

> These [Cuvier's] ingenious insights [determining the limits of stratified layers] . . . add a positive implication for the study of fossils. The comparisons among the remains of diverse formations were multiplied infinitely with the general result that they demonstrated that species do not pass from one formation to another but that they are circumscribed within the limits that correspond to a great extent to the divisions between the strata. . . . For my part at least I know up to this point within the classes that I studied in particular [fish] that no species can be found in two geological formations at the same time.[16]

Two years later, in 1835, reporting his visit to Gideon Mantell's marine fossils museum in Brighton to the readers of the *American Journal of Science*, Agassiz reasserts his hypothesis that marine fossils remain conserved by water in their proper geological layer, for there he saw a marine fossil perfectly preserved and monumentalizing with uncanny precision the very moment when the fish died. In Mantell's museum Agassiz saw displayed, as if on a photographic plate, a fish fossilized while swallowing—with its mouth wide open—

accurately capturing the moment at which the era of life it belonged to was terminated: "In some specimens the mouth is open, as if in the act of swallowing. . . . This fact is of considerable importance, as it proves that the animal perished by some sudden evolution of mineral matter, which encased the body before the putrefactive process had commenced, and enabled it to resist the pressure of many hundred feet of chalk deposited over it."[17] Because strata are thus firmly delineated—the geological environment preserves fish in their originary condition—the study of marine fossils becomes necessary for deciphering the archaeology of the most unarticulated, and so most inhuman, life.

During the winter of 1846–1847, when Thoreau is working hard on the second draft of *A Week*,[18] Agassiz is in Boston giving a series of talks at the Lowell Institute on the "Plan of Creation in the Animal Kingdom," which he will finally publish in 1848 as a handbook for American colleges.[19] There polygenesis, immobility, and the hierarchal fixity of both species and individual identities are intensified, dismissing inter-species connections as a dilettantish fictionalization and allowing for no variations within genera, since "*varieties*" are always anomalous and "seldom endure beyond the causes which occasion them."[20] Moreover, Agassiz's "American" book essentialized all varieties into a "type" that would only interest an antiquarian:

> For each of these groups, whether larger or smaller, we involuntarily picture in our minds an image, made up of the traits which characterize the group. This ideal image is called a Type, a term which there will be frequent occasion to employ, in our general remarks on the Animal Kingdom. This image may correspond to some one member of the group; but it is rare that any one species embodies all our ideas of the class, family, or genus to which it belongs. Thus, we have a general idea of a bird; but this idea does not correspond to any particular bird, or any particular character of a bird. . . . And yet every one has a distinct ideal notion of a bird, a fish, a quadruped, &c. It is common however, to speak of the animal which embodies most fully the characters of a group, as the type of that group. Thus, we might

perhaps regard an eagle as the type of a bird, the duck as the type of a swimming-bird, and the mallard as the type of a duck.[21]

In Agassiz's circular procedure of ideating reality, individuals come to represent multiplicities (the eagle represents birds), because multiplicities (general ideas) don't represent individuals accurately. Thus, because the idea of a "bird" doesn't necessarily lead us to imagine an eagle (the "general idea of a bird . . . does not correspond to any particular bird"), the eagle (the strongest, the fastest) should be taken as representative of nonswimming birds in general, a gesture of substitution with significant political consequences. For when the "ideal strongest" is claimed as a "typical" representative of all, then everything that is fragile, low, or mixed becomes anomalous, hence contingent ("seldom endur[ing] beyond the causes which occasion" them), something that can be dismissed as irrelevant. Following his earlier ideas concerning revolutionary and polygenetic gaps among beings, Agassiz's "American" book opens his life science toward the biopolitical imagination of the preferred ideal.

The final version of the "discourse on fish" that Thoreau offers in *A Week* clearly opposes the Cuvier-Agassiz theory of catastrophe, affirming, as I argue, the value of what is continuous, anomalous, fragile, empirical, and analogous.[22]

The continuous. Thoreau's discourse on fish explicitly insists on the continuity of life: "fishes . . . are . . . forms and phases of the life in nature universally dispersed" (W, 25). To suggest that fish that can be observed now *are* phases of a life that has been is in fact to disclaim life's discontinuity. It is to affirm, as Thoreau argues elsewhere in *A Week*, that far from being the sedimentation of eroded life, fossils are animated by the life integral to all other, "later" life forms, testifying that life is incessant: "The eyes of the oldest fossil remains, they tell us, indicate that the same laws of light prevailed then as now" (W, 157). Hence, if past life is mixed into different, current life forms, then there is no life deposited into a severed history of fossils, as was advocated by Agassiz. And because all life survives in what comes after it, uninterrupted by catastrophes, Thoreau can, in a stunning passage from *A Week*, radically redefine the nature of fossils:

> It concerns us to attend to the nature of fishes, since they are not phenomena confined to certain localities only, but forms and phases of the life in nature universally dispersed. The countless shoals which annually coast the shores of Europe and America are not so interesting to the student of nature, as the more fertile law itself, which deposits their spawn on the tops of mountains, and on the interior plains; the fish principle in nature, from which it results that they may be found in water in so many places, in greater or less numbers . . . The seeds of the life of fishes are everywhere disseminated, whether the winds waft them, or the waters float them, or the deep earth holds them; wherever a pond is dug, straightway it is stocked with this vivacious race. They have a lease of nature, and it is not yet out. (W, 25–26)

Whereas in Agassiz's paleontology fossils are confined to their specific stratum by the more or less dense milieu they inhabit, Thoreau's earth is less dedicated to maintaining zones of belonging than to releasing and mixing. In it nothing is ever confined to a "certain locality only," but instead elements, fossils, and beings merge; "fossils" are "everywhere disseminated" by winds, water, or earth: "on the tops of mountains," "on the interior plains," "by the seaside, or by the lakes and rivers, or on the prairie," so that "even in clouds and in melted metals we detect their semblance" (W, 26). Fish are thus not fossilized, because, far from being petrified traces, their remnants are absorbed into the flow of terrestrial mobility and turned by it into seeds dispersed in nature, confusing geological classifications, revising the past in all present organizations. Animated relics animate, functioning as transmitters of life, embodying nature's "fertile law." In fact, this "fertile law"—the vitalization of marine fossils and their dispersion throughout strata—is the law of all life, which is why Thoreau calls it the "fish principle in nature": all life is a transmutation of seeds called "fossils"; it is formed, transformed, and then disseminated into new seeds without ever being interrupted by death.

Additionally, to claim that everything mixes into everything else does not only advocate life's linear continuity, it does not only insist on the "progression" of the tree of life; it also tacitly disclaims any possibility of the genealogical ordering of beings. It suggests that not only do fish precede certain other life forms (birds, humans), but because water doesn't confine them, they can also be contemporaneous with life that comes *after* them: by acting as seeds, they undergo mutations and display incomprehensible similarities. In fact, a note for *A Week*, written in 1844, seven years before Thoreau reads Darwin—but left out of the final version perhaps because it was too openly polemical—explicitly claims such disoriented chronologies of life and radically resists Agassiz's creationist polygenesis that degraded fish into a life form that was separated from especially created man. Thoreau claims that the difference between fish and man is impossible to determine:

> How many young finny contemporaries of various character and destiny, form and habits, we have even in this water- . . . We shall be some time friends I trust and know each other better. Distrust is too prevalent now. We are so much alike? Have so many faculties in common!- I have not yet met with the philosopher who could in a quite conclusive undoubtful way-show me the and if no then how any-difference, between man and a fish. We are so much alike. (J, 2, 108)

All the characteristics Agassiz points out to distinguish fish from humans—their unarticulated body as opposed to humankind's complex corporeal organizations (the fish's "head is not detached at all from the trunk" but "is a simple extension of it"), and their life verging on the inorganic (fishes have "dual members" not "yet organs")[23]—are for Thoreau inconclusive. They betray a fear of connecting with what seems different (Thoreau identifies it as the "distrust" of relations) and disregard the substantial sameness of a life that generates unarticulated as well as articulated beings, that minimizes the precise delineation of species boundaries. Rendering life continuous and relational—in short, as one—Thoreau's democratic ichthyology also cancels any ranking of living beings: the "highest" life is of the "lowest," human is fish but, by the same token, the lowest is of the highest, fish is human. This is not to say that all life is degraded but, oppositely that all life—even its "lowest" forms—is elevated into what must be dignified by the same utmost care: "How much could a really tolerant patient humane and truly great and natural man make of them [fish], if he should try? For they are to be understood, surely, by no other method than that of sympathy. It is easy to say what they are not-to us-i.e. what we are not to them-but what we might and ought to be is another affair" (J, 2, 108).

The anomalous. Additionally, the inarticulate nature of fish—the fact that fish are not "yet organic," but has "dual" members—which produces Agassiz's diagnosis that they are the "lowest" life far surpassed by more stable forms, is welcomed by Thoreau. The unformed or, alternatively, multiform nature of fish life, reinforced in *A Week* by his assertion that fish are "slightly amphibious" beings (W, 29), signals the multiple, hence, mobile nature of all living forms. Fish's amphibious nature, always branching in at least two directions—an amphibian lives in different elements (for instance, water and air) because its form of life is not quite "determined" but diploid (from Greek *amphy* + *bios*)—clues us into the insight that life can be without being confined to a single form, or else, that animation doesn't necessarily end in the self-stabilization of a body closed on itself. In other words, the "non-dwelling," "migratory," or "dancing" characteristics of fish life (W, 25, 29), which *A Week* systematically reaffirms—so that even when they seem to be "stationary" like the bream, fish "keep up a constant sculling or waving motion with their fins" (W, 27)—demonstrate that life never settles but is always in a process of formation. The amphibiously imagined boat on which the Thoreau brothers glided up and down the Concord and Merrimack rivers—"If rightly made a boat would be a sort of amphibious animal, a creature of two elements, related by one half its structure to some swift and shapely fish, and by the other to some strong-winged and graceful bird" (W, 16)—was thus not simply an exercise in allegory on the part of a philosophical imagination. It was also an embodiment of the idea that, materially, all life is a force of endless digression, producing the potential for strange attachments.

Thoreau witnesses such attachments everywhere: the cork, a "true product of the running stream," mixes with the vegetal, suddenly emerging "amid the weeds and sands" (W, 30); the American

Stone-Suckers transgress the boundary between animate and inanimate, developing an assemblage between their body and stones: they appear to eat stones ("collect these stones, of the size of a hen's egg, with their mouths"), play with them (they "fashion [the stones] into circles with their tails"), and climb rather than swim ("ascend falls by clinging to the stones" [W, 33]); the Common Eel, has a strange "inorganic" body, made of earth and found on dry land ("a slimy squirming creature, informed of mud . . . it too occurs in picture, left after the deluge, in many a meadow high and dry" [W, 32]). Fish thus become trees or reshape themselves into stones, signaling that life is composed of divergences. In contrast to Agassiz's understanding of life as invariable ("*varieties* . . . seldom endure beyond the causes which occasion them"), in Thoreau, then, there are nothing but ad hoc causes related through a network of unpredictable occasions.

Following the insights of his "fish discourse" and influenced by Edward Tuckerman's biology, Thoreau will expand on the multifariousness of life in the direction of radical transformability. For even though he reads Darwin's *The Voyage of the Beagle*[24] in 1851, and takes extensive notes from the book, his 1852 meditations on mutation signal that the philosophy of life he began to fashion with the "fish discourse" was more profoundly advanced by Tuckerman, a Bostonian naturalist and Amherst college professor of Botany, than by Darwin. Thoreau reads Tuckerman's *An Enumeration of North American Lichenes*, originally published in 1845—and dedicated to Thoreau's professor, Thaddeus Harris—in February 1852. Reflecting on how not just classification but also perception can remain true to the flow of life ("How can we simply follow nature, and where must art come in?"), the preface to the book—from which Thoreau also took extensive notes—clearly privileges the singular as opposed to the categorical or taxonomical. Although Tuckerman recognizes that the "tendency to an individual and separate nature [of] all natural bodies" is what allows for the scientific separation of families and classes, he also claims that separateness is never fixed but subverted by affinities.[25] In a cogent passage he thus activates Coleridge's "affinity" rather than the scientific concept to suggest, rather subversively—because such a claim disables the classifying he is about to undertake—that all life is connected, hence harmonious, but consists of unpredictable or "promiscuous" relations: "we know of a law of nature which concerns all plants, and seems to indicate a harmony which will never admit of complete expression in a system. 'Do not all,' says Coleridge, 'press and swell under one attraction, and live together in promiscuous harmony, in the immediate neighbourhood of myriad others that in the system of thy understanding are distant as the poles?"[26] In Tuckerman's understanding, affinities are attractions of functions or structures that expel being to relate it, often, to what is incongruous or disparate, thus generating anomalousness ("promiscuous harmony"). Tuckerman detects those affinities at work not only within the same families or kinships ("immediate neighborhood") that Darwin recognized,[27] but through and across them, relating what is remote and questioning the existence of stable identities. The idea of affinities as infinite relations thus suggested to Tuckerman that life is made of "diverging or converging" phenomena, each point of divergence constituting a "deviation" that will multiply throughout the network: "the series of nature and of plants is not simple, neither as respects all plants, nor as respects particular groups; . . . deviations observable from particular types do not form a single, but more, series."[28]

Thoreau drew radical but logical consequences from Tuckerman's suggestion that life is composed of diverging relations, receiving it as confirming his idea that forms are migratory, as formulated in the discourse on fish:

> Tuckerman says that Fries "states formally the quaquaversal affinity of plants," and hence rejects once more the notion of a single series in nature. He declares species "unica in natura fixe circumscripta idea," and hence all superior sections are more or less indefinite." Just as true is this of man.- even of an individual man. He is not to be referred to or classed with any company. He is truly singular-and ab-normal ever. (J, 4, 334–335)

Importantly, multiple serialization doesn't here mean separate creations, but infinite ramification through the activity of affinities ("quaquaversal affinity of plants"). It allows Thoreau to emphasize his own understanding of life: if not only the boundaries between genera or families but also the forms delineating individuals are destabilized by mobile affinities ("as true is this of . . . an individual man"); if each individual is contrived by slightly different affinities that ramify into an indefinite number of other beings, then each individual is not only infinite (transcending its form into others) but also irreducibly specific, precarious, and precious (specific because contrived of a unique combination of affinities; precarious because forms are unstable; precious because each form exists for a moment only, always on the way into other forms). And since all living beings are such peculiar, unclassifiable singularities they are all "abnormal ever." But universal abnormality is not a normative term identifying pathological deviations of a "type," instead it is the general logic of life that ordinarily proceeds by manufacturing the atypical.[29]

Thoreau remains so convinced of the incomparable, and hence unclassifiable, nature of every living being that as late as October 13, 1860, after reading Darwin's *On the Origin of Species*, he will still claim the only order in nature to be transformative "disorder," for nature doesn't proceed by "system or arrangement":

> You are not distracted from the thing to the system or arrangement. In the true natural order the order or system is not insisted on. Each is first, and each last. That which presents itself to us this moment occupies the whole of the present and rests on the very topmost point of the sphere, under the zenith. The species and individuals of all the natural kingdoms ask our attention and admiration in a round robin. We make straight lines, putting a captain at their head and a lieutenant at their tails, with sergeants and corporals all along the line and a flourish of trumpets near the beginning, insisting on a particular uniformity where Nature has made curves to which belongs their own sphere. (Jn, 14, 119–120)

Instead of being a progressive evolutionary ordering that would proceed along "straight" lines to privilege humankind ("captain at their head"), life proceeds by "curves" that circle or come back to produce a changeability that disables uniformity. Thus, to understand life—which in curving upon itself always takes strange and unpredictable routes—is to understand each singularity that occupies the "present" in the uniqueness of its phenomenality. And because everything is atypical and at variance with everything else—producing its own "music"—nothing can be identified as standard or "mainstream"; or else, there are no deviations, because there is nothing but deviance; there are only incessant, albeit slow, efforts at mutation, a straying that disorients all boundaries and forms.

The right that Thoreau accords the atypical—suggesting that no being should be marginalized on the basis of its difference—represents the radical originality of his philosophy of life. For other nineteenth-century naturalists—from Geoffroy Saint-Hilaire to Darwin—did indeed recognize variability but only as the more or less complex divergences of an individual against the backdrop of a "specific type," which safeguarded relatively stable kinship and species boundaries.[30] But in Thoreau, such "specific types" are missing, for not even species boundaries are stable. Thus, only a week after reading Tuckerman, Thoreau suggests that the distinction between animal and vegetable is porous, since various animals, such as deer and oxen, routinely become trees:

> It is interesting to meet an ox with handsomely spreading horns. There is a great variety of sizes and forms though one horn commonly matches the other. I am willing to turn out for those that spread their branches wide. Large and spreading horns methinks indicate a certain vegetable force & naturalization in the wearer-it softens & eases off the distinction between the animal & vegetable- The unhorned animals and the trees- I should say that the horned animals approached nearer to the vegetable- The deer that run in the woods as the moose for instance carry perfect trees on their heads. . . . Oxen which are de-animalized to some extent approach nearer to the vegetable perchance than bulls & cows—& hence their bulky bodies & large & spreading horns. Nothing more natural than that the

> deer should appear with a tree growing out of his head. Thus is the animal allied to the vegetable kingdom & passes into it by insensible degrees. (J, 4, 352)

Just as animals and plants are caught in a process of vital mixing, "softening" the boundaries among life forms, so humans mix with plants, their brain serving as a rhizome that grows not just thoughts but also grass and flowers. That is, at least, how Thoreau understood the function of his own brain, for he explained how he would put botanical specimens he found during his walks in a "special shelf" built inside his hat, so that they could be close to his brain, because it was his brain that "helped to keep the specimens moist," and so to nourish them.[31] Thus, when Thoreau suggests in *A Week* that all life is multiply articulated like the life of fish; or when, reaffirming that insight, he argues in *Walden* that, like any other life in nature, human life is "various" and of "several constitutions" (Wa, 10); or when he suggests in the *Journal* that humans, like lichens, ramify and mix to the point of becoming unclassifiable, he cancels de facto all "backgrounds" (congruities and kinships) and instead suggests that there is nothing essentially human, nothing that separates humans from other living beings, nothing that ranks them as the "highest" living beings. Because the human is a site of infinite becomings, hosting everything human eyes see, what the human is will have to be decided each time anew: "Man is but the place where I stand & the prospect (thence) hence is infinite. it is not a chamber of mirrors which reflects me-when I reflect myself-I find that there is other than me. man is a past phenomenon to philosophy" (J, 4, 420). At the time when anthropology is emerging to position humankind at the center of epistemological concerns and to organize those concerns into what will become known as the "humanities," Thoreau declares the datedness of such a project, clearly foreseeing what Michel Foucault will diagnose a hundred years later: that the faces of humankind are variable signposts in a transformative process in which an "I" finds itself strangely missing, dislodged into trees, birds, and animals. Thoreau finds out that humankind is a series of faces of others, like Foucault's unstable figure in the sand, representing the approaching end of anthropocentric thinking: "one can certainly wager that man would be erased, like a face drawn in sand at the edge of the sea."[32]

The fragile. Because forms are precarious, understanding life requires not only that one render transparent the moment of transformation but also that one grasp how it is that a form endures at all. The discourse on fish presents such a transparency by systematically registering how living forms emerge. Thus the bream appears in the summer "assiduously" (W, 26), the minnow "instantly" (W, 27), the red chivin "suddenly" (W, 30), the pickerel "slowly," and the sucker "mysteriously" (W, 32). But more than referencing the styles of life's appearance, Thoreau's discourse attends to the moment forms dissolve to be figured otherwise and to the disorientation of boundaries between beings that such dissolutions enact. Thus, most spawns—anomalously living "attached to the weeds" (W, 27)—are swallowed by minnows before ever being articulated into a fish ("The spawn is exposed to so many dangers, that a very small proportion can ever become fishes" [W, 27]); the shiner "glides its life through" until in summer it "almost dissolve[s] by the . . . heats" (W, 31); American Stone-Suckers migrate out of their forms into what they anomalously love ("as they are not seen on their way down the streams, it is thought by fishermen that they never return, but waste away and die, clinging to rocks and stumps of trees for an indefinite period" [W, 33]); the Horned Pout retains its life even when it is disfigured ("they are extremely tenacious of life, opening and shutting their mouths for half an hour after their heads have been cut off" [W, 31]); whereas the pickerel, strangest of all, can under certain circumstances be half formed and half liquefied into another body ("I have caught one which had swallowed a brother pickerel half as large as itself, with the tail still visible in its mouth, while the head was already digested in its stomach" [W, 31]).

Thoreau's examples don't simply identify fragmentation and disfiguration as strategies of life; more radically, they suggest that dissolutions are often not a result of the strong overpowering the weak, involuntarily performing the natural selection that Darwin described. When suckers collectively mutate into the stones that attract them; or when spawns build their nest "so near the shore, in shallow water, that they are left dry in a few days, as the river goes down" (W, 27) and the whole generation of larvae is extinguished, those species have not been overpowered by stones or water; rather, it is their own life that leads them to dissolution. This is to say that Thoreau's observations describe life unguarded against its own motions, as if ill-attuned to its interests and so inherently unprotected. Too fragile to act against its disintegration, life turns that disintegration into a pervasive phenomenon. That is so even when such disintegration appears unnecessary (not caused by sickness or sudden change of conditions), or amounts, as Thoreau suggests regarding suckers, to a pure "wasting away," for ultimately it happens to all the fish Thoreau mentions and it happens systematically, summer after summer. And yet, this disintegration is not the cessation of animation. Reinforcing his claim about life's continuity, and contrary to Agassiz's catastrophism, Thoreau's examples suggest that, rather than being interrupted, life is only reorganized into other bodies as its forms dissolve (half of a pickerel becomes its brother's body, while the other half remains its own; spawns disseminate into trees), as if dying beings were carried on by other living ones. In short, for Thoreau, life's continuity is not interrupted by its fragility, an insight he confirms and generalizes when observing a sparrow in 1845:

> It is a marvel how the birds contrive to survive in this world- These tender sparrows that flit from bush to bush this evening- though it is so late do not seem improvident, to have found a roost for the night. They must succeed by weakness and reliance, for they are not bold and enterprising as their mode of life would seem to require, but very weak and tender creatures. I have seen a little chipping sparrow come too early in the spring shivering on an apple twig . . . and it had no voice to intercede with nature but peeped as helpless as an infant and was ready to yield up its spirit and die without any effort- And yet this was no new spring in the revolution of the seasons. (J, 2, 125)

As so many of Thoreau's examples assert, life doesn't acquire what its conditions "seem to require," as if its forms can't adapt to the conditions it generates (sparrows are thus not adjusted to habitual New England cold, nor suckers to stones or shiners to heat), an incapacity that appears to run counter to life's interest in self-prolongation. However, if despite its delicacy life does continue and beings do survive, it is perhaps because fragility is not contrary to life's continuation, but sufficient for it. Hence, Thoreau's distinctly non-Darwinian idea: survival is enabled by what is unable and nonfitting; life "succeeds by weakness" and tenderness, not by "boldness and enterprise"; its way is to yield, not to aggress and dominate; it is the way of survival, marvelous for being the contrivance of an exhausted fragility "ready to yield up its spirit and die" effortlessly. Thoreau generalizes that proposition to include humans a year later, while still revising his "fish discourse": "I am astonished at the singular pertinacity and endurance of our lives. The miracle is that what is-is-precisely that-that our particular life succeeds so far. That every man can get a living and so few can do any more So much I can accomplish ere strength and health are gone and yet this much suffices" (J, 2, 328). Years after health appears to be diminished and strength consumed, when the body feels it no longer has any possibilities at its disposal, exhaustion stirs and impels ("accomplishes much") to regenerate life. That is what is so astonishing about the fact of existence.

It is against the background of these reflections on life's fragility that the "discourse on fish" ends by offering the radical principle of life's inconsistency and reliance on illness:

> As if consistency were the secret of health, while the poor inconsistent aspirant man, seeking to live a pure life, feeding on air, divided against himself, cannot stand, but pines and dies after a life of sickness, on beds of down. The unwise are accustomed to speak as if some were not sick; but methinks the difference between men in respect to health is not great enough to lay much stress upon. Some are reputed sick and some are not. It often happens that the sicker man is the nurse to the sounder. (W, 36)

If the idea of pure health represents an untruthful way of looking at life (only the "unwise" make such a distinction), neglecting the fact that all life is sick to a degree and so animated by illness and tiredness, then insistence on physical strength and health is fictional and hence ideological. In opposition to the discourses of empowerment and health that were about to appear in Continental philosophy, science, and politics, leading to catastrophic historical developments, Thoreau thus claims that because all life is feeble, shivering helplessly, and because feebleness suffices and survives, it is the frail, the helpless, and the delicate that animates and creates.

The empirical. Life conceived of as multifarious and relational requires an approach different from that offered by scientific strategies. Thus, against Agassiz's ideational practices that generate the typological, Thoreau's discourse on fish promotes a radical empiricism. It is radical because it refuses to mediate, even if provisionally, the mobility of life by means of ideal classifications, eschewing taxonomies that would assimilate what is specific to the singular.[33] By 1851 Thoreau will come to state explicitly that to honor the singular, this radical empiricism must by definition do away with generalized scientific theories: "Ah give me pure mind-pure thought. Let me not be in haste to detect the universal law, let me see more clearly a particular instance. . . . Dissolve one nebula-& so destroy the nebular system & hypothesis" (J, 4, 223). Radical empiricism abandons systems and hypotheses to follow what Walls calls " 'mingled and imperfect' particulars of [the] world."[34] Because it treats every being as incomparable in its singularity ("What slight but important distinctions between one creature & another" [J, 4, 208]), radical empiricism—unlike a classical empiricism, such as that of Bacon—doesn't even allow for generalizations by induction. Hence, Thoreau dismisses such inductive protocols in "The Natural History of Massachusetts": "we cannot know truth by contrivance and method; the Baconian is as false as any other."[35]

And although Thoreau did use generic scientific names in his discourse on fish, his correspondence with James Eliot Cabot—Agassiz's assistant, through whom Thoreau and Agassiz conversed briefly if indirectly—showed a resistance to do so. In the spring of 1847, while at Walden Pond revising *A Week*, Thoreau also collects specimens of fish "for Agassiz" but ships them to Cabot, all the time voicing his disagreement with Agassiz's typologies, something Cabot simply registers as Thoreau's naivety. In a letter to Cabot of May 8, 1847, at the risk of having his descriptions identified as "impertinent and unscientific," Thoreau suggests that no fish can accurately be called "pickerel" because one found at the brook bites differently from one found near his house: "*Pickerel.* Besides the common, fishermen distinguish the Brook, or Grass Pickerel, which bites differently" (C, 293). The same refusal to substantiate through identifications is registered in Thoreau's description of pouts, breams, and trout: "*Pouts.* Those in the pond are of different appearance from those that I have sent. *Breams.* Some more green, others more brown. . . . *Trout.* Of different appearance in different brooks in this neighborhood. . . . *Red-finned Minnows,* of which I sent you a dozen alive" (C, 294). Three weeks later, replying to Cabot's report that the twelve minnows Thoreau sent to Agassiz were unknown, Thoreau explicitly questions scientific nomenclature: "Do you mean to say that the twelve banded minnows which I sent are undescribed, or only one? What are the scientific names of those minnows which have any?" (C, 303).[36] Provocatively, Thoreau's questions suggest that in making concessions to generality, scientific identifications move away from the empirical and produce epistemological fallacies.[37]

As late as October 9, 1860—a telling date, since it is sometimes presumed that in later life Thoreau became more supportive of the strategies of contemporary scientific research—he dismisses the idea that life should be killed to reach the stability necessary for scientific observation to propose a generalized classification: "This haste to kill a bird or quadruped, and make a skeleton of it, which many young men and some old men exhibit, reminds me of the fable of the man who killed the hen that laid golden eggs, and so got no more gold. It is a perfectly parallel case. Such is the knowledge you may get from anatomy as compared with the knowledge you get from the living creature" (Jn, 14, 109).[38] Importantly for my argument, the same critique of scientific taxonomies was made regarding ichthyology on November 30, 1858:

> Though science may sometimes compare herself to a child picking up pebbles on the seashore, that is a rare mood with her;

> ordinarily her practical belief is that it is only a few pebbles which are *not* known, weighed and measured. A new species of fish signifies hardly more than a new name. See what is contributed in the scientific reports. One counts the fin-rays, another measures the intestines, a third daguerreotypes a scale, etc., etc.; otherwise there's nothing to be said. as if all but this were done, and these were very rich and generous contributions to science. Her votaries may be seen wandering along the shore of the ocean of Truth, with their backs toward it, ready to seize on the shells which are cast up. You would say that the scientific bodies were terribly put to it for objects and subjects. A dead specimen of an animal, if it is only well preserved in alcohol, is just as good for science as a living one preserved in its native element. (Jn, 11, 359–360)

Thoreau doesn't voice the skeptic's concern that truth is inaccessible; rather, he states that truth can't be attained by the scientific protocols of generalizing through classifying, since in maintaining a distance from the particulars of life classifications amount to the *poiesis* of fiction. This is to say that all science—and Agassiz's typology is an extreme example—offers a figuration of what does not exist (scientists are said not even to see the truth, having "their backs" turned to it) but is only imagined. Science, then, just like the poetry Thoreau criticizes, is an aesthetisized idealism. And since ideation is ingrained in both poetics and scientific epistemologies, for Thoreau the question will not be how to make poetics scientific or how to poeticize science, as is sometimes suggested, but instead how to generate a new poetry (predicated not on prosody) and a new science (predicated not on taxonomies), what Philip Cafaro understands as "knowledge supplementary to science."[39]

In the same 1858 *Journal* passage that critiqued scientific classifications, Thoreau—still loyal to his early claim from the "Natural History of Massachusetts" that scientific knowledge should be replaced by "direct intercourse"—finally offered a precise description of what is required and enacted by a perception that "circumnavigates naming":[40]

> In my account of the bream, I cannot go a hair's breadth beyond the mere statement that it exists,—the miracle of its existence, my contemporary and neighbor, yet so different from me! I can only poise my thought there by its side, and try to think like a bream for a moment. . . . I only see the bream in its orbit, as I see a star, but I care not to measure its distance or weight. The bream, appreciated, floats in the pond, as the centre of the system, another image of God. Its life no man can explain more than he can his own. I want you to perceive the mystery of the bream. (Jn, II, 358–359)

The difference between two contemporaneous beings sharing the same universal life is here expected to be bridged by a strange exercise of self-transformation on the part of the perceiver, leading, as Walls puts it, to his "total immersion to the point of saturation."[41] In Thoreau's account, long and patient observation of the object of knowledge doesn't generate personal knowledge of the observed that would subsequently be generalized (the bream remains inexplicable and so mysterious), but, to the contrary, impressions that undo the observer. Impressions, as Gillian Beer notes when discussing Darwin, "are a ferment, sharing the body, keeping the mind agog, multiplying positions."[42] Thoreau's "impressions" of the bream are intense enough to keep his mind agog so that he crosses over to the bream's "side," attempting to mutate into a fish ("I try to think like a bream").[43] What Thoreau imagines as new scientific knowledge, then, is a perception that metamorphoses the scientist.

The analogous. Analogy would be a linguistic and logical practice commensurate with such a process, but it was resisted by nineteenth-century science. As Agassiz's ideational strategies suggest, early nineteenth-century sciences of life welcomed homology, a form of reasoning that emphasized similarities between different phenomena before reducing them to the homogeneity of a type. Homology was thus a crucial discursive vehicle for achieving "total and satisfying congruity"[44] on the basis of which stable denominations and classifications were made. As Mary B. Hesse argues, not only were Cuvier's or Agassiz's fossil theories centered on it, but even Goethe's morphology subscribed to it: "When homologies were first described in detail by Goethe, Cuvier, Saint-Hilaire, and their successors, it was commonly suggested that community of type among organisms is based on an 'ideal type,' an archetype, or natural plan."[45]

When Thoreau explicates his 1851 dream "of a return to the primitive analogical & derivative senses of words" (J, 4, 46), he effectively welcomes analogy for the same reasons that science dismisses it. In opposition to homology, analogy brings differences into relation without gathering them into identity, thus preserving the transformative. Those are the qualities Thoreau explicitly recognizes in analogy, again in 1851, while insisting that not just scientific but all thinking should be predicated on it: "Improve the opportunity to draw analogies. There are innumerable avenues to a perception of the truth. Improve the suggestion of each object however humble-however slight & transient the provocation. . . . It is not in vain that the mind turns aside this way or that. Follow its leading-apply it whither it inclines to go" (J, 4, 41). In obeying analogical thinking, Thoreau suggests, the mind refrains from imposing criteria of unity and instead follows the heterogeneity of "the transient," extending in diverse directions ("this way or that").

But the specificity of analogical thinking, leading to its being discredited by scientists, is that in observing similarities (morphological or otherwise) among diverse phenomena, it claims them to be causally connected even though such connection is not perceptible. This connecting of different phenomena is neither allegorical (for there is no "one-to-one correspondence of object and meaning"), nor metaphorical (different entities are not considered simply similar, and "resistance as well as accord" persists),[46] nor emblematic (one phenomena doesn't represent another). As Hesse explains, in analogy—hence its strangeness—similes serve as forms of "analogical inference" connecting co-occurring phenomena whose causal relations are not clearly visible.[47] Analogy is thus not simply a rhetorical device meant to suggest imagined connections but a strategy of reasoning that uses language to claim actual relations. For instance, in Thoreau's discussion of the bream, the bream's color, analogous to a "brilliant coin," is claimed to be a material "concentration" of light, that traverses "floating pads and flowers," materially relating them all to the "sandy bottom" of the river. Similarly, in his "discourse on fish," the observable chromatic similarity among phenomena as different as metal, fish, and clouds results in a morphological analogy that suggests a causal relation in analogous coloration: "fishes too, as well as birds and clouds, derive their armor from the [copper] mine" (W, 30). The chivin's skin is not simply *like* copper; rather, the two entities are causally connected as fish *derive* their armor from the copper banks, as if metal brings the fish skin into being. By registering phenomenal similarities (for instance, the similarity in color between skin and copper), analogy concludes the causal bonds (it presumes that the sameness of color must be generated by the same cause). In that way it existentially bridges beings. Moving from what is visible (fish, copper banks) to what is obscure and unknown (the existential link between the fish and the banks), analogy claims an invisible agency (of the copper on the fish) and so casts doubt on the distinction between words and bodies. Analogy is imagined as a language capable of evoking things, which means, as Gillian Beer argues, that it aligns itself with magic:

> The shifty, revelatory quality of analogy aligns it to magic. It claims a special virtue at once incandescent and homely for its achieved congruities. A *living*, not simply an imputed, relation between unlikes is claimed by such discourse. The power of analogy to transform the homely into the transcendent or the lesser into the greater intensity draws on a formulation of experience learnt by Christians in the sacraments. Analogy requires transformation and implicitly claims transubstantiation.[48]

I suggest that Thoreau precisely celebrates analogy as a sorcerer's tool, representing it as demonological in his "Natural History of Massachusetts," a text he writes while mourning John in 1842: "He has something demoniacal in him, who can discern a law or couple two facts. We do not learn by inference and deduction and the application of mathematics to philosophy, but by direct intercourse."[49] A connection claimed (a "coupling" of facts) is not predicated on methodically organized observation or logical reasoning ("interference"), but on the immediacy of seeing through what is here into what is not perceptible, thus shortening the distance between stating and enacting causations. And since the "facts" Thoreau wants to couple are never simply imagined but are, as Walls underlines, always "*vital*," to couple them is to generate, just as sorcerers do, real vital relations (what Beer refers to as a "living, not simply an imputed, relation").[50] If, then, Thoreau insists that "all perception of truth is the detection of an analogy" (J, 4, 46), it is because analogy is a practice of reasoning adequate to the vitalism he imagines: a force of stating, and so generating, continuous connections and mutations. By demonstrating that sense can still be detected in the collapse of the categories and methodologies supposed to generate it, analogy shocks reason into awe, and renders utterable and material what is illicit, suspect, and barely thinkable: metal fish, a flower related to a snake, or, by extension, a brother who is a bird.

Stones

The idea that fossils are not monuments separated by catastrophes from current geological conditions was something Thoreau learned from Charles Lyell. Lyell's *Principles of Geology* was published in London between 1830 and 1833, and Thoreau read it in 1840. Judging by the notes he took through October 1840, Thoreau found the issues Lyell raises to be of particular importance not just for his scientific thinking but also for his own ontology. For they enabled Thoreau to give coherence to his own vitalism by expanding life into what is commonly understood as inorganic. Drawing radical consequences from Lyell's geology—consequences that Lyell only gestures toward but never explicitly claims, and that Thoreau finds confirmed in the structure of Native American shell middens—Thoreau came to include the earth and the elements among living beings partaking in universal life. But his interest in the earth transcended the boundaries of geology, for his belief that even stones are animated triggered a radical critique of monumental architecture and, in relation to it, Western burial rites, leading him, finally, to propose treating the dead in a quite different manner.

The central proposition from Lyell that Thoreau registers is that geological changes testify not to the discontinuity but to continuity of earthly life. His *Journal* notes Aristotle's remark quoted by Lyell: "time never fails and the universe is eternal" (J, 1, 187). For Lyell, if time never fails, if it is uninterrupted, then the causes of distant changes are still working, still changing us. In an explicit rebuttal of Cuvier, he suggests that all past transformations, far from being separated by disruptive catastrophes, are connected into a continuity of life by forces that shape current geological changes: "We hear of sudden and violent revolutions of the globe. . . . We are also told of general catastrophes, . . . of the sudden annihilation of whole races of animals and plants. . . . In our attempt to unravel these difficult questions, we shall adopt a different course, restricting ourselves to the . . . operations of existing causes," to prove that the earth is "capable of modifications."[1] The earth was thus imagined as a continuity of mutability, a principle Thoreau registers by remarking that "we discover the causes of all past change in the present . . . order of the universe" (J, 1, 191). Even if it does so imperceptibly slowly, the earth mutates, and those mutations are never sufficiently interrupted to constitute death, which is why, according to Thoreau's understanding of the consequences of Lyell's geology, the slow motion of geological phenomena marks precisely the perseverance and continuity of life. As Thoreau puts it: "the longer the lever the less perceptible its motion. It is the slowest pulsation which is the most *vital*" (J, 1, 190). Thoreau's understanding of the continuity of geological processes as the continuity of life does not amount to a complete misunderstanding of Lyell, as Lyell himself suggested that geological principles could be expanded to the biological when he claimed that "numerous proofs were discovered of the tranquil deposition of sedimentary matter and the slow development of organic life."[2] Yet, while Lyell compares the deposit of matter to the development of organic life without identifying the two concepts (his formulation is "matter and life"), Thoreau's *Journal* entry, written a week after reading Lyell's theory, gestures toward a biological interpretation of geology, describing the first causes of geological changes as living: "pulse [of the first cause] had beat but once-is now beating" (J, 1, 191). Thoreau continues to outline his own geological theory in numerous later *Journal* entries, intensifying identification of the geological with the biological. Thus, a *Journal* entry from October 2, 1843, remarks that the "simple but slow life" of a stone contains "very few fibers . . . very little organism" (J, 2, 472), whereas the famous ending of *Walden's* chapter "Spring" claims even more radically that a stone's life is continuous with, and indistinguishable from, the organized life of vegetation.

The last pages of "Spring" to which I am referring are a version of an 1848 *Journal* entry. In both instances Thoreau reports on his observation of "clay and sand . . . flowing down the sides of a deep cut on the rail road" through which he walked (J, 2, 382), which to him seemed similar to the process Lyell described, of rock transitioning from granite to clay slates. Yet, the differences between two versions are significant. Whereas in the *Journal* the "interlocking" and "overflowing" of streams of clay is observed to look "like some mythological vegetation" (J, 2, 383), *Walden's* version abandons the simile to claim that the interlocking of clay and sand actually forms vegetation. In *Walden* streams of clay mix with water, so that clay obeys the "laws of nutrition" and engenders hybrid formations of sandy-grass and clay trees: "Innumerable little streams overlap . . . one with another, exhibiting a sort of hybrid product which obeys half way the law of currents, and half way that of vegetation. As it flows it takes the forms of sappy leaves or vines. . . . It is a truly *grotesque* vegetation . . . a sort of architectural foliage . . . destined perhaps . . . to become a puzzle to future geologists" (Wa, 305). The clay of 1848, only reminiscent of "mythological vegetation," had thus by 1854 become a true, albeit grotesque, vegetation, generated from moisture and still in the process of transforming itself from the "vitals of the globe" into the "vitals of the animal body" (Wa, 306). In *Walden* the moist clay becomes self-organizing—"it is wonderful how . . . the sand organizes itself as it flows" (Wa, 307)—functioning as a "moist thick *lobe*, a word especially applicable to the liver and lungs" (Wa, 306); functioning, that is, as the breath of the earth, turning its giant body into a living being, enabling Thoreau to proclaim his axiom "there is nothing inorganic" (Wa, 308). Here, then, is the question: if, following Lyell, Thoreau tended from 1840 to extend the biological into the geological, what happened in 1848 and after that encouraged him, at the end of "Spring," to claim their unquestionable identity and dismiss the existence of inorganic matter?

In 1851 Thoreau read Arnold Guyot's *The Earth and Man* (1849) and took extensive notes from it. A Swiss geologist educated in Berlin, where he had met Humboldt, Guyot was a colleague of Agassiz at the College of Neuchâtel; following Neuchâtel's suspension, it was Agassiz who helped Guyot come to the United States and settle in Cambridge, Massachusetts, arranging for him a series of lectures on physical geography that Guyot delivered at the Lowell Institute in 1848. And although it is commonly recognized that Guyot was influenced by both Humboldt and Agassiz, it is less often noted that *The Earth and Man*—based on his Lowell Institute lectures—was in fact an open critique of both those influences, voicing differences among them similar to those that separated Nuttall from Audubon or Tuckerman from Linnaeus.

Guyot's physical geography—much like Nuttall's ornithology or Tuckerman's botany—was predicated on the idea that sentient affinities move and organize the earth. Introducing the vitalistic orientation of his physical geography, which "forbids [him] to employ . . . the word 'geology,'" Guyot described the earth to Bostonians as a living being:

> Physical geography, therefore, ought to be, not only the description of our earth, but the physical science of the globe, or the science of the general phenomena of the present life *of the globe in reference to their connection and their mutual dependence.* This is the geography of Humboldt and of Ritter. But I speak *of the life of the globe*, of the *physiology* of the great terrestrial forms! . . . This nature, represented as *dead*, and contrasted in common language with *living* nature, because it has not the same life with the animal or the plant, is it then bereft of all life? . . . Yes, gentlemen, it is indeed life. . . . It is life . . . this general life, this physical and chemical life, belongs to all matter.[3]

On Guyot's account the living earth transforms its affective body much faster than Lyell proposed; it works at a physiological rather than geological speed—which, as Guyot explained, "forbids [him] to employ . . . word 'geology'"—rendering its changes clearly perceptible as the outcome of every motion of water or wind. The earth's life makes the expression lines of its surface mutate all the time, whence the necessity to talk about physiology instead of geology, as Humboldt or Lyell suggested.

It is Guyot's vitalism that, I am proposing, encouraged Thoreau to intensify his already vitalistically oriented geology and to repeat in a *Journal* entry from 1854 that there is indeed nothing inorganic: "There is nothing inorganic- This earth is not then a mere fragment of dead history-strata upon strata like the leaves of a book- an object for a museum & an Antiquarian but living poetry like the leaves of a tree——not a fossil earth—but a living specimen" (J, 7, 268).[4] As always, Thoreau means it literally. For the conclusion that the earth is a plastic being ("The very earth . . . is plastic" [ibid.]), which initially relied on Thoreau's observation of the actually flowing clay, was also supported by a series of experiments he conducted on Walden Pond. The experiments demonstrated that the pond reacted to atmospheric changes and, moreover, that the pond water, as if it were an animal, had a clear presentiment of approaching storms. Thus, only a week after observing the living clay and claiming the nonexistence of anything inorganic, experimenting with the pond water resulted in a *Journal* entry that declared the whole earth to be not simply animated but, moreover, imbued with feelings:

> Who would have suspected so large & cold & thick-skinned a thing to be so sensitive- Yet is has its law . . . as surely as the buds expand in the spring- For the earth is all alive & covered with feelers of sensation.—papillae. The hardest & largest rock- the broadest ocean-is as sensitive to atmospheric changes as the globule of mercury in its tube. Though you may perceive no difference in the weather-the pond does- So the alligator & the turtle with quakings of the earth come out of the mud. (J, 7, 285–286)

In negative terms, the idea that water is sensuous like the turtle swimming in it doesn't mean that water is capable of suffering the same feeling as an animal, because differences in intensity of sensation and reflexivity still remain; to write that water and earth have "feelers of sensation" like a turtle is not to claim that harming a turtle is the same as swimming in the pond. But it does suggest in positive terms that both turtle and water partake of the same life; that nothing is dead. In Thoreau's world a sensitive water-animal bathes in sensate water that touches the sensate land. Life pulsates through each of them.

Lyell's geological assessment that on a vital earth "geological monuments"—rocks, shells, bones—are less commemoration of past changes in the "regions inaccessible to us" than "evidence" that those changes "*may be in progress*" now[5] influenced not only Thoreau's geology but also his understanding of the monuments made by humans, specifically in the form of monumental and cemetery architecture. The transition from geology to history and culture was only logical, for if, as Lyell's geology suggested, stones are not monuments at all but living relics (past change still in progress); if they are, to use Jane Bennett's phrasing, "swirls of matter, energy and incipience that hold themselves together long enough to vie with the strivings of other objects,"[6] then to dislocate them from the living relations in which they change, as is always the case with monuments, is to transform them into a dead thing. By enacting such relocations, which turn stones into dead monuments that are paradoxically asked to commemorate life, monumental architecture intervenes in some fundamental way in the circulation of life.

The insight that the way we memorialize has more to do with the interruption than with the continuation of life, is, I am suggesting, what motivates Thoreau's systematic, indeed obsessive, critique of monuments found in texts from *A Week* to *Walden* and even to "Autumnal Tints." Such a critique is voiced paradigmatically in remarks on his visit to an old graveyard overlooking the Merrimack, which I quote here in the manuscript version not entirely included in *A Week:*

> we noticed there two large masses of granite more than a foot thick and rudely squared, lying flat on the ground over the remains of the first pastor and his wife. It is remarkable that the dead lie everywhere under stones——That when a man's spirit takes its flight, the parishers (variant: his neighbors) should come together and cart some huge block from its place in the ledge and place it over his body as a composter, or for fear he should rise before his time—Each under his own stone. "Strata jacent passim suo quaeque sub" lapide corpora. . . . These condemned, these damned bodies. Think of the living men that walk on this globe, and then think of the dead bodies that lie in graves beneath them, carefully pact away in chests as if ready for a start! Whose idea was that to put them there? Is there any race of beetle-bugs that disposes of its dead thus?"[7]

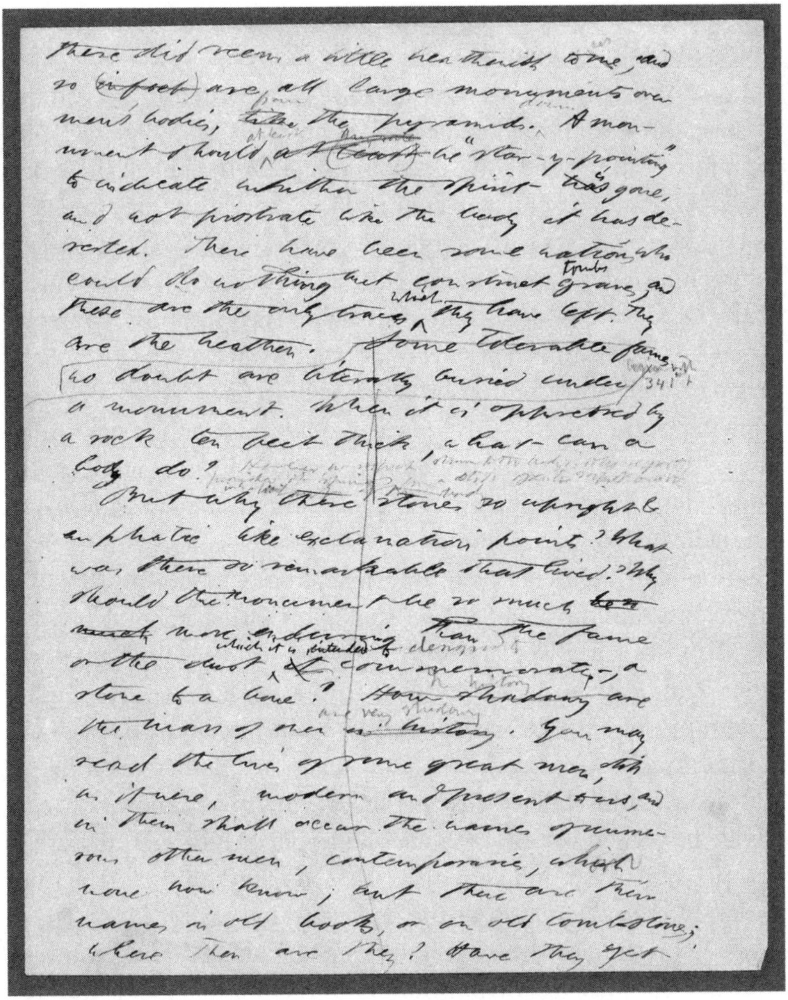

there did seem a little heathenish to me, and
so (in fact) are all large monuments over
men's bodies, like the pyramids. A monument should at least be "star-y-pointing"
to indicate whither the spirit has gone,
and not prostrate like the body it has deserted. There have been some nations who
could do nothing but construct tombs, and
these are the only traces which they have left. They
are the heathen. Some tolerable fames
no doubt are literally buried under
a monument. When it is oppressed by
a rock ten feet thick, what can a
body do?
But why these stones so upright &
emphatic like exclamation points? What
was there so remarkable that lived? Why
should the monument be so much too
much more enduring than the fame
or the dust which it is intended to commemorate, a
stone to a bone? How shadowy are
the mass of men in history. You may
read the lives of some great men, still
as it were, modern and present to us, and
in them shall occur the names of numerous other men, contemporaries, which
none now know; but there are their
names in old books, or on old tombstones;
where then are they? Have they yet

"These damned bodies," excerpt. Henry David Thoreau, Houghton Library, Harvard University, Cambridge, Massachusetts, MH 15, J. Courtesy Houghton Library, Harvard University.

Thoreau worries that the heaviness of the granite dislocated from its habitat will oppress the corpses, as if blocking their decay into the circulation of natural life. The stone extracted from the living earth to be transformed into a monument performs on those it memorializes the same abuse as has been done to it: it excludes the dead from the vitality of nature.

Other intellectual developments in Cambridge, Massachusetts, would intensify that initial geologically motivated critique of monumental and cemetery architecture and transform it into a critique of burial rites. Among those developments were Jacob Bigelow's philosophy of life and the consequences it had for the burial rites. Bigelow, yet another vitalist related to Harvard University, to whose work Thoreau refers throughout the *Journal*, explained the relation between his vitalism and burial rites in the address "On the Burial of the Dead" before the Boston Society for the Promotion of Useful Knowledge. For him the subject, far from preoccupying an isolated group of intellectuals only, had "within the last few years . . . occupied a greater share than formerly of the public attention in our own vicinity."[8] The philosophical idea that was mobilizing public attention and finding its embodiment in the Rural Cemetery Movement, whose beginnings are commonly related to the consecration of the Mount Auburn Cemetery in Cambridge, Massachusetts, in 1831, was precisely vitalism. As Stanley French put it, starting from the premise that all matter is alive, the main philosophical thesis of the movement held death to be subservient to life because "in the mighty system of the universe," what died was considered included in the eternal "circle of creation."[9] Bigelow made that point explicit in his address when he claimed that the burial rites of Western humanity should be revised according to the philosophy of life—hence death—practiced by plants and animals, which "do not become extinct nor useless after death; they offer themselves as the materials from which other living frames are to be constructed."[10] Humans, in contrast, who isolates their dead from the living earth, act unethically: "Although [man] is prodigal of the lives of other classes, and sometimes sacrifices a hundred inferior bodies, to procure himself a single repast, yet he regards with scrupulous anxiety the destination of his own remains."[11] Driven by human exceptionalism, which renders human remains superior to those of other living beings, people construct "subterranean vaults and the walls of brick" and thus "divide the clay of humanity from that of the rest of creation."[12] On Bigelow's understanding, Mount Auburn was envisaged as embodying a critique of such exceptionalism, for there the dead would be laid in "the quiet verdure of the field," where only a light emblem would mark the grave to "sooth" the survivors rather than to isolate human from the circulation of nonhuman clay.[13] The fusion of human with nonhuman "clay" is a value that Thoreau's critique of burial rites proposes, perhaps even more explicitly than Bigelow, when he insists that "the farmer who has skimmed his farm should at least leave his body to nature, to be ploughed in and in some measure restore its fertility"; or when, developing the similar idea that "fertility is increased by . . . decay," he proposes that the earth be regarded not as "a graveyard or merely a necropolis" but as a field with the "sees of life" (J, 8, 327).

Consequently, when *A Week* includes in its critique of cemetery architecture and burial rites an unfavorable comparison between Mount Auburn and the catacombs or other Continental cemeteries that Bigelow disparaged, it will not be because Thoreau disagrees with Bigelow regarding the central issues implied in new ways of burying the dead. Rather it will be because by the time he composes *A Week* Thoreau has seen more clearly than Bigelow that Mount Auburn too was becoming everything Bigelow hoped it would have eschewed. He is therefore explicit: "these [stones] did seem a little heathenish to us; and so are all large monuments over man's bodies, from the pyramids down. . . . I have not taste for such collections as they have at the Catacombs, Père Lachaise, Mount Auburn, and this Dunstable graveyard. For my part I had rather walk where there is no carrion" (W, 170, 458). Thoreau's comparison of Mount Auburn to Egyptian grave architecture is not hasty. For if the gravestones at Mount Auburn minimized Christian symbolism, it was only because by the late 1840s the cemetery had become the site of an Egyptian revival, and Egyptian emblems of death were everywhere: pyramids, obelisks, lotus flowers, and sphinxes appeared on a massive scale from the entrance of the cemetery to the individual graves. As Stanley French remarks: "There were objections to the propriety of using ancient Egyptian burial architecture . . . but as one apologist explained: 'Egyptian architecture is essentially the architecture of the grave.'"[14] The cemetery that was designed to do away with the architecture of the grave became the central site of funerary monuments.

Obviously contradicting Bigelow's prediction that Mount Auburn would do away with "monumental fortresses . . . piled over their [the

Mount Auburn Cemetery, obelisk. Courtesy David Wills.

Mount Auburn Cemetery, urns. Courtesy David Wills.

Mount Auburn Cemetery, sphinx. Courtesy David Wills.

dead] decaying bones,"[15] the most immediate effect of erecting Egyptian style monuments was the individualization of graves by means of fencing.

Thus, "by the late 1840s people started to complain that the cemeteries were becoming seas of fences. . . . According to the opinion of a foreign couple who visited Mount Auburn in the 1850s: 'the elegant iron rails, which divide the different small lots, are neither ornamental, nor . . . reverential for the place. . . . the idea of private property, carried even into the realm of the dead . . . has something unnaturally strange."[16] The graves reaffirmed the value of isolation and individuation, featuring the grave as private property, allowing both the dead and living visitors to feel at home there. It was as if even after death their inhabitants would still own a "home," a closed-off space to represent the enduring concretization of their personhood, similar to what Benjamin observed in so many European cemeteries. Commenting on David Octavius Hill's photographic portraits taken at the Greyfriars Churchyard, Edinburgh, Benjamin said: "Many of Hill's portraits originated in the Edinburgh Greyfriars cemetery—nothing is more characteristic of this early period, except maybe the way the models were at home there. And indeed this cemetery, according to one of Hill's pictures, is itself like an interior, a separate closed-off space where the gravestones, propped against gable walls, rise from the grass, hollowed out like chimneys, with inscriptions inside instead of tongues of flames."[17]

And if such cemetery architecture offers what Eduardo Cadava calls "a 'literalization' of the conditions of living," suggesting that "we live as if we were always in a cemetery,"[18] then Thoreau's critique of the Mount Auburn graves proposes to revise the way we die in the context of revising the way we live. In trying to institute the value of transforming and becoming, Thoreau sought to vitalize both our mortalized life and our mortalized death.

In this way, when Thoreau says that we shouldn't "waste time on funerals," and instead allow "a speedy resurrection of the body in some other form, in corn for fodder, in wood for fuel, in grain or flowers for use or beauty,"[19] he is not calling on us to forget the dead. On the contrary, he wants to say that it is precisely our ways of memorializing that generate forgetting. It is precisely because our burial rites exclude the dead from what is going on and materially taking place, finally rendering them nothing but dead, that for Thoreau make all efforts to preserve corpses border on sacrilege: "God says plainly, for even I hear him, lay not up this matter, but disperse it. Go into a museum and look at the mummy and consider whether the man that did that work of embalming has not got something to answer for."[20] Because the human remains hosted by graves, catacombs, or museums are not, as Thoreau puts it, "dispersed," and therefore are not, as Lyell would have it, a "change still in progress" and are still affecting living beings, they become "Old Mortality . . .

David Octavius Hill and Robert Adamson, Dennistoun Monument, Greyfriars Churchyard, Edinburgh. Calotype known as "The Artist and the Gravedigger." Collection of the Scottish National Portrait Gallery, Edinburgh. Courtesy National Galleries of Scotland.

rather [than] Immortality," "defaced and leaning monuments," a storyless void hollowed out by what is utterly gone (Wa, 269).

But when Thoreau critiques so intensely the human effort to preserve human corpses, he is not speaking simply about the struggle on the part of the survivors against the material decomposition of the dead, and not even about what Derrida so accurately called "the old humanist and metaphysical theme . . . (the) burial (place) [as] the proper(ty) of (the) man" meant to "wrest [the] singularity" of the dead from nature.[21] He is also questioning the very premise of that metaphysics, that is, the priority of spirit over matter that burial rites are supposed to embody. In so doing he articulates yet another objection to idealism. For as Derrida convincingly argued in his discussion of Hegel, Western/Christian burial rights are organized around the idea that the burial and mourning are what "transforms the living into consciousness. . . . In embalming it, in shrouding it, in enclosing it in bands of material, of language, and of writing, in putting up the stele, this operation raises the corpse to the universality of spirit. The spirit extricates itself from the corpse's decomposition, sets itself free from that decomposition, and rises, thanks to burial."[22] When Thoreau radically reverses this logic to say that he doesn't understand why a monument—the sign of a successfully executed ideation of matter into spirit—would be "so much more enduring than the fame which it is design to perpetuate," he is in fact saying that pure spirit—spirit that has lost its concreteness—is good for nothing, and that its operation of the transubstantiation of matter into idea, far from enlivening the dead into universal spirit, is an act that brings "pure death." Those are precisely the terms of a short discourse on museums that he recorded in his *Journal* on September 24, 1843, in many ways anticipating the critique of graveyards in *A Week*. There, he explicitly wrote that embalmed bodies—bodies under pyramids and obelisks, bodies in brick-walled graves, like exhibits in museums—do not preserve life even in spiritual form, as Hegel had it, but instead, preserve only death:

> I hate museums. . . . They are catacombs of Nature. They are preserved death. . . . I know not whether I muse most at the bodies stuffed with cotton and sawdust-or those stuffed with bowels and fleshy fibre. The life that is in a single green weed is of more worth than all this death. They are very much like the written history of the world. (J, 1, 465)

In what way is a weed more historical than the archive or museum, and what kind of traces does a leaf chart so that its surface can become much like the written history of the world? A single green weed writes history, because the shapes preceding and generating its leaves—the seed, the stem—are not distant or severed from it but instead generate their life, which the leaves then vitalize. This is not a metaphor pointing to organicism. Thoreau doesn't want to say that one form smoothly becomes another, following the neat narrative order of organic development. Rather, the metaphor of the weed functions counter to organicism to suggest a radical disturbance of the narrative ordering of phenomena, for what it accentuates is that the transition whereby one organic form vanishes in the arrival of another in fact doesn't happen to what lives at all. The weed tells the story not of linearity but of simultaneity: through the living link of the stem the seed nourishes the life of the leaves that nourish it; the leaves feed the life of what preceded them. This idea is less trivial that it might seem, for it in fact says that causes neither simply generate consequences nor continue to inhabit them; instead, less intuitively, they are regenerated by what they produce. The weed thus tells a different ontological story—one that reveals the logic of life—according to which consequences come to cause what caused them. If only life has capacity to commemorate, as Thoreau says, it is then because, in this nonlinear way, it can make what comes after keep causing, gathering anew, and recollecting what preceded it. Archives, museums, and monuments, in contrast, are precisely consequences that can't affect their causes and therefore can't recollect: "the monument of death will outlast the memory of the dead. The pyramids do not tell the tale that was confided to them. . . . Why look in the dark for light? Strictly speaking, the historical societies have not recovered one fact from oblivion" (W, 154). If monuments testify to anything, it is to oblivion. Such is the final paradox of all cemeteries and museums according to Thoreau: far from being sites of power—of eternal memory—cemeteries locate memory's disconcerting fragility; they constitute embodied oblivion.

Thoreau's critique of memorial architecture as well as his insistence that no historical fact can be recovered from the darkness of the archive is not to be understood as a call to abandon the past and history, commemoration and archivization. Instead, it is his call to revise what an archive is and what it means to commemorate. He specified what such a proper archive might be, one that speaks still,

The Whaleback Shell Midden, Damariscotta, Maine.

in yet another fragment that didn't make it into *A Week:* "Where is the proper Herbarium—the true cabinet of shells—and Museum of skeletons—but in the meadow where the flower bloomed, by the seaside where the tide cast up the fish, and on the hills where the beast laid down its life."[23] In what follows I argue that to state that the earth is the true museum of shells is not a trivial pronouncement but a precise reference to a practice of archivization that Thoreau learned from his research on Native American shell middens around Concord, Massachusetts, sites where history and geology, natural and artificial, came together.

Early American naturalists typically considered shell heap formations, which were spread throughout northeastern America, to be of natural origin, even though they were mixed with animal bones and earth, as well as Native American artifacts, from pottery to beaded jewelry and arrowheads. According to geological logic, such traces of nonaquatic life forms shouldn't have mixed with marine remnants. Thus, in the 1760s and 1770s John and William Bartram, to whom, as Andrew L. Christenson remarks, we owe "the first extended discussions of North American shell heaps," made a series of observations of the heaps that lined the St. John's River in Florida and, while recognizing the presence of Native American material culture, claimed the heaps to be of natural origin.[24] John Bartram "seems to have considered such mounds to be natural formations," whereas William "found shell heaps with conflicting evidence of their origin: 'heaps or mounds of sea-shell, either formerly brought there by the Indians . . . or which were perhaps thrown up in the ridges.' "[25] Thanks to increasing interest in geology, by the late 1830s the origin of shell heaps in general and in New England in particular had become the "subject of public debate," with most geologists still suggesting their natural origin. Emerson's brother-in-law, famous New England geologist Charles Thomas Jackson, whom Thoreau read too, was the first to attempt a scientific explanation of the deposits at the Whaleback Shell Midden in Damariscotta, Maine, where even human bones were found but where "there were apparently no traditions connecting the Indians with the deposits."[26]

Jackson acknowledged the presence of "arrow heads, bone stilettos, and human bones," all of which were found in the shell beds, and agreed that those finds "encourage conjectures that the shells were heaped up there by the ancient Indian tribes who formerly frequented the spot," but nevertheless he concluded that because of their "regular stratiform position and the perfection of the shells" the Native American–origin theory should be opposed.[27] To Jackson, who attributed regularity to nature and irregularity to humans, Native American middens appeared to be geological. It would take American geologists another twenty years to claim, beginning with P. A. Chadbourne in 1859, the unquestionable human origin of the deposits. But even as late as 1868, Jeffries Wyman, who understood them to be of human origin, would also claim their accidental nature, calling them accumulated "rubbish"—an ad hoc monument—comparable to the waste one finds on the streets of big cities and which, despite being unintentional, can still clue us into the "modes of life" of former inhabitants.[28]

When Thoreau, in 1837, starts describing Clamshell Hill—a freshwater mussel midden on the Sudbury River—as a Native American–made site that intentionally confused geology and history he is, as Christenson argues, "nearly two decades ahead of most every other naturalist in the region," becoming by the late 1850s "one of the most knowledgeable persons about shell middens in New England."[29] In his "Natural History of Massachusetts," written in the wake of John's death, he had thus described the midden as interlacing strata of natural and artificial combined into a new type of memorial: "That common mussel . . . left in the spring by the muskrat upon rocks and stumps, appears to have been an important article of food with the Indians. In one place where they are said to have feasted, they are found in large quantifies, at an elevation of thirty feet above the river, filling the soil to the depth of a foot, and mingled with ashes and Indian remains."[30] He remains unfazed by the fact that shells are found well above water level, nor does he interpret that as the effect of mere geological accumulations. Rather, he suggests that the layers of shells, in which ashes and remains are interposed, function as a deliberate geology in which the earth is made to tell the tale of burials, feasts, divinations, and ordinary life as it was shared with fish and animals, birds and vegetation.[31]

Thoreau realized just how deliberate this geology was in researching the same midden in 1860 after a heavy rain had "washed the bank considerably" (Jn, 14, 58), exposing new strata. Then he found pieces of Native American pottery ornamented with a type of earth-writing, "not much more red than the earth itself" and so integrated into the earth that to put them together to reconstruct the large vessel they were broken from amounted not to an archaeological but to a "geographical puzzle" (Jn, 14, 59). Mixed with arrowheads and ornamented pottery he also found "many small pieces of bones in the soil of this bank, probably of animals the Indians ate," and on the other part of the bank, "a delicate stone tool . . . of a soft slate-stone . . . used to open clams with" (Jn, 14, 59). Vestiges, in other words, of hunter and hunted, hunting tools and hunted animals, animal bones and the pots in which they might have been cooked, clam and clam-opener combined together with sand, dirt, and clay to create a new earthly entity. As he moves through this landscape Thoreau also sees "under this spot and under the layer of shells a manifest hollowness in the ground, not yet filled up" (Jn, 14, 59), which suggests to him that the mound is a concretion of past communal life stacked in a geological stratum above which other layers might be mixed. Faced with the earth stratified in zones that gathered cultural and natural objects, Thoreau witnesses what Peter Nabokov encountered when researching Navajo practices of archivization, namely that the "tribe's past" was referred to as "*atk'idaa* ('on top of each other')," and was marked by the " 'stacking' of former experiences or events, which could draw in, according to Kenneth J. Pratt, everything from material 'particular to an individual's recall or personal opinion' to separate clan migration narratives to acquisitions of the major 'sources of healing power.' "[32] Thoreau's realization that "Clamshell hill" is made of "temporal blockings" stacked in a new earth, confirms the "preeminence of topography over chronology," as Nabokov calls it, in Native American "historicity in general,"[33] which is not to suggest that in being inscribed in the earth Native American history collapses into a murky mythology but instead, that linear temporality is demoted in favor of different temporal moments that are simultaneously embodied and visible at a certain concrete physical site. Thoreau takes in the fact that for Native Americans, as Nabokov puts it, "history always took place," that it occurs as embodied in the earth, which is why, as Alfonso Ortiz notes, the history of many Native American tribes has "no meaning apart from where they occur."[34] "The full-bodied role of non-built environments in Indian history"—rock accumulations, mountains, ravines, canyon, forests—are thus "more than painted canvas backdrops for human events";[35] because they provide an ambiance that still hosts the living they also function as archives affecting and so constituting what is taking place in the present. That is why, as Sam and Janet Bingham put it, "a wise person can look at the stones and mountains and read stories older than the first living thing . . . The land still remembers the [first people] . . . and keeps . . . the things they left behind."[36]

Thus, when Thoreau writes in *A Week* that stones "teach us lessons . . . ; verily there are 'sermons in stones, and books in the running streams' " (W, 248);[37] or when, again in *A Week*, he opines that a smart person would know how to "decipher . . . time's inscriptions" on our cliffs (W, 250); or when he suggests in *Walden* that his philosophy comes out of his experience of the place, for he had "travelled a good deal in Concord" (Wa, 4), he is neither romanticizing and hence aestheticizing the American landscape, nor voicing his parochialism or regionalism. Instead, he is conveying a rigorous lesson in history learned from Native American geo-history: history happens by occupying space and *taking place;* it is embodied in concrete sites. Landscapes are the annals of history and the earth is a living archive from which nothing disappears, but the "land . . . keeps" everything "many people and many tribes [who] have come and gone . . . left behind." Everything, that is, takes place precisely according to the ethical premise of Thoreau's perpetual grief, which he entrusts to nature but desires for humans.[38]

The idea that history is embodied by being stacked in strata of the earth shapes Thoreau's understanding of time generally. In *Walden* he famously depicts time as a mark "notch[ed] . . . on [his] stick" (Wa, 17), carved out in space to suggest that temporal intervals are contemporaneous sites rather than chronologically ordered points. His discussion of modes of temporality in *A Week* similarly argues that when we regulate history in a linear way—which Aristotle defined for Western humanity as the succession of discreet "nows"—events of the past and future are entrusted to memory and imagination, rendering time limitless. By this Thoreau means that a remembered—and therefore by definition ideated or dematerialized—moment doesn't really have boundaries but can instead last as long as we allow it to last, such that the elongation of minute historical "nows" generates not just an elongation but the unhinging of history into endlessness. That is the sense of his statement that historical time or "the age of the world is great enough for our imaginations, even according to the Mosaic account, without borrowing any years from the geologist" (W, 324). Remembered history becomes older than the earth. However, as a result of the potentially endless duration of each remembered now that immediately precedes our present, the past that preceded that present is made to retreat into obscure remoteness. In contrast, when, as a result of the prolongation of the remembered moment its past is rendered obscure, the flow of the remembered into its own past is obstructed, making it difficult for us to read even Native American worlds that were alive on the banks of the Merrimack "not so far back" (W, 324). In attempting to remember linear time, history not only generates forgetting, but by the same token also separates what is closer from what is further removed in time into discreet temporal units.

Our understanding of history changes radically if we decide not to understand time in a progressivist or narrative way, but instead, like the Native Americans, as a contemporaneity of embodied sites, clustered in a spatial vicinity. That is what Thoreau proposes in *A Week:* "From Adam and Eve. . . . to——America. It is a wearisome while.—And yet, the lives of but sixty old women, such as live under the hill, say of a century each, strung together, are sufficient to reach over the whole ground. Taking hold of hands they would span the interval from Eve to my own mother. A respectable tea-party merely,—whose gossip would be Universal History" (W, 325). Temporal units are here materialized in the bodies of women whose linear or generational succession is turned into a simultaneity. Yet because each of the female bodies that concretize a "now" is also holding on to the two other bodies co-present with it, it is far from a discreet unit positioned in delineated present; each "now" is traversed by a doubly oriented temporality, constantly decomposing into its past by touching it, while already inhabiting, by also touching it, its future. Each present moment simultaneously retreats and advances. The category of the present thus dissolves into gradations that enable rather than disable—as is the case with linear time—the flow of now into what was and what is about to be.

The charts Thoreau drew up between 1860 and 1862, known as his "Kalendar," were, I am suggesting, his final effort to unify all his previous thinking about time, from the early interest in chronic or circular time understood as an embodied sphere on which a poet can move from the future into the past, to his understanding of time as spatially stratified, according to the logic of Native American middens. In reductive terms, the charts tabulate the natural and meteorological phenomena that Thoreau calls "general," as they appear, intensify, or disappear throughout the year (the most extensive charts of general phenomena are for April, May, June, October, and November, even though Thoreau did make lists of natural phenomena for other months also).[39] The more extensive charts are typically drawn on two double-sided sheets, with most of their information written in black ink, but in some cases, as Peck speculates, "apparently tentative, notations, appear in pencil."[40] However, over the grid drawn in pencil delineating charts into squares of equal size—thus registering that in the logic of social calendric time a day is in fact an abstract category always signifying the same duration—Thoreau drew another grid in ink, with squares of irregular size. For instance, on the April chart, the size of the ink-boxes varies so that some are smaller than regular pencil-boxes, while some expand over four or more regular-sized squares, as if Thoreau were signaling that the "communal" way of measuring time by dividing it into days of equal duration was nothing but a fallacious form of ideation, since the amount and duration of phenomena occurring makes some days empirically shorter than a calendar day (in the chart for April 17, "everything very shortened"), while enabling other days to linger for a period longer than a calendar day.

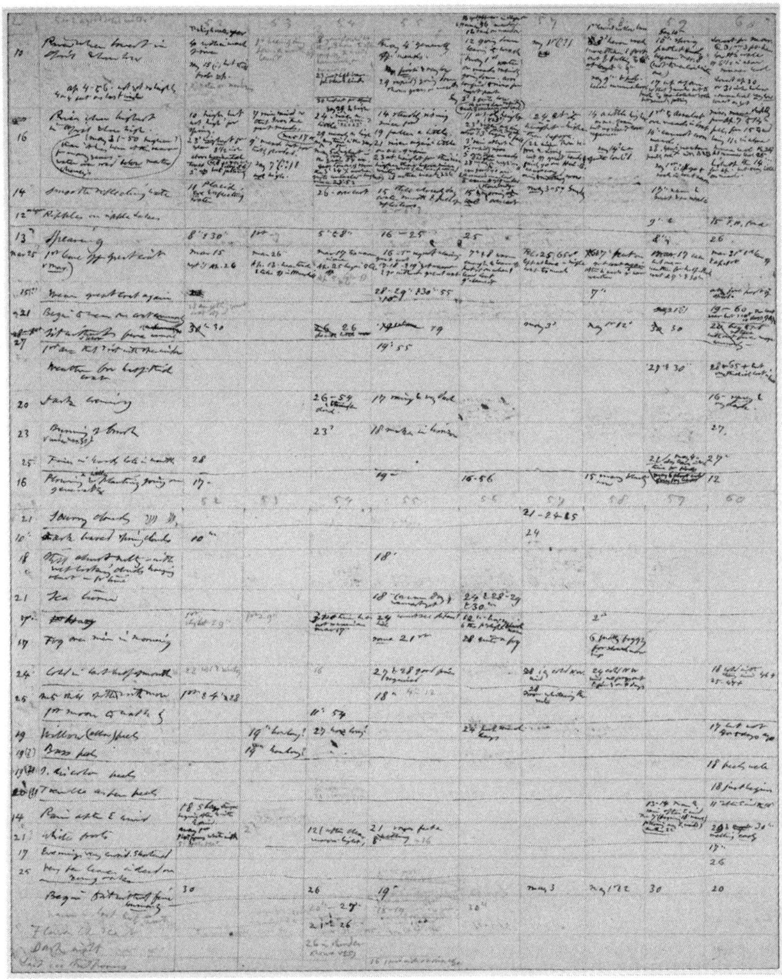

Calendars, General Phenomena for April, Tabulation. Nature notes, charts and tables: autograph manuscript, 1851–1860. The Pierpont Morgan Library, New York. MA 610. Photograph The Pierpont Morgan Library.

Similarly, in each of the individual boxes delineating calendar days, Thoreau predictably inserted numbers signifying the dates of the month, only to then destabilize this numerical category of calendric time also, for phenomena registered in the boxes are sometimes characteristic of a part of a day only ("dark evening" in the April chart, for instance), while at other times they reference minute temporal

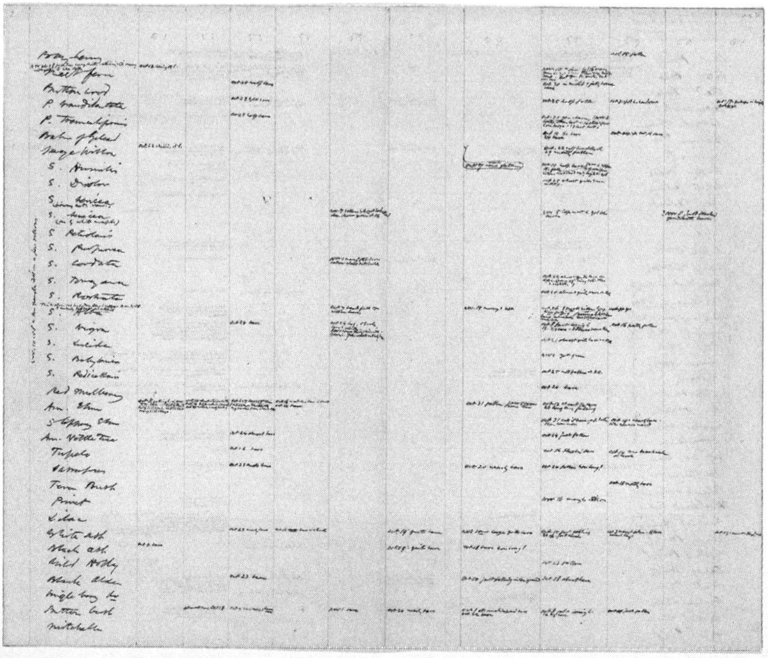

Calendars, General Phenomena for April, Autograph Notes of Thoreau on the Leafing of Trees and Shrubs. Nature notes, charts and tables: autograph manuscript, 1851–1860. The Pierpont Morgan Library, New York. MA 610. Photograph The Pierpont Morgan Library.

units (for instance, April 25, 1856: "7′ 28″ warm enough"), as if Thoreau were suggesting that dating also is fallacious, since to measure the duration of phenomena is to register their mutations as they unfold in seconds or minutes.

But if Thoreau's charts are unstable, with dividing lines curved unlike any common calendar, it is because they reflect the lived duration of the phenomena they register. This is to say that Thoreau is attempting to produce time charts that would accurately register the empirical as opposed to the abstracted or categorical time of calendars.

He wants to register the moment of change of slight instances he calls "general phenomena" to suggest, time and again, that all general phenomena are minute, their processual minuteness being what

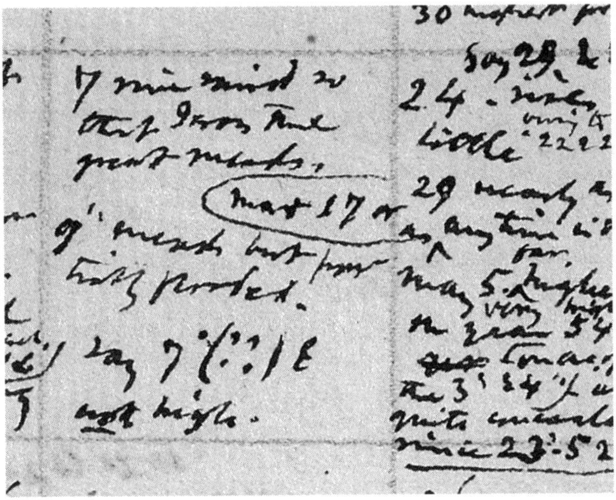

Calendars, General Phenomena for April, detail. Nature notes, charts and tables: autograph manuscript, 1851–1860. The Pierpont Morgan Library, New York. MA 610. Photograph The Pierpont Morgan Library.

alters and contrives the empirical. Thus, in some cases Thoreau notes how a slight difference in intensity changes the same phenomenon (as, for instance, when the "smooth reflecting water" of April 14, 1851, is compared to the "placid reflecting water" of April 11, 1852), whereas in other instances he records how a phenomenon may be affected by remaining at the same intensity (as for instance, on April 15, 1855, when he writes "still cloudy day" but adds below it in pencil: "5" or 6"?," wondering what exactly the duration of "still" is, when does something that "still is" like it was begins or ends). And because the duration of the "now" of such transitioning phenomena varies, being sometimes very slow, some phenomena are registered on the very grid separating two days—taking place simultaneously in their past and their future. Others are circled and transported into another day, appearing in a different temporal strata, just as fossils do in Thoreau's geology, and just as shells do in Native American archaeology, suggesting that as in geological time, the historical time of all phenomena allows them to move from one temporal zone to another without obeying the logic of succession. As if to assert this point, Thoreau ends up radically confusing temporal strata from the

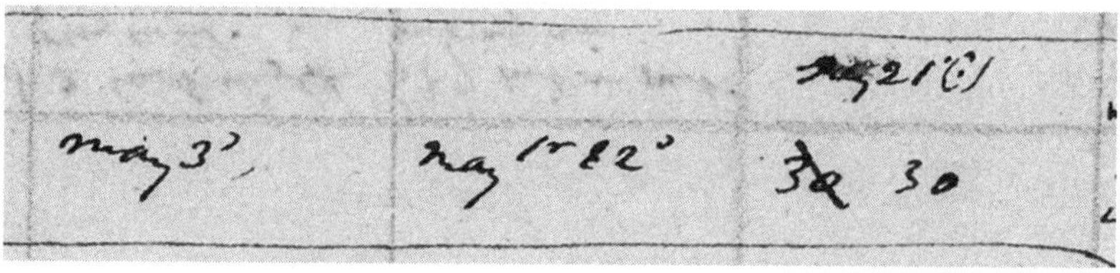

Calendars, Phenomena for April, detail. Nature notes, charts and tables: autograph manuscript, 1851–1860. The Pierpont Morgan Library, New York. MA 610. Photograph The Pierpont Morgan Library.

middle to the end of the April chart, as when April 25, 1851, appears before April 16 of the same year, or when April 19, 1851, appears three times in a row, or when, in the lower right corner of the chart, May phenomena overlap into April.

For if May 3 in Thoreau's calendar arrives before April 30, it is because time—far from being an ideal succession of progressive numbers allowing us to enumerate and so order change—is generated by the material and concrete duration of phenomena themselves, the unevenness of which scrambles temporal order. Thus, the shifts and transportations of phenomena from one day into another; the variable duration of days, the variable duration of hours, the disorderly appearance of dates in Thoreau's calendars, far from testifying to his "struggle to determine which natural phenomena are central and discrete enough to serve as viable categories," as Peck suggests,[41] demonstrate rather his struggle to overcome all categories. Thoreau's calendars are neither about fixed temporal categories nor about clearly bounded discreet phenomena, but about gradations and intensities that can by definition be registered only outside conventional structures and paradigms, for they occur precisely between categories, between days, between hours, among colors, in the midst of relations, and move from one temporal zone to another while being simultaneously in several.

Photographs

In 1851 Thoreau read Robert Hunt's *The Poetry of Science* (1848), a massive book in which Hunt—a chemist and mineralogist himself—endeavored to connect contemporary research in geology, physics, and chemistry to the arts. The lengthy notes that Thoreau took from that book attest to the fact that it was Hunt's discussion of light, chemical radiation, and actinism, advanced in chapter eight, that affected him most. In that chapter Hunt offered a sustained scientific explanation of the transformism he defended throughout the book. Very much along the lines of Thoreau's own understanding of life, Hunt suggested that "transmutation is a natural operation" working not just in structured but also in unorganized phenomena, reshaping everything from a grain of dust to rocks, animals, and stars. This global shape-shifting is generated by light, which is necessary not just for maintaining the processes of vegetal life but also for the generation of all forms. "Mixing up of light and matter," Hunt argues, causes a "striking change" in matter itself; it even enacts the "crystalizing forms of nonorganic matter" and by manufacturing a form out of the inanimate and unformed proves wrong those who refuse to believe that what is material is also transformable.[1] On Hunt's understanding, contemporary chemistry owed this discovery to Nicéphore Niépce, the inventor of photography. In a series of experiments with light that Niépce conducted at Châlon-sur-Saône, France, which were crucial for the invention of photography, he proved (in a sentence that Thoreau quotes from Hunt, both in Thoreau's essay "Walking" and in the *Journal*) that "no substance can be exposed to the sun's rays without undergoing a chemical change."[2] When Niépce demonstrated that even "horn silver blackened when exposed to the light," he in fact proved, on Hunt's understanding, that creative light is operative everywhere, engraving forms and working changes in phenomena like metals and stones.

But this creative power also has its down side, because the informing force of light can also de-create what it has created. On Hunt's understanding, that is what Niépce determined when he established that the more a surface is exposed to sunlight, the more distinct the form sunlight has engraved on it will become, so that if not counterbalanced by another force, the sunlight would continue sharpening forms until finally severing them from their background. As Hunt understands it, then, sunlight wouldn't just create forms but, if intensified, would also be able to peel them off matter, as if skinning it and annulling what it had manufactured. This disturbing power doesn't affect tissues alone but also geological formations:

> We now know that it is impossible to expose any body, simple or compound, to the sun's rays without its being influenced by this chemical and molecular disturbing power. To take our examples from inorganic nature, the granite rock which presents its uplifted head in firmness to the driving storm, the stones which genius has framed into forms of architectural beauty, or the metal which is intended to commemorate the great acts of man . . . are all alike destructively acted upon during the hours of sunshine, and, but for provisions of nature no less wonderful, would soon perish.[3]

In generating, sharpening, and then engraving forms on a silver plate, the heliographic principle of making images thus followed the logic of the sunlight. Heliography is what Niépce experimented with but what Louis Daguerre advanced, giving it the famous name of "daguerreotype." As Daguerre explains in an article extensively quoted by Hunt, in renaming heliography as daguerreotype, he didn't mean to suggest that he had discovered a new process of picture making that wasn't based on "sun-drawing," but only that he had found a way to control and intensify the effects of sunlight. As he put it, he had "perfected" Niépce's procedures, enabling them to "render the effects of nature sufficiently correctly." The "effects of nature" that his daguerreotype managed to duplicate were phenomena observable in Spain, Italy, and Africa, territories "particularly suited" for understanding the process of form-making, thanks to the intensity sunlight assumes there. In such places, Daguerre argued, natural sunlight cuts forms so clearly and so rapidly that the eye has difficulty perceiving the shift of shadows, seeing only shadowless objects clearly formed against the bright background. By inventing a process of rapid generation of sharp images by dramatically increasing the intensity of the light to which the silver plate is exposed, the daguerreotype thus discerned and repeated the natural operation. But, as Daguerre adds, because it manages to repeat what nature does—that is, enacting shadowless form creation—the daguerreotype becomes more than "merely an instrument which serves to draw Nature; on the contrary it is a chemical and physical process which gives [nature] the power to reproduce herself."[4] An apparatus capable of individuating formless matter, repeating the process of natural life, the daguerreotype becomes a technological version of nature. That is why the daguerreotype, as Oliver Wendell Holmes will soon phrase it, is more than art. It is an ontological machine that in generating a form, only to then sever it from the matter to which it was initially attached and archive it on paper, renders that matter irrelevant: "form is cheap and transportable. We have got the fruit of creation now, and need not trouble ourselves with the core. Every conceivable object of Nature and Art will soon scale off its surface for us. Men will hunt . . . objects, as they hunt the cattle in South America, for their skins, and leave the carcasses as of little worth."[5] Photography, then, is even better than nature. For not only, like sunlight, does it peel from matter the form it creates, thus distilling the very "fruit of creation" into pure essence; also, more conveniently than nature, it preserves these pure essences by entombing them in the photograph, while leaving the carcasses of the photographed, skinned bodies to their worthless destiny. In the new world announced by the arrival of daguerreotype, matter—bodies and objects—finally becomes mere waste, being instead replaced with pure forms; it is nothing less than Platonism realized. It doesn't disturb Holmes's logic that the photograph into which these pure forms are deposited and archived is also a body, since the endless technological reproducibility of photographs—on which Benjamin focuses his well-known essay—can endlessly multiply the photographed forms, enabling their "transportation." In easily and endlessly reproducing pure forms, the very essence of creation, daguerreotype renders them cheap.

Unlike Daguerre, who claimed the daguerreotype to be merely an improvement on Niépce's discovery, Hunt argued that Niépce's process differed in essence from Daguerre's. For heliography was only part of Niépce's photographic interest, the other part being a process called "actinism."

As Hunt reconstructs it, during the years when Niépce experimented with heliography, he also experimented with exposure of silver plates to the less intense light of the moon and determined that moonlight affects objects differently than does sunlight. It doesn't act on them directly, by burning them, as it were, but by "radiating" into them heat of low intensity. Once infused into the object this gentle radiation would rise to its surface, where it causes chemical changes that annul those generated by the sunlight; it would spread along the lines of the form inscribed by sunlight, to reconnect the molecules that had been separated, and in so doing blur and even erase the inscribed form. That process was what Niépce called "actinisim." In discovering actinism, Niépce in fact discovered that light operates through two "antagonistic"[6] modalities: the photometric modality (which engraves and sharpens forms, damaging their matter during hours of intense sunlight); and "actinic" or chemical modality, where "we find that the radiations which produce chemical change are more refrangible . . . the maximum of this power [being] found where no trace of light under ordinary conditions exists."[7] In conditions that the human eye registers as shadowy or downright dark, calm light remakes damaged matter chemically. As Hunt puts it in the passage that Thoreau finds important enough to quote it in "Walking," Niépce thus "showed that all bodies which underwent this change during daylight possessed the power of restoring themselves to their original conditions during the hours of night, when this excitement was no longer influencing them."[8] He discovered that light "actinically" annuls its own destructive daytime photometric acts, thus affirming what Thoreau always intuitively sensed: nature recovers what it destroys, and what is done can be undone.

In contrast to Daguerre, Hunt argues that Niépce wasn't interested in intensifying light and shortening the time of exposure to generate more distinct, shadowless forms. As a result, Daguerre didn't "improve" something at which Niépce had failed. Instead of sharpening forms, Niépce was concerned with a complex process designed to generate images that were simultaneously heliographic and actinic, registering not just the forms inscribed by sunbeams but also the way nocturnal darkness blurs them to restore matter that the day has damaged. He wanted both the "writing of the sun" (heliography) and its erasure, trace and disappearance. He intended to name that complex process "photography," because the literal meaning of that word refers not only to the photometric modality of light but also to the operation of light in general, and therefore includes its actinic activity.[9] A photograph was thus supposed to be not just a snapshot of death—a fixed form peeled off matter—but also a record of life's restoring power.

On Hunt's account, Niépce came close to inventing this restorative, vitalistic photography when he determined that "paper prepared with the iodine of platinum, which, being impressed with an image by heliographic power . . . restores itself in the dark . . . to its former state . . . and recovers its sensibility."[10] Hunt is here talking about Niépce's experiments from 1816, more successfully executed after 1820, which resulted in the discovery that the actinic force of light is best transmitted through bitumen. In *Researches on Light* (1844) Hunt published Niépce's own description of this process, which explained how he would place a silver plate, covered with a film of bitumen, in a camera obscura directed at an object and exposed to light. As Niépce explains, after interrupting the exposure and taking the plate out of the camera the "forms of the . . . picture remain still invisible," as if no form of the photographed object had been inscribed on the plate. Only after starting to remove the film of bitumen would the photographer begin "to perceive the images of the objects . . . gradually unfolding their forms." But the objects were in fact only vaguely formed, because, thanks to the long exposure—reported by Hunt to be twelve hours—the photographed object was illuminated differently. As the sun moved from east to west it generated differently positioned shadows, which influenced the varnish of the plate otherwise than the sunlight, causing it to be

Nicéphore Niépce, View from the Window at Le Gras, 1826, Getty Version. Courtesy the Gernsheim Collection, Harry Ransom Center, University of Texas at Austin. Photograph Harry Ransom Center.

solidified and brighter where it was affected by intense light, and blurred, darker, and less solid "on the parts upon which the shadows have fallen."[11] The silver matter registered both the damaging sunshine and the restorative actinism of shadowy light.

That is what Niépce communicated to the Royal Society of London in December 1827, explaining how five photographs he had brought with him to England were obtained. Hunt lists and describes all five but pays special attention to one that became known as the "first photograph," a view of a courtyard seen from an upper window of Niépce's attic at Le Gras, France: "it is less distinct than the former, the outlines of the black portions are bordered by a white fringe, whiter than the adjacent parts. The surfaces on which the pictures appear are metallic, but are blackened."[12]

Niépce's pictures did not sever the object from its shadows by means of sharp contrast but incorporated shadows into the motion of light falling on the object while also registering that motion. For Hunt it was a process that photographed the movement of the living world. Niépce wanted his images to both perform and record the process of life, which is why, contradicting Daguerre and Arago, Hunt argues that Niépce's shadowy pictures were intentionally, not accidentally, blurred: "Niépce was acquainted with the method of making the shadows and lights of his pictures."[13] In showing simultaneously different views of the object (the variety of ways in which it was illuminated and shadowed), and in the process gathering the transformation or history of what was photographed on the single site of the photograph, Niépce rendered different moments of the object's history concurrently. He turned the photograph into a physical site harboring not only the moment taken out of time, but also time itself. If Niépce's photograph is often compared to an archaeological or even geological site—to a record of things and phenomena dispersed through time, rather than being thought of as the impression of a thing "fixed in time" that photography will become—it is because it is just that: a multiplicity of images stacked one on another, a co-presence of moments stratified much like the shells of the geological-archaeological Native American middens.[14]

The idea that the whole of natural life is photographic fashioned Hunt's vitalistic understanding of geology and organic life. His reasoning—on which much of Thoreau's argument advanced in "Walking" is predicated—was quite logical, even if it led to bizarre conclusions. Hunt's idea could be summarized in the following way: if the whole of life is photographic, then each phenomenon, from stones to animals and humans, is a photograph. Hunt explained what that meant, taking stones as his example. Thanks to their slow transformation (Hunt here has in mind Lyell's hypothesis that stones too metamorphose, but at the pace of geological time), stones are the natural phenomena with the longest continuous exposure to sunlight and moonlight, and so they are the clearest examples of all being's photographic nature. A stone behaves like Niépce's silver plate on an infinite scale—at least from the point of view of a finite human life—as if positioned in the camera obscura of the earth. As beings move around it, their shapes and shadows are inscribed on it by sunlight and then blurred by the moonlight. The forms of beings captured by a stone are thus detectable on it as more or less blurred spots, as if it were indeed one of Niépce's photographs (one of the many cases Hunt discusses is the "building of Stonehenge," where figures of "flying monsters, griffins and others, which now live only in the meaningless embellishments of heraldry" appear as images of a past reality).[15] Because stones preserve the blurred images of past life, Hunt argues—in a sentence that Thoreau quotes in his *Journal* of February 27, 1851, and then discusses extensively in "Walking"—that stones are "a faint and shadowy knowledge of a previous stage of organic existence." As Amanda Jo Goldstein puts it, a stone is therefore a "present . . . shot through with old and distant things," hence the photograph of what isn't.[16] Yet, because stones are also the living "process of communion between man of the present, and the creations of a former world" so that "the forces which have worked still labour"[17] in them—which is how Hunt formulates Lyell's uniformist belief that the former causes of life actively function in all rock formations—Hunt will additionally conclude that "the phenomena which [those forces] have produced will be repeated."[18] Hence, stones are not just photographs of life but also living photographs: at some point the forces of life will revisit the shadows of their previous forms of existence, reviving them as bodies again through the restorative photographic or actinic process.

Such is Thoreau's understanding of Hunt's fantastic hypothesis in "Walking": "Some expressions of truth are reminiscent,—others merely *sensible*, as the phrase is,—others prophetic. . . . The geologist has discovered that the figures of serpents, griffins, flying dragons, . . . 'indicate a faint and shadowy knowledge of a previous state of . . . existence. . . . I confess that I am partial to these wild fancies, which transcend the order of time and development. They are the sublimest recreation of the intellect" (Wk, 180). Whereas for Hunt the idea that figures of beings captured by stone-photographs will be embodied into life in different climatic circumstances is scientific, for Thoreau, more reasonably, it is a wild fancy. Yet, he insists that that wild fancy not be dismissed as trivial. To take seriously the idea of revived forms—that is, the idea of causation that defies time—and of mixtures that transcend the order of hierarchical development of species is to accept the wildest thought proposed by Homeric mythology: that what is sensibly present embodies a past that will yet come to be; that what is remembered will occur. Because it violates all the logical and categorical orders we know of, taking such a wild idea seriously would require a "recreation" of the intellect, its fashioning otherwise, its emancipation it into thoughts currently classified as unthinkable; it would require that thinking go wild. Such a moment of hubris—the moment in which all determinations are destabilized—is wildness itself, the moment when one is overwhelmed with life, becoming "most alive," for "the most alive is the wildest" (Wk, 175). "Wildness" is thus not just a key word in Thoreau's new vitalist ontology, predicated on the premise that "life consists with wildness" (Wk, 175), and accordingly resides in transcending orders and in delineations of all sorts. It is also required by Thoreau as a performative, which shifts our way of thinking closer to the "raw material of life" (Wk, 175).

Every once in a while Thoreau would take a walk in the night.[19] But in the summer of 1851 his interest in night walking changed (J, 3, 110). As Harding reconstructs, throughout that summer Thoreau undertook systematic full-moon walks as though generated by an aesthetic experience of the sublime: "In June 1851 [Thoreau] was suddenly struck by the different look at the sandbank in the Deep Cut near Walden Pond at night and decided it was necessary to see his world by moonlight as well as sunlight."[20] However, the walk of June 9 that Harding here references was part of a broader experimentation with moonlit walks (there was an extensively recorded night walk on June 6, for instance), and it was clearly organized around the night's capacity to heal beings rather than any experience of the beautiful. In contrast to Harding's aesthetic understanding, I suggest that Thoreau's investment in night walks was a response to his reading of Hunt on Niépce and his new understanding of the literal capacity of moonlight chemically to insert restorative actinic forces into creaturely life. If Thoreau starts walking in the night after reading Hunt, it is to witness how damage can heal.

For that reason Thoreau's recordings of his night walks regularly mark a progression from intense to actinic light, from destruction to recovery. Starting from where there was still "too much white light-like the pale remains of day light," he would walk to where there was a "yellow gloomy . . . light-that . . . would be like a candle light by day" (J, 3, 249), continuing down to the "lower field," stopping at the spot illuminated by the palest light, a place of pure actinic restoration: "I now descend round the corner of the grain field . . . in to a lower field . . . & find my self in a colder damp & misty atmosphere . . . I seem to be nearer to the origin of things- There is something creative & primal in the cool mist— . . . fertility the origin of things- An atmosphere which has forgotten the sun-where the ancient principle of moisture prevails" (J, 3, 250). At that point everything comes together in Thoreau's head: ancient materialism is confirmed by the contemporary sciences of life. In complete accordance with Hunt's account, the dark swamp becomes the site where actinic forces work through matter to restore it, counteracting the power of sunlight. Simultaneously, in complete accordance with Thales, the site of restoration also becomes the moist and damp site of fertility out of which things originate. The actinic principle of recovery is thus revealed as coincident with the materialist principle of generation. To originate things is to recover them; life begins through convalescence. There is no other beginning, no origin; rather, everything starts from the middle. That is why, as Thoreau later explicitly states, we always "find ourselves in a world that is already planted, but is also still being planted as at first."[21]

Throughout "Walking" and *Walden* one finds Thoreau summoning the powers of nocturnal life—"not even does the moon shine every night, but gives place to darkness"—pointing to how a type of actinism wonderfully restores "granite rocks, and stone structures, and statues of metal" (Wk, 183), to his belief "in the night in which the corn grows" (Wk, 174), or his growing "like corn in the night" (Wa, 111). In such cases he is not voicing an aesthetic joy generated by some deep ecological immersion in the forces of the mystical earth. Rather, he maintains that life proceeds through a loosening of sharp distinctions, which enables healing mixtures. Thus, when he says "I speak out of the night," he means that he speaks from the fragile yet persevering node of healing forces.[22]

Swamps, Leaves, Galls: A Treatise on Decay

As THOREAU is walking through the swampy terrain of Walden Woods on June 13, 1851, he reaches a point in the night where "unmixed moonlight" falling on the east side of the wood blended into the appearing "red" sunlight falling on its western side (J, 3, 260), and so witnesses the moment at which forces of life—recovery, creation, and formation—fuse in the heart of the swamp. Swamp then becomes for Thoreau the site in which actinic recovery and Thalesian generation literally coincide, and on December 31, 1851, still insisting on the correctness of ancient materialism, he declares that wetlands are the Thalesian "principle of all things" (J, 4, 232). Thus, when he calls the most "dismal swamp" a "sacred place" it is not because he is aestheticizing it into a "sublime enchanting scenes of primitive nature" as swamps earlier appeared to William Bartram, quoted in *Walden.*[1] Instead, it is because the geologically unstable sites—liquid earth with no stabilized form, allowing all forms to come to be—that traversed forests around Concord functioned as generators and conduits of life. Identified by Thoreau as the "laboratory of continents" in which matter is in the process of individuation—as Gilbert Simondon would later put it, along very Thoreauvian lines, "the swamp . . . is prior to the form"[2]—swamps become workshops of life, sites in which all strange or "cunning" mixtures are enacted, confirming the findings of his early ichthyology, namely, that life doesn't obey categorical divides but is instead anomalous from the point of view of any scientific classification.

For instance, entering barefooted into the "soft open sphagnous centre of the [Beck Stow's] swamp" to investigate the effects of rising water, Thoreau sees "cunning little cranberries," some of them "lying high and dry on the . . . tops of the sphagnum" becoming "almost parasitic, resting wholly . . . in water instead of air," while others "lay very snug in the moss," tangling "their vines into moss" (Jn, 9, 38–39). Swamp, then, unfixes the habits of living beings; in allowing cranberries to turn into parasites or to become moss, to live in the water rather than in the air, it sets them free to change their shrub-nature. The anomalous transformations of life forms into new specific differences suggest to Thoreau that life continues not because beings slowly adapt but because the nourishing swamp milieu enables them to swiftly become, as when cranberries rapidly become parasites, or, more astonishingly, when on the burned ground of the cedar swamp cinders bloom into "starlike . . . fertile flowers" (Jn, 10, 400). The swamp embodies the incessant continuity of mutation, and at any given point in time swamp life embodies the current, but temporary, configuration of life as it was "a thousand years ago" (Jn, 9, 44).

When in 1856 Thoreau comes to identify the swamp as a "planetary matter . . . terrene, titanic matter extant in my day" (Jn, 9, 45), he means precisely that the swamp is the oldest material life still in the process of differentiation and transformation, or restorative creation that always was. He confirms as much in October 1857, when in the Lee farm swamp he finds "two kind of ferns" mixed in with grapevines and witch-hazels, growing "half-chokingly" under fallen leaves. The "moist, trembling, fragile greenness" of the ferns' "persistent vitality" (Jn, 10, 149) convinces Thoreau that the swamp is the site of literal immortality: "Even in [the ferns] I feel an argument for immortality. Death is so far from being universal. . . . [in] the concentrated greenness of the swamp . . . winter and death are ignored" (J, 10, 150–151). Life survives in conditions meant to outpower and crush it, as in delicate ferns, which go on living thanks to mixing with roots of witch-hazels. If in the same entry Thoreau calls the ferns he observes "outlandish vegetation," it is, then, because they move counter intuitively, outside themselves into witch-hazels, leaving their "land" to maintain itself.[3] And if the swamp life embodied in outlandish ferns becomes for Thoreau the "marrow of nature" (Wk, 242)—something like the essence of life—then by extension all life must be thought of as outlandish, literally outside the land. As Laura Dassow Walls puts it, slippery and shifting swamps "make for uneasy ground,"[4] preventing roots and fixed forms from implanting themselves. Life, then, doesn't have its proper territory, but flourishes in deterritorialization, in shifting terrestrial conditions.

Rodney James Giblett is thus right to argue that Thoreau's understanding of swamps as seats of life "subverts many aspects of standard swamp-speak," most importantly "the miasmatic theory" of swamps as sites of disease."[5] Although swamps are certainly not represented as the sites of miasmic contagion, for Thoreau they raise otherwise the question of disease and decay. Because moss and fungi at the center of a swamp both decay and behave as living seeds, hosting cranberries—that is, functioning as a life force generative of other organisms—they raise a centrally important question for understanding life, namely, the question of decay, leading Thoreau to inquire into the role of decaying processes in generation.

The decaying processes of fungi preoccupy Thoreau as much as the budding of flowers, as his *Journal* testifies, even if he always finds them repulsive ("that offensive mustiness of decaying fungi" [J, 6, 304]), sometimes being "impressed by them . . . like carbuncles & ulcers" (J, 6, 36), at other times being awed by the rapidity of their "dissolution" and wondering whether the "mass of man who ignore them" aren't right to do so (J, 6, 36). If he doesn't ignore them, it is because he believes that understanding how fungi proliferate and dissolve affords a "glimpse" (J, 6, 36) into how life in general operates. That is what is suggested in his late claim about needing to decipher the "cryptogamia" of life (J, 6, 36).

The cryptogam references the twenty-fourth class of the now-discredited Linnaean taxonomy of plants, including fungi, ferns, mosses, lichens, and algae, that is plants whose life—since they don't reproduce by seeds or flowers but by spores—appeared to a previous botany as secretive. Hence their name: "cryptogam" means "hidden marriage" (from Greek *kryptos*, hidden and *gameein*, to marry) and references a form of generation that remains obscure, life that conceals itself. Thus, while flowering plants (phanerogamae) generously reveal their process of generation and decay (*phaneros* is Greek for "visible"), manifesting their linear order (decay results from halting generation), cryptogams, classified by Linnaeus as the lowest form of life, obscure the order and causal relationship among such processes.[6] Despite his remark that observing fungi is "stooping too low" (J, 6, 36), Thoreau nevertheless continues such observations, believing precisely that to understand them is to finally coerce into visibility what there is of life's logic that still remains hidden.

Even though, as I argue here, Thoreau's thinking about the relationship between generative and decaying processes eventually assumes a philosophical and scientific dimension—influenced by Harvard vitalists' take on the question of pathological changes and decomposition—he began to record his ideas in a very general way following John's death. On March 14, 1842, fungi and moss are intuitively understood as emblems of the decay that overtakes the graves: "By the mediation of a thousand little mosses and fungi-the most unsightly objects become radiant of beauty. There seems to be two sides to this world presented us at different times-as we see things in growth or dissolution" (J, 1, 372). But unlike that remark, which registers growth and decay in successive order, an observation made only several days later, concerning fungi that are both decaying and growing on a decayed tree, affords Thoreau a more complex insight into the temporal order between generation and dissolution. It astonishes him so much that he decides to dedicate himself to the "botany of fungi":

> Nature has her russet hues as well as green- Indeed our eye splits on every object, and we can as well take one path as the other . . . I may see a part of an object or the whole- I will not be imposed on and think nature is old, because the season is advance[d] I will study the botany of the mosses and fungi on the decayed- and remember that decayed wood is not old, but has just begun to be what it is. (J, 1, 381)

Here the separation of "russet hues" from "green" is identified as the habit of the eye, not as objectively generated. Our mind imposes its narrative logic onto natural phenomena, splitting the simultaneity of life processes into successive events, choosing to see first one and then the other part of an object, whereas the "object as a whole" is neither green nor russet, but both "green and grey," (J, 1, 381), concurrently growing and decaying. Fungi thus reveal that life's temporality is not linear, for if what is about to be (disease, decay) traverses what now is, then life's temporal zones are strangely simultaneous, occurring in a dynamic in which what is late organizes what is early, and consequences shape their causes. The botany of fungi, then, is concerned with this overlapping of health and decay.

The question of temporal overlap is rendered even more complex by the capacity of fungi to challenge somatic boundaries without ceasing to be animated. On placing in his cellar a "giant parasol fungus" he found in 1853, Thoreau watches how it "went on rapidly melting and wasting away from the edges upward, spreading as it dissolved, till it was shaped like a dish cover" (J, 6, 220). In Thoreau's cellar the fungus decays and speedily "wastes away" into a new form without any discontinuity, its dissolution functioning precisely as the unformed life connecting forms. By flowering into many forms, it reveals to Thoreau not just what Ben Woodard calls the fungal "unhinged spatiality of life"—an astonishing disclosure that not all life needs a frame—but also the capacity of life to attach itself to a new shape as it moves along.[7] The existence of the dissolving mushroom is so fragile that it comes closer to some delicate quality than a being, and Thoreau is fascinated precisely by its fragility, its incapacity to maintain a shape while yet not dying: "it was so delicate & fragile that its whole cap trembled on the least touch" (J, 6, 218). Even if it does appear close to the simple gelatinous spit that Francis Ponge calls a mollusk, that spit that must nevertheless be regarded as a "precious reality," a "mighty energy" capable of recreating itself.[8] Thoreau generalizes the insight offered by the dissolving parasol fungus into all life—insight suggesting that decomposition animates itself into new forms—and comes to propose, in the same *Journal* entry, that this fungal transitional life is the oldest life on earth, a life that emerges out of swamps but is indistinguishable, as life, from the life we still inhabit, "all[ying] our age to former periods such as geology reveals." If then he says that the "simplest and most lumpish fungus has a peculiar interest for us," or similarly that it is better to "learn

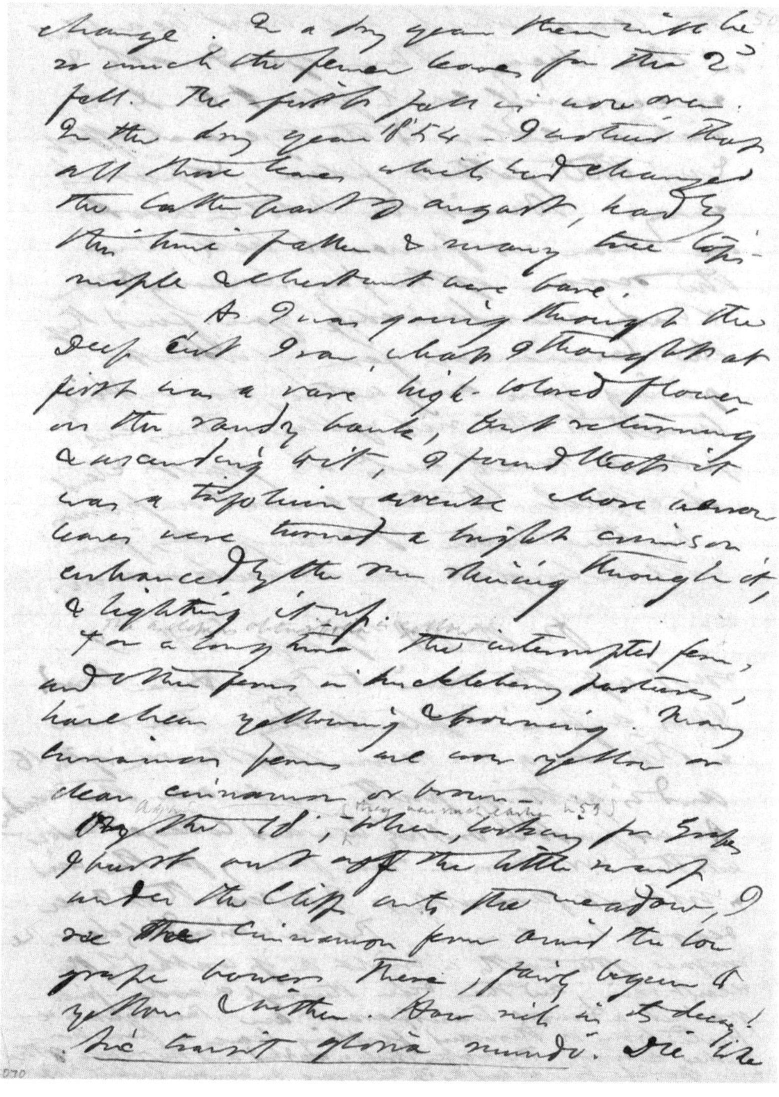

change. In a dry year there will be so much the fewer leaves for the 2d fall. The first fall is now over. In the dry year 1854 I noticed that all those leaves which had changed the latter part of August, had by this time fallen & many tree tops – maple & chestnut were bare.

As I was going through the Deep Cut I saw what I thought at first was a rare high-colored flower on the sandy bank, but returning & ascending it, I found that it was a trifolium arvense whose narrow leaves were turned a bright crimson enhanced by the sun shining through it & lighting it up.

For a long time the interrupted fern, and other ferns in huckleberry pastures, have been yellowing & browning. Many cinnamon ferns are now yellow or clear cinnamon or brown.

On the 1st, when looking for grapes I burst out of the little swamp under the Cliffs into the meadow, I see ~~the~~ cinnamon ferns amid the low grape bowers there, fairly begun to yellow & wither. How rich is its decay! Sic transit gloria mundi. See the

The fall of the leaf; autograph manuscript leaf, [ca. 1854–1858]. The Pierpont Morgan Library, New York. Gift; Mr. and Mrs. Donald Hewat; 2001. MA 5070. Photograph The Pierpont Morgan Library.

fungi" in swamps than study colors in art class, it is because fungi suggest the continuous existence of a "vegetative force" animating all beings and disclosing the logic of vital transformations (J, 6, 218).

Leaves, however, play a more important role in Thoreau's thinking on generation and decay than do fungi. As early as 1837, following Candolle and Sprengel's botany,[9] Thoreau became interested in the process of crystallization in fallen leaves stricken by frost, while in 1840 he elevated leaves into the pattern of all life, arguing that all beings assume variations of leaf-form, and that a leaf records "the print of nature's footstep" (J, 1, 475). Thoreau will believe throughout his life that leaves provide an entry to understanding the alphabet of life, so much so that, as some scholars argue, toward the end of his life began work on a book on the "fall of the leaf" and its meaning.

Yet, exactly what kind of message that alphabet wrote will change. For instance, in late 1843, decaying fallen leaves are separated from the healthy processes of natural life: "It is remarkable how . . . all things else, recor[d] only life . . . never death and decay. We are obliged often to contradict ourselves to express these as the tree requires the vital energy to push off its dead leaves" (J, 1, 483). However, Thoreau eventually abandons the presumption that fallen leaves are not part of the language of life and as of 1850 at least, the botany of (fallen) leaves will fuse with the botany of fungi into a more generalized botany of decay that continues to preoccupy him. Thus the late lecture, "Autumnal Tints," explicitly declares his preference for decaying processes over processes of growth: "We are all the richer for their decay. I am more interested in this crop than in the English grass alone or in the corn."[10]

Here are some exemplary claims of that botany, chosen randomly from the Journal in 1853:

> June 3: "Saw a large moth or butterfly exactly like a decayed withered leaf—a rotten yellowish or buff" (J, 6, 176).
>
> August 7: "Their turning or ripening looks like decay" (J, 6, 287).
>
> November 30: "I find that those large triangular . . . shoulders on this root are the bases of leaf-stalk-which have rotted

off-but toward the upper-end of the root are still seen decaying" (J, 7, 182).

November 30: "All these leaves are lightly rolled up. . . . There is a vast amount of decaying vegetable matter at the bottom of the river" (J, 7, 183).

There is, of course, nothing unusual in recognizing decay as part of natural processes. Aristotle had introduced the idea for the Western naturalist mind when in "De Anima" he argued that plants belong in the realm of living beings by virtue of possessing nutritive force, a vital capacity that leads them through "growth, maturity and decay."[11] But, for Aristotle these processes are successive, correlative, and mutually exclusive. Thus growth is the healthy development that increases the body in a quantitative way ("everything that grows increases its bulk"),[12] leading it to maturity (the time during which the quantity of the body's matter is stable), and ending in "growth's correlative," decay, which decreases the quantity of the being.[13] This correlative difference in the quantity of the living body has also its qualitative side ("but change of quality and change of quantity are also due to the soul").[14] Qualitatively speaking, growth is the process of increased composition and organization of the body that intensifies its form, rendering life and matter as indistinguishable as the "wax and the shape given to it by the stamp,"[15] while in contrast, decay is the process of "disintegration," signaling that the nutritive soul has left the body: "the soul holds the body together; . . . when the soul departs the body disintegrates and decays."[16] According to this linear logic, decay names what happens to matter after life abandons it; it is a part of nature—insofar as nature also includes the inorganic and inanimate—but it isn't part of what lives. And it is attributed to living beings only to mark the moment when they turn into what is not living.

Western naturalists won't disturb the Aristotelian linearity of health and disease, growth and decay until well into the eighteenth century. It will be maintained by even the eighteenth-century radical vitalist John Turberville Needham, who is important for my argument not just for believing, similarly to Thoreau, that the "productive force in Nature resides in all substances and things" and is of a "vegetative nature,"[17] but also because he undertook the first important research on the pathology of plants. Explaining why wheat galls and rye blight cause "the grain [to] crumble . . . into a black Powder," he suggested that the mortification generated by pathological changes was nothing but "exuberant Matter breaking forth," causing "the Vegetative force to withdraw" and resulting in "decomposition."[18] Just as in Aristotle, then, decay is what happens after "vegetative" life abandons matter.

This "afterness" is certainly claimed in Candolle and Sprengel's *Elements of the Philosophy of Plants*, the book that first prompted Thoreau to write about the crystallization of leaves. Candolle and Sprengel discuss the vegetal health and decay of plants precisely by focusing on their leaves. Leaves assumed a central role in their botany, because they were identified with a plant's breath, judged to be its respiratory organs or the power that keeps it alive: "leaves are the organs, which by exaltation and absorption . . . maintain the proper composition of the plant."[19] That a plant remains composed by breathing through its leaves raises for Candolle and Sprengel the crucial question of the status of fallen leaves: are they still the part of a plant (albeit decomposed), that is, do they still live, or is the decomposition they embody already on the side of death? The problem is less bizarre than it might appear at first glance, for in Aristotle's ontology of life that is precisely what happens: the breath (soul, vital force) leaves matter but continues to live. Through its various adaptations in Christianity and the Middle Ages, that idea in fact became one of most domesticated presumptions of the Western mind. What makes that logic strange in the context of botany is its concretized instantiation: if the leaf is the breath of the plant, can't that breath leave the plant—fall off—and continue to live (which would explain why fallen leaves keep changing color after they fall, as if they were still animated)? And even if Candolle and Sprengel don't discuss hues of leaves in any detail comparable to Thoreau, they do differentiate among them quite carefully, starting with "vital" green, moving to purple and blue—which manifest a "higher degree of vital activity in leaves" due to the "excess of Hydrogen"—and ending with variants of red and yellow characteristic of decaying and falling leaves. In their view the transformation of coloration in fallen leaves is not a sign of life left in them: "Decayed and falling leaves are yellow and red, because the oxygen remains in them after the vital activity is gone."[20]

Thoreau's preoccupation with decaying leaves produces different insights. It is precisely fallen leaves that support the argument that vital processes are indistinguishable from decaying. Thus on November 19, 1850, clearly contradicting Candolle and Sprengel, Thoreau will declare that fallen leaves still live: "I find on breaking off a shrub-oak leaf a little life at the foot of the leaf-stalk so that a part of the green comes off" (J, 3, 145). Similarly, on October 10, 1851: "When freshly fallen with their forms & their veins still distinct they have a certain life in them still" (J, 4, 138). Life, then, lingers in what decays—as if the "breath" didn't quite leave matter—even if it is barely perceptible: "red cedar & white show no life except on the closest inspection" (J, 7 135). And the fragile protraction of life in decaying leaves becomes for Thoreau an emblem of the immortality of all vegetal life. As he discovers that decaying vegetal bodies are not decompositions of lifeless matter, but instead remain in life and grow "fresh radical greenness" in themselves, Thoreau will come to identify vegetal life as the "essential vitality" (J, 7, 166):[21] "I saw the other day a dead limb which the wind or some other cause had broken nearly off, which had lost none of its leaves" (J, 3, 142); "The dry fields have for a long time been spotted with the small radical leaves of the . . . life everlasting" (J, 7, 168). The minute variations of color that Thoreau registers with such precision and even obsession throughout his *Journal*—from intense to pale red, to intense yellow to pale brown to "grey smoke"—thus read as the annals of this survival of delicate life in what appears to be inanimate, a survival that annuls the Aristotelian chronological ordering of healthy and pathological, of growth and decay.

However, as a consequence of cancelling Aristotle's successive orders, it becomes more difficult for Thoreau to separate healthy and pathological processes. And even if "Autumnal Tints" is focused on "exhibit[ing] the process of seasonal ripening, from late August through late October," as William Rossi accurately observes, the *Journal* nevertheless testifies to the fact that for Thoreau, decay came to traverse the whole of life and not just its autumnal phase, just as health came to traverse all life and not just that of the spring.[22] There will thus be decaying leaves in July ("The heart leaves are . . . turned dark-but the less decayed part in the centres still green" (J, 6, 256); or in August: "the pontederia leaves are already half of them turned brown. . . . Why should they decay so soon? . . . The fall has come to them" [J, 7, 4]). Reciprocally, there will be blossoming in October ("[Witch-hazel's] blossoms smell like the spring. . . . Suggesting amid all these signs of Autumn—falling leaves & frost—that the life of nature—by which she eternally flourishes, is untouched" [J, 4, 135]), and there will be fresh life in December ("I found this afternoon cold as it is, and there has been snow in the neighborhood, some sprouts . . . covered still with handsome fresh red & green leaves. . . . There were also some minute birches only a year old their leaves still freshly yellow-and some young wild apple trees apparently still growing their leaves as green and tender as in summer" [J, 3, 145]).

The simultaneity of generation and decay, however, radically confuses the ordering of seasons. For if ripening and decay, emblematic of fall and winter, are in fact not specific to any season, as Thoreau thought they were in 1842 when he said after John died "I die in the annual decay of nature" (J, 1, 368); if, instead "each season is . . . drawn out & lingers in certain localities," in every other season (J, 3, 146), then the difference between seasons stops being a qualitative difference and becomes one of intensity. Thus, seasons can be more or less generative or more or less "dry" ("plants have two states . . . the green & the dry" [J, 4, 138]), but there will be no season of death, no interruption of generation: "the autumn then is in deed a spring. All the year is a spring" (J, 4, 136). By the same token, and even though the freshness of spring becomes something like a "meta-season" traversing all the others, there will be no health untouched by decay. In that sense when Thoreau says "the day is the epitome of the year" (J, 7, 8), he means that every moment in the year—every moment in life—simultaneously blossoms and deteriorates, is traversed by little deaths and recoveries as it keeps going on.

Many critics who comment on Thoreau's preoccupation with vegetal foliage have found its sources in Ovid's, Linnaeus's or Goethe's thinking on metamorphoses, and understood it to be a metaphor of human transformation.[23] Accordingly, as Rossi correctly argues, Thoreau's "Autumnal Tints" is typically considered to be "the least 'scientific' in any conventionally accepted sense"[24] of all his late writings on natural life. Similarly, the rare scholars who discuss Thoreau's botany of fungi and moss suggest that his knowledge of botany was "minimal." Lichenologist Reginald Heber Howe Jr. judged that "Thoreau's observation on lichens showed only a slight knowledge of the species."[25] In the case of fungi, Ray Angelo explained Thoreau's unfamiliarity with the species by the fact that fungi were a "difficult plant group" to grasp, for they "resisted the study owing to the absence of good manuals."[26] But although Thoreau certainly wasn't averse to improving his botanical knowledge about fungi, lichens, or leaves, his primary interest in them, I argue, was in using them to glimpse into how life works. Appearing at the threshold of life and death—decomposing, transforming, exchanging—they provided an insight into pathological and decaying processes that could reveal much about the logic of life.

Besides, if Thoreau didn't look to botany for an answer to the question of what life is—Angelo notes that Thoreau's collection of books on botany was quite modest[27]—it is because botany wasn't able to answer the question. For, as Canguilhem puts it "eighteenth-century naturalists such as Comte Buffon and Carolus Linnaeus . . . describe[d] and classif[ied] life forms without ever defining what they meant by 'alive.' "[28] As a consequence, "in the seventeenth and eighteenth centuries, the study of life as such was pursued by physicians rather than naturalists, and it was natural for them to associate life with its normal mode, 'health.' From the mid seventeenth century onward, then, the study of life became the subject of physiology (narrowly construed)."[29] This is to say that seventeenth- and eighteenth-century medicine synecdochically translated the Aristotelian exclusion of decay from life to a relationship between health and disease. Mechanists and vitalists, solidists and humoralists identified life with health and opposed it to disease imagined as a force external to and opposed to life. The separation of physiology from pathology that followed merely reflected the stance that life, as what encompasses all "normal" processes of "vital phenomena," is opposed to disease in its various "abnormal, pathological and morbid" modes.[30] And even though the vitalists especially claimed a continuity between the state of health and disease, the two states nevertheless remained heterogeneous, for, as Canguilhem argues, "the continuity of the middle stages d[id] not rule out the diversity of the extremes."[31] Thus, the identification of health with life and the insistence on the heterogeneity between normal and pathological established "health" or "normality" as a norm for the living, which immediately started to function as a value; to say "life is valuable," came dangerously close to meaning "healthy life is valuable," for "this normal or physiological state [was] no longer simply a disposition which can be revealed and explained as a fact, but a manifestation of an attachment to some value."[32] And, as Canguilhem puts it, when normality is established as a value, the *"ideal of perfection soars over"* it, opening the door to a monstrous ethics and politics that regards the imperfect, sick, or "abnormal" as worthless.[33]

Even though there were eighteenth-century physicians who formulated a radically opposing view on the status of the pathological—most notably Scottish physician John Brown (1735–1788)—it was only at the turn of the century that Xavier Bichat's *Research on Life and Death* (1800) radically changed the mainstream scientific understanding of life precisely by arguing that there is no normality heterogeneous to the pathological. Bichat's immediately influential *Research* established that no regularity and stability—nothing that might be called a state of perfect health, for instance—characterized life, but "instability and irregularity are the essential characteristics of vital phenomena, such that forcing them into a rigid framework of metrical relations distorts their nature."[34] Bichat rejected the presumption that normalcy or health establishes the stable center of a scale along which excess or deficiency are measured by their distance from that center, what Canguilhem calls the "framework of metrical relations." Instead, he argued that the only constancy of life is its variability; that it is through variations inherent to it that life mutates into the pathological, thus blurring any clear distinction between the normal and the pathological. With Bichat disease becomes a mode of health within life, so that, as Foucault puts it, "the idea of a disease attacking life must be replaced by the much denser notion of *pathological life*. Morbid phenomena are to be understood on the basis of the same text of life, and not as a nosological essence . . . [disease] is no longer an event or a nature imported from the outside; it is life."[35] To understand and study life, then, is to study pathological changes. This is, of course, not to say that "pathological life" doesn't allow for the category of health, but rather that "health" becomes an unstable "margin of tolerance for the inconstancies" of life. Health is no longer regarded as normative, because it is no longer regarded as normal: "being healthy and being normal are not altogether equivalent since the pathological is one kind of normal."[36] Thus, with the arrival of the idea of "pathological life," the healthy stops functioning as a norm, and a space opens for an ethics and politics of inclusion of the supposedly degenerate, precisely on the basis of its normalcy.[37]

The first American edition of Bichat's *Research* appeared in Philadelphia in 1809, exerting an immediate influence on physicians at Harvard and the University of Pennsylvania medical school, which only intensified with time, so that in early 1820 the Philadelphian *Journal of Foreign Medical Science and Literature* was publishing lengthy extracts from French medical books and journals.[38] In 1822 George Hayward, fellow of the Massachusetts Medical Society, translated Bichat's *General Anatomy*, to be followed in 1823 by the whole translation of Beclard's *Additions to Bichat's General Anatomy* and in 1827 by a much improved new translation of the *Research*. That intellectual atmosphere inspired many American physicians to study in Paris. As John Duffy details, "some 105 American physicians studied in Paris during the 1820s and another 222 in the 1830s. The list of those studying medicine in France in the 1830s is a virtual catalogue of the outstanding names in American medicine, men such as Jacob Bigelow, James Jackson, Jr., Oliver Wendell Holmes, John Collins Warren, Valentine Mott. . . . All told, almost 700 American physicians spent some time in France in the years from 1820 to 1860."[39] And while all those physicians brought back to America medical practices predicated on a new understanding of life and the pathological, it was, as Duffy argues, Jacob Bigelow who "first . . . express[ed] this new outlook," giving it a theoretical formulation in an address on "self-limited diseases" that was delivered in 1835 at the Massachusetts Medical Society.[40]

Duffy is certainly correct that Bigelow's address expressed the new French understanding of the pathological. Having no doubt been exposed to it in his early studies in Philadelphia, as well as during his later trip to Europe, Bigelow deals with it explicitly and references Bichat and Laenneck as the "most distinguished lights" of modern medicine."[41] Moreover, as Joan Burbick points out in her fine discussion of him, Bigelow calls for Americans to follow the method "employed by Louis in his extensive pathological researches, and now adopted by his most distinguished contemporaries in France."[42] Yet Bigelow's address was much more than a simple expression of French views in an American context. His earlier empirical research into the medical properties of "vegetables of the United States," published as *American Medical Botany* (1817–1820), regarded vegetable life as possessing the power to heal pathological processes in the larger economy of vegetal, animal, and human life, hence as an instance of nature's power to regulate its own pathological life. Building on this research, Bigelow's address generalized his views concerning pathological processes in human life across all natural life. Far from being a strictly medical treatise, then, Bigelow's address offered a philosophy of the universal economy of life.

For Bigelow, just as for Bichat, pathological processes are not external but immanent to life. Observing that in some cases disease is cured even when the patient forgets to take medicine, or simultaneously takes different medications with opposite effects, he concluded that some diseases can't be affected by an artificially generated chemical pharmacon but are instead self-regulated or, as he puts it "self-limited." Recognizing, then, that some diseases can't be treated from the outside, he concluded by extension that such diseases are not external to the living body but are indistinguishable from life, that they appear following life's own logic. Pathological processes are immanent to a life that limits pathological processes. Bigelow then generalized life's setting of limits for pathological processes to its universal feature, and found it operative not just in the cases of disease he classified as self-limited but also in cases of incurable disease.

Moreover, he found the distinction between self-limited and incurable diseases to be untenable—existing for the "convenience" of classification only—believing in a "general sense" that the category of the incurable falls into that of "self-limited" diseases.[43] The particular and "convenient" sense of the word "incurable" is medical. Physicians use it to designate pathological processes that end by dispersing the form in which they have occurred, coinciding with the death of an individual. Ascribing absolute, incomparable value to each human being, Bigelow then goes on to call for the epistemological reform of medicine with the aim of bringing about more efficient ways of extending the limits incurable diseases impose on the duration of human life. But the "general sense" of the incurable that Bigelow talks about, which makes it a subset of the self-limited is, as he puts it, "philosophical," taking into account life as a whole.[44] According to such a philosophy, pathological processes that occurr in all living beings—hence in nature—do not interrupt life but are instead absorbed by and refigured in it, for nature possesses a "spontaneous principle of restoration," without which all life "would long ago have been extinct."[45] Thus, on the one hand, for Bigelow as for Thoreau, life always survives; yet on the other hand it never enters a state of pure health but instead remains traversed by pathological processes. Hence the title *Nature in Disease*, which Bigelow gave to his 1854 collection of essays, in which the address on "self-limited" diseases was included.

In diagnosing nature as always sick—Bigelow's way of expressing what Foucault called the "discovery of pathological life"—Bigelow posited natural life as the absence of perfection or health, and thereby opened up the possibility of a politics that would take from life the clue that there is no standard of normalcy against which one could measure discrepancies as degenerative processes worthy of exclusion. And even though, as Oliver Wendell Holmes later testified, Bigelow's ideas regarding the pathological exerted a tremendous influence on "American medical practice," eventually entering the mainstream, they were still being resisted some ten years later, both by influential physicians and lay audiences, less for scientific reasons than for the progressive politics they implied.[46] Thus, in a tacit polemic against Bigelow and in open disagreement with French developments, Elisha Bartlett, chair of Materia Medica at the University of Maryland and College of Physicians and Surgeons in New York City, enthusiastically argued in *An Essay on the Philosophy of Medical Science* (1844) that our knowledge of "morbid processes cannot be inferred or deduced from our knowledge of their healthy processes. *Pathology is not founded upon physiology . . . The one does not flow from the other.*" Instead, for him, health and normalcy exist in an isolated way until attacked by the pathological as something utterly foreign to them.[47] Melville's *Confidence-Man* gives us a glimpse into how an antebellum lay audience, similarly believing in "pure health" and a standard of perfection embodied by natural life, dismissed Bigelow's ideas on religious and ethical grounds: "*Nature in Disease?* As if nature, divine nature, were aught but health . . . nature is health; for health is good, and nature cannot work ill. As little can she work error."[48] What Melville suggests here is not only that for ordinary antebellum Americans health functioned—just as Canguilhem argued that it had for mid-nineteenth-century Europeans—as a norm of normalcy in contrast to which disease was a degeneration from the norm, hence an error. It also recognizes in Bigelow a theological sacrilege: that God, insofar as he creates out of and in accordance with his nature, is himself diseased and erroneous. This "sacrilege" that rebels against excluding the sick from the normal is precisely what Bigelow's and Thoreau's theory of life in disease proposed.

In May 1851 Thoreau began to read Bigelow's *American Medical Botany*—predicated on the understanding of life developed in "On Self-limited Diseases"—at the same time as his interest in decaying processes (falling leaves, decaying fungi, and moss) intensifies. However, as shown by the notes he took from Bigelow over the next several months, he shifted Bigelow's emphasis from observing plants that heal life to plants in need of healing. From May to August he records a variety of vegetal pathological phenomena: the late budding of blue vervain, vines anomalously connected with trees in the swamp, a wingless female cricket, "everlasting Tansy . . . already getting stale," "Utricularia on the North branch . . . flowered upright [but] . . . rooted yellow . . . no bladders nor inflated leaves" (J, 4, 6, 11, 13, 26). And, as a climax to those observations, he asserts on

August 31, while observing a vegetal gall, his own generalized pathology of life: "Is not disease the rule of existence? There is not a lily pad floating on the river but has has been riddled by insects- Almost every shrub and tree has its gall—oftentimes esteemed its chief ornament-and hardly to be distinguished from the fruit. If misery loves company-misery has company enough- Now at midsummer find me a perfect leaf-or fruit" (J, 4, 26). Thoreau's early remarks on the fragility of life, his early refusal to accept anything as anomalous, which developed under the influence of Tuckerman, and his already acquired insights into decaying leaves are all summoned by the brute materiality of the gall as the crystallization of life's generalized decrepitude. There is nothing purely pastoral ever, not even in midsummer, for there is nothing that doesn't rot as it buds and therefore certainly nothing that can be called "perfect." As the pervasive presence of galls on trees in the summer of 1851 suggested to him, life proceeds through imperfections and suffering that inhabit the core of its generative power.

For the Greeks and Romans, who ascribed them predominantly to oak trees, galls were precisely that: the fusion of sadness and life that rushes through the forms it has created and so de-creates them. Indeed, the word is derived from the Latin *galla* and its older root, the Greek *kēkis*, meaning "something gushing forth, ooze or outgrowth and links with the production of poison or bile . . . there is a connection between galls and gall, or bile," the humors of which produce melancholia.[49] For Greeks and Romans, then, the pathological growths on oaks were created by a life so abundant that it overflows, and in generously giving its excess, sickens what it creates. As an oozing that ruptures a form, galls were thus understood to incarnate the literal coincidence of illness and life's crying over it, to embody nature's melancholy.

As H. Peter Loewer remarks, Thoreau kept "documenting occurrence [of galls] and their forms throughout the *Journal*.[50] For the continuous interest in galls that followed Thoreau's first observation that so many trees are marred by morbid outgrowths only confirmed for him that life lives not just by growing forms but also by growing out of forms. Just as flowers burst from buds, so pathological formations bloom in it:

Phylloxera; Surface of a leaf covered with galls, vintage engraved illustration. Trousset Encyclopedia, 1886–1891. Courtesy CanStock Photo.

> It is remarkable that a mere gall-which at first we are inclined to regard as something abnormal should be made so beautiful as if it were the *flower* of the tree . . . -we see that beauty exhibited in a gall, which was meant to have bloomed in a flower-unchecked- . . . How oft it chances that the apparent fruit of a shrub-its apple-is merely a gall or blight! . . . So many plants never ripen their fruit. (J, 6, 168)

Or, more precisely, galls sometimes become the only fruit that a plant ripens: "The gall on the leaves of the slippery elm is like fruit" (J, 6, 250). Galls are alternative flowers, because, like them, they burgeon and blossom until they burst out of the body they thus reform ("This gall had completely checked the extension of the twig" [J, 6, 37]). As Foucault puts it in the context of Bichat, "disease assumes the figure of a great organic vegetation, which has its own forms of sprouting, its own ways of taking root, and it own privileged regions of growth."[51]

Oak Apple or Oak Gall, of Oak or *Quercus* sp., vintage engraved illustration. Dictionary of Words and Things—Larive and Fleury—1895. Courtesy CanStock Photo.

The widespread and regular coincidence of flowering and "galling" ("how regularly these phenomena appear" [J, 6, 200]) thus eventually leads Thoreau to detect the logic of vegetal pathology to be operative in all life. The whole of nature is identified as a gall: "Saw some green galls on a golden-rod . . . completely changing the destiny of the plant. . . . It suggest that nature is a kind of gall" (J, 6, 283).

The idea that all organic life grows pathologically—continuously "completely changing its destiny"—does point to its contingent, nonteleological nature, but it does not mark its deficiency or signal a constitutive lack that it is unable to overcome. Nor is it a sign of life's "autoimmunity" (a process whereby life uses its defense mechanisms to attack itself and so destroys itself), just as it is not the strategy of "immunization" by which life absorbs what destroys it in order to become resilient to it.[52] Rather, pathological growth is life's constitutive "rule," precisely what, together with nonpathological processes, makes it: "Disease is not the accident of the individual, nor even of the generation but of life itself. In some form & to some degree or other it is one of the permanent conditions of life" (J, 4, 34–35). That disease is life's permanent condition implies that it never destroys life, which would be the case according to the autoimmunitarian scenario. Entrusted to itself, then, life wouldn't go on dismantling forms by sickening them until all differences were reduced to a unity tantamount to death. What keeps this bursting of forms in check is life's other tendency to constantly differentiate by individuating. But neither is this tendency unchecked, for it never ends in pure health separate from "the mortal risk that runs through it," which would constitute life's success according to the immunitarian logic.[53] In Thoreau, great health is always cracked. Never dead yet never healthy, life is made of two forces that never reconcile yet remain inseparable. But the constitutional impossibility of their reconciliation—suggested by Thoreau's remark that disease is a permanent condition of life—means only that from the very beginning, at the fantasmatic origin of life, life was already twofold and torn. It was sick from the very beginning and so was from the beginning in recovery.

Coda: Antigone's Birds

In the "Monday" chapter of *A Week*, where references to geology and Aristotle mix with reflections on memory, tombstones around Concord, and funeral rites, Thoreau inserts discussion of Antigone "concerning the burial of the dead body," suggesting that her way of burying the dead is in fact what he had in mind when he was talking about necessity of revising such methods (W, 135). That might seem strange, because, ever since Hegel standardized Antigone as a figure obsessed with proper burial, she has come to represent for us precisely the logic of burial rites to which, I have argued, Thoreau was radically opposed.

In burying her brother Polynices—who fought against Thebes—against Creon's newly imposed law that an enemy of the city must not be buried, Antigone, on Hegel's account, performs a "positive ethical action" that derives its ethical status from obedience to the "perfect divine law." That law orders the family to bury their dead, which interrupts both the political and the natural. This law is considered divine, because it concerns a dead person who is no longer a citizen and who, for the state, becomes "only an unreal . . . shadow," entrusted to what is behind the real world ("the divine right and law has for its content and power the individual who is beyond the real world, yet . . . not without power").[1] Antigone's action becomes positively ethical rather than, for instance, political, precisely because its object, expelled from the state into the world of shadows, is not a political being. But her action is positively political also, because it is not guided by any love or passion ("the ethical connection between the members of the Family is not that of feeling, or the relationship of love"),[2] or fashioned by the subjective interest of the doer, being motivated instead by the "duty of the member of the family." Antigone's story is tragic precisely because to act as a purely ethical being—unconcerned with the political, and emotionally indifferent—she must transgress the law of the city, and so finds herself accursed as unethical by the civic and political forces.

But to maintain itself as ethical by obeying the divine, Antigone's act of burial, on Hegel's account, doesn't interrupt the civic only. It must also interrupt the operation of the natural. That is what the burial does: it isolates the corpse from exposure to the "forces of abstract material elements." As Hegel puts it, "the family keeps away from the dead this dishonouring of him by unconscious appetites and abstract entities, and puts its own action in their place."[3] That is to say, instead of entrusting the dead body "solely to Nature," which is "something irrational," the family consciously performs a burial, and through that conscious act asserts the "right of consciousness" over nature.[4] Thus, by dispelling the natural and maintaining something conscious, the burial enables the dead, who have left the "accidents of life," to be transformed into the "calm of simple universality."

Thoreau also celebrates Antigone as a figure who interrupts the political, and guided by her "resolute and noble spirit," dares to transgress the laws of the state (W, 134). But in Thoreau the higher laws that Antigone follows in resisting the state do not regulate what is "beyond the real world," as was the case for Hegel. In complete opposition to Hegel, the higher laws Antigone obeys are earthly in Thoreau's account; they do not interrupt but instead affirm the natural. For in Thoreau, Antigone responds to the call to "be awake, though it be stormy, and maintain ourselves on this earth and in this life, as we may . . . [to] see if we cannot stay here where He has put us, on his own conditions" (W, 134). To maintain ourselves on this earth also means to maintain as earthly the way we die. For Thoreau it is precisely the manner in which Antigone buries her brother that demonstrates that "God's conditions" or Hegel's "divine laws" are of the earth. Far from performing a complex rite that imposes the "right of consciousness" over the earth, as Hegel would have it, according to Thoreau, Antigone performs an act so slight that it verges on the natural. This is how Thoreau describes Antigone's burial act: "Antigone has resolved to sprinkle sand on the dead body of her brother, Polynices, notwithstanding the edict of king Creon" (W, 134). Far from "wedding the blood-relation to the bosom of the earth"[5] as Hegel represents it, Antigone less buries the body in the earth than sprinkles it with sand, in fact barely covering it, for wind could easily expose it again. Nor does she leave signs or marks of any "human deed" in the sand, which Hegel maintained was necessary to extract the dead from the natural. That is why, for Thoreau, Antigone's act becomes the paradigm of new burial rites in accord with the natural he has in mind when he urges reform of Christian funerary customs.

Thoreau's reading, which understands Antigone's burial act as a sprinkling of sand over the dead body and more proximate to how nature would cover it than how humans would, is in fact closer to Sophocles's text than is Hegel's rendering. In Sophocles, Antigone's act is difficult to differentiate from the natural, for it verges on non-burial. In response to Creon's question as to what man has dared perform the act of burial against his law prohibiting it, the Guard who has brought the news that the burial took place replies: "I do not know. For at the spot there was no stroke made by a pickaxe, no impression of a mattock. The ground was hard and dry, unbroken and unmarked by cartwheels; there was no sign who the culprit had been . . . the body was out of sight—not entombed, but lightly covered with dust [sand, in Thoreau's rendering], as if by one who was avoiding pollution."[6] There are no signs, traces, or marks of the funeral, to testify to conscious and a human denaturalizing act. The body is here, as Carol Jacobs puts it in her astonishing reading of Antigone, to which mine is greatly indebted, "out of sight and yet not out of sight, out of sight and yet 'not entombed, but lightly covered with dust.' The body is buried and yet not buried, actually, with just the thinnest veil of dust upon it. Antigone's gesture does not really make the child above the ground belong to those below. . . . Is this then the gesture that will . . . save [Polynices] from Nature and natural death and deliver him to the universal as Hegel had described it?"[7]

In fact, the indiscernibility between the human act of sprinkling sand over a body and a natural operation whereby wind covers it with dust is the central issue raised by the scene that occurs after Creon has ordered Polynices's body unsanded and left on the surface of the earth to rot. The guard appointed to watch that the body remains uncovered, reports to Creon as follows:

> When we arrived there . . . we wiped away all the dust that covered the corpse, stripped the damp body well, and sat on top of a hill to windward. . . . And then suddenly a whirlwind raised from the earth a dust-storm, a trouble in the sky, and filled the plain, spoiling all the foliage of its woodland; and the wide air was choked with it. We shut our eyes to keep out the god-sent plague. And when, after a long time, this had cleared, the girl was seen.[8]

As Jacobs puts it, commenting on these lines, "that same earth, which Antigone had left without a trace, that same matter so dear to Antigone . . . has been raised by a whirlwind to trouble the sky. The whirlwind . . . fills the plain with dust. . . . This is a dust so overwhelming it covers all the foliage and surely, therefore . . . also the damp corpse of Polynices."[9] If, in the first act of sanding the body the act was performed in the manner of nature, in this second occurrence—especially since nobody has actually seen Antigone do anything with the body—it is impossible to decide between Antigone and nature, "both arriving" to quote Jacobs again, "coincidentally and performing the same rite of scattering the dust that neither buries nor leaves unburied the decomposing figure of the male."[10] In the coincidence of natural and human, it becomes hard to assert what constitutes burial, just as Thoreau imagined for human burials.

But the confusion of human and natural in that scene is intensified by the fact that once the whirlwind settles and the guards clear their eyes, they in fact don't quite see the "girl." What they see instead is a girl in the process of becoming a bird: "And when, after a long time, this had cleared, the girl was seen; and she uttered a piercing cry, the shrill note of a bird, as it cries when it sees, in its empty nest, the bed bereft of nestlings."[11] Antigone's becoming a lamenting bird-mother is not Sophocles's imprecise simile whereby he transfers to Antigone the feminine capacity of giving birth and makes her lament the loss of her brother as if he were her son. Moreover,

Antigone's becoming-bird is not a simile at all, for she doesn't cry like a bird but utters the cry of a bird; her becoming-bird thus has the status of a real transformation, which, as Jacobs writes, "undefines the human as either origin or product."[12] What, then, does this Antigone-bird mourn, releasing a cry verging on a screech? What she cries over is what she herself sees as she opens her eyes after the sand storm, what she herself is becoming: something that is and isn't, neither hidden (dead), nor visible (alive). At the exact intersection of life and death, of natural and human (bird and woman), articulate and inarticulate (utterance and shrill cry), Antigone becomes, in Jacobs, the mother of "that which has not and cannot maintain its figure."[13] That is not to say that Antigone becomes the mother of the dead, but rather the mother of "dispersal" of any claim to "completeness of shape" and "final figuration."[14] In Sophocles then, Antigone does what nature does in Thoreau: she disperses what was whole into new relations, becoming a coincidence of de-creation and making.

Part III

Walden Pond, Concord, Massachusetts, 1845: Epistemology of Change

Introduction: On Embodied Knowledge and the Deliberation of the Crow

In the first part of the book I discussed Thoreau's early efforts to emancipate perception from representation and imagination, indeed to empty the mind from all content in order to experience beings and things in their "suchness," as they appear in the world from which the human gaze has been absented. But my concerns there were not epistemological. Instead, I charted ways in which these early epistemological concerns were mobilized by John's death and put into the service of grieving.

I now wish to turn to the epistemological concerns that I there left aside. In this part I claim that with *Walden*, Thoreau's desire to recover the suchness of things results in the articulation of a complex anti-Kantian epistemology, with carefully crafted protocols of knowing and the difficult comportment they required. I thus argue that what at first, in early *Journals* and in *A Week*, might have appeared as a naive idea—the idea of self-suspension and stilling and emptying the mind—is elevated in *Walden* into what Stanley Cavell calls a rigorous philosophical doctrine, one that radically revises Western concepts of subjectivity. To reconstruct this epistemology I assume the role of an archaeologist of Thoreau's thought, charting his resistance to the Western understanding of subjectivity (as self-posited reflexive mind representing the external and material world), truth (understood as the adequacy of subjective representation to the represented object) and logic (understood, in the wake of Aristotle, as the grammatical distribution of categories that navigate thinking). While suggesting that Thoreau's critique of dominant Western epistemologies (with the important exemption of Plotinus, whose influence I also discuss) was supported by philosophies of the East—specifically those he finds in *The Bhagavad-Gita* and *The Sānkhya Kárika*—I also argue that Thoreau followed them only to the extent that they fit into his own understanding of mind and truth. In other words, he appropriated from the *Gita* and *The Sānkhya* only those aspects of their teachings that could strengthen his own theories while making them appear less eccentric. Thus, taking his distance from Western traditions while profoundly altering the Eastern, Thoreau, in my reading, emerges as an astonishingly original—and, as I suggest in a little coda on Walter Benjamin, modern—philosopher, offering a materialist epistemology in which the self is successfully suspended and the suchness of the world restored.

My understanding of Thoreau's epistemology as quintessentially materialistic, as invested in accessing things as they are without mediation by the mind, is in contradistinction to many influential readings of Thoreau. There he emerges as an idealist who posits a self that searches for truth in what is commonly, but vaguely, described as the sphere of the "spiritual" or "divine." In those readings Thoreau is a "spiritual seeker." Perry Miller, for instance, proclaimed that even Thoreau's notorious interest in nature does not signify an interest in the material. Reading Thoreau through the lens of puritanism, Miller argued that he never saw nature as nature, but instead, in accordance with puritan typology, as a figure of the divine, as the material embodiment of the spiritual. On that understanding Thoreau's empiricism was a pretense and his perceptive investment in the natural something strictly instrumental, providing a passage to the spiritual world, which found only imperfect expression in material terms. The empirical was understood to be a passage to the world of immaterial essences, both the essence of the mind and the essence of the divine. Once the ideality of the mind, whether finite or infinite, was revealed, the natural could be left behind: "[Thoreau] strove to transcend not only experience but all potential experience; had he achieved what he intended, he would have become pure act, and his beloved nature could have been consigned to oblivion."[1] Following Miller, his professor at Harvard, Sherman Paul saw the Walden experiment as a similar "inward exploration." On Paul's understanding, Thoreau goes "outward" only "to take the higher leap." For Paul, nature, and everything material, is only a surface—a base and corruptible, changeable existence behind which lurks what Thoreau is really after, the permanence of essential spirit. Thoreau's trajectory advances from the material to the ideal, from the natural to the spiritual, from disease to health: "[Thoreau] began with surface before he spoke of depth, with the transient rather than the permanent, with complexity before simplicity, disease before health."[2] Following Paul, indeed declaring his book to be "probably the single most important work of Thoreau scholarship," Walter Benn Michaels, as I've already suggested, understood *Walden* to be a version of an "almost Cartesian process of peeling away until we reach that point of ontological certainty where we can say 'This is, and no mistake.' " As Michaels specifies, the certainty Thoreau finds is just like Descartes's *cogito*, that of a being that is thought; he therefore emerges as an idealist immersed in a dualist ontology where "the human and the natural are conceived as standing in implicit opposition to each other."[3] Readings suggesting that Thoreau needs materiality, exteriority, and otherness only to sublate it, and through this sublation posit the self, continued into first decade of the twenty-first century, culminating perhaps, with Alfred I. Tauber's comparison of Thoreau's self with Fichte's self-posited "I," or Reid, Furtak, and Ellsworth's assessment that Thoreau is an idealist Kantian invested in conceptual mediation.[4]

My Thoreau couldn't be further from those interpretations. He is closer to readings that have argued that he sought a profound revision of everything we mean when we talk about personhood, and that such a revision required an ontology that is deeply invested in the question of the material. Here I have in mind Cavell's 1972 claim that Thoreau wanted to turn self-reflexivity into the "impersonality" of a "self's standing beyond self-consciousness," sensing the "thing-in-itself"; or Lawrence Buell's claim that Thoreau wants to enact the "radical relinquishment" of "individual autonomy itself," to cancel not just mental but "even bodily apartness from [his] environment";

or Sharon Cameron's suggestion that in Thoreau not only the object, but also the origin of thought resides outside the self, so that thinking is not within the mind but without it. I also include here H. Daniel Peck's idea that Thoreau's thought is relational but that the relationality in question is not between subject and object but among objects; or Russell Goodman's assertion that Thoreau is a thinker of the body, not in the sense that he thinks about the body, but rather with the body, to generate an embodied instead of a spiritual knowing.[5] In what follows I take seriously those remarks with the aim of describing the complex epistemology that, as I see it, Thoreau generated. My analyses are guided by the belief that what these careful readers of Thoreau registered as his wish to relinquish the self or his desire to fuse with his environment, must be translated from a metaphorical language, whereby Thoreau would only imagine his fusion with the natural, into an epistemological discourse that answers precise questions: why exactly would a self wish to fuse with the natural? Leaving aside the vague and dubious quasi-ecological answer that such a fusion was mobilized by Thoreau's affinity with nature, a more rigorous response is required, one that explains what is an epistemological gain of such fusion; what kind of knowing would self-cancellation generate; what kind of self would be capable of enacting its own cancellation; and, finally, *how exactly*, by what practices such a cancellation of the self or such a fusion with the natural might occur?

But how is this rigorous epistemology that I detect in Thoreau related to mourning, why do I call it an "epistemology of mourning"?

In "The Difficulty of Reality and the Difficulty of Philosophy," Cora Diamond analyzes epistemological responses to experiences in which "we take something in reality to be resistant to our thinking it, or possibly to be painful in its inexplicability. . . . *We take things so.*"[6] She illustrates what those things might be with two examples, both related to how we take the suffering and death of others, how we cope with loss. The first example is a statement from Elizabeth Costello, a character in J. M. Coetzee's *The Lives of Animals.* Explaining the intensity of her concern with the suffering of animals, with the massive proportions of the slaughter, Costello declares that when she thinks about killed animals, she doesn't simply imagine them as dead and then, on the basis of that representation, draw a conclusion that it is unjust to treat animals so. Instead, she says, for an instant, she *becomes* a dead animal, "for an instant at a time . . . I know what it is like to be a corpse. That knowledge repels me. It fills me with terror."[7]

Diamond takes her second example from Ted Hughes's poem, "Six Young Men." The speaker of that poem looks at a photograph, taken in 1914, of six smiling men, all of whom will be dead six months later. The last stanza of the poem summarizes the epistemological crisis that the sight of a happy yet dead man causes in the observer:

> That man's not more alive whom you confront
> And shake by the hand, see hale, hear speak loud,
> Than any of these six celluloid smiles are,
> Nor prehistoric or fabulous beast more dead;
> No thought so vivid as their smoking blood:
> To regard this photograph might well dement,
> Such contradictory permanent horrors here
> Smile from the single exposure and shoulder out
> One's own body from its instant and heat.

Both examples are about "experiences of the mind's not being able to encompass something which it encounters,"[8] something—the pain and death of others, for instance—that is so intense that it instantaneously dismantles the analytical categories and concepts we typically use to mediate reality, filling reason with terror, leading it not toward the safety of abstract thinking but to the awe of embodied knowing. Such an experience of the unspeakable Diamond calls a "difficulty of reality."

Philosophy responds to this difficulty inadequately, by avoidance and deflection, producing the dualistic divide between mind and body. By "deflection," Diamond means "what happens when we are moved from the appreciation . . . of a difficulty of reality to a philosophical or moral problem apparently in the vicinity."[9] This is to say that philosophy responds to the difficulty that "pushes us beyond what we can think," toward the suffering flesh and embodied knowledge that can easily unhinge the mind, by transforming it into a thinkable philosophical problem; philosophy idealizes and so distances us from reality. To philosophize about reality is to initiate "coming apart of thought and reality," to enact a magical metamorphosis of pain into a rigorous argument, aimed at protecting the thought's protocols and its togetherness as guarantors of the coherence of our persons. But, as Diamond points out, what is injured in this separation of thought and reality is "flesh and blood." For when the mind, believing in the insurmountable difference between thought and matter, withdraws into itself to think through a death of the body, the body is abandoned as an unthinking being, left to itself, on the outside, to cope with its own experience of grief, which doesn't make it into words and concepts.

I want to suggest that Thoreau shares with Hughes and Costello a refusal to deflect loss and pain into the safety of rigorous argument, even if they can unhinge us; or alternatively, that he accepts the risk of unseating "ordinary modes of thinking" embedded in attending to the "difficulty of reality." That refusal to protect the mind by means of idealism, while leaving the body to reality is, precisely what Thoreau voices in a famous passage from *Walden*, inviting us to "crave only reality." To crave nothing but reality, as Thoreau specifies there, would require nothing less than the radical abandonment of everything we know; it would require us to break through all the layers of idealism, through "the mud and slush of opinion, and prejudice, and tradition, and delusion, and appearance . . . through church and state, through poetry and philosophy and religion, till we come to a hard bottom and rocks in place, which we can call *reality*, and say, This is, and no mistake" (Wa, 97–98). To abandon the opinions, judgments, values, traditions, and the cultural or communal world we inhabit; to abandon all our philosophy, religion, law, and literature, is, in fact, to abandon everything, to start thinking from scratch, in a different way, in a real way, in a way that would enable thought to settle in the real, in its own body and so to become embodied. The first effects of such a fall into the real might be horrifying, for, as Thoreau suggests, when the mind leaves how it is taught to think, it is the body that falls and bruises on the hard rock. But it is from such hurting that new thinking—in which the body thinks—will emerge;

hence, "be it life or death, we crave only reality. If we are really dying, let us hear the rattle in our throats and feel cold in the extremities" (Wa, 98).

Thus, when Thoreau cancels our ordinary modes of thinking or when he insists on shrinking the distance between mind and body so that something like embodied knowing can appear, he is not expressing a vague romantic aesthetic, a desire for a drifting mind. Rather, he is answering to a pressing epistemological crisis, a crisis generated precisely by the unspeakable loss that reality presses on us. Thoreau's response to the pressure of such loss—a pressure that is nothing other than grief—may transgress proper philosophical reasoning, but it nevertheless forms an epistemology, I argue, insofar as it responds to what is unbearable in reality by bearing with it, not by deflecting it.

Just as crows do. Crows fascinated Thoreau throughout his life, because they are birds that defy patterns and orderliness, being perhaps "our only large birds" that "hover and circle about . . . in an irregular and straggling manner" (Jn, 14, 101). They do so not only to passively endure the conditions that befall them, but to come to know them. Such is the case of the crow that astonishes Thoreau by consistently trying to fly "directly forward against the wind," even if it is always "blown upward and backward within gunshot" (Jn, 14, 102). But the crow is not simply a creature that is incapable of adjustments. Faced with the wind that blows it away and unwilling to take an easy approach to reality by flying with the wind to its back, the crow changes strategy; it slows down "and . . . only advances directly forward at last by stooping very low within a few feet of the ground" (Jn, 14, 102). It risks coming nearer to the ground, where it will be trapped into walking as opposed to swift flying, to achieve one thing only: to advance by facing "very deliberately" what is blowing it away. In what follows I tell the story of Thoreau's crow-epistemology, which devises ways of thinking that allow that thinking to face, in order to know, what blows it away.

Toward Things as They Are

"Portions," the last essay from Cavell's *The Senses of Walden*, suggests that the sense of loss fashions Thoreau's whole philosophical project. The Thoreau that Cavell attends to is always saying that all our experiences are desolate: "From our own experience we draw or project our definitions of reality; . . . only the experience we learn from, and know best is our failure. . . . We were to be freed from superstition; instead the frozen hopes and fears which attached to rumored dictates of revelation have now attached themselves to the rumored dictates of experience."[1] In Cavell's interpretation, Thoreau feels that our experience is transformed into "rumored" representations of it. Cushioned by rumors, we distance ourselves from reality: "we crave only reality; but since 'we know not where we are' and only 'esteem truth remote' . . . we despair of ourselves and let our despair dictate what we call reality."[2] Our despair comes to be our reality, and so our reality gets lost. This condition complicates our experience of loss, for if everything we have ever lost is lost to reality, which is also lost to us, then our loss is doubled. To recover loss would first require learning how to find the reality in which our losses have been deposited. That is why Thoreau is so obsessed with learning "what finding is, what it means that we are looking for something we have lost."[3] But to learn the meaning of finding—which would in turn offer us knowledge of reality—amounts to revisiting and relearning what knowing is. Hence, the central preoccupation of Thoreau's writing, on Cavell's account, becomes epistemological. Moreover, in Cavell's reading, this epistemological commitment becomes so central that Cavell identifies it as Thoreau's "work of systematic philosophy,"[4] fundamentally different from the enterprises of both Emerson and the European romantics, who, unlike Thoreau, followed the path outlined by Descartes and, especially, Kant.

Cavell suggests that the importance of Kant for understanding Thoreau's project is in fact signaled by Emerson himself. "The Transcendentalist" states that the New England transcendentalists found their philosophical orientation in Kant's idealistic claim that "an idea of Reason is an Idea to be aspired to." To come close to Reason is an idealistic aspiration, because materialists, Emerson understands Kant to be saying, follow a path opposite to the Idea of Reason, a path found by the senses. Idealists, in contrast, "perceive that the senses are not final, and say, The senses give us representations of things, but what are the things themselves, they cannot tell."[5] But the idea that the senses don't give us things as they are, is, as Cavell interprets it, Emerson's fundamental misunderstanding of Kant: "The directly un-Kantian moment in Emerson's essay is his notion that the senses are the scene of illusions. This at a stroke misconceives the undertaking of the *Critique of Pure Reason*, an essential half of which was exactly to show that *things* (as we know them) *are as the senses represent them.*"[6] A thing in itself, then, is given through senses, as sensation, and the main question for Kant becomes how to access sensation, how to come to know sensing. For him, access to sensations as "objects of our knowledge" doesn't just happen—we don't just sense in a wild and unstructured way—but instead "requires a transcendental . . . preparation," the following of certain protocols. However, the protocols humans follow when they want to know a thing are not something that they have to negotiate each time, but instead, something given to them before any experience has taken place. When we approach a thing in order to know it, we thereby subject it to "these *a priori* conditions" of knowing, which we "understand as knowledge." Only things that obey "*a priori* conditions" of knowing become for Kant "what we understand as an object." An object is thus not just any thing but a thing subjected to us in such a way that it "answers to our conditions for knowing anything." Objects are what in reality, among things, are submissive to the workings of our epistemology; as Cavell puts it: "if something does not answer to our conditions of knowledge . . . it is not what we understand as an object. A thing which we cannot know is not a thing."[7] This "not a thing"—

this nothing—is precisely what Kant calls the "thing-in-itself," understanding it as anything that in "escap[ing] the conditions of knowledge" remains untrue for us. But to declare things false or senseless because they don't fit our categories—and so appear as illogical or obscure—is not only to neglect a great deal of the world but also to completely disregard that "our primary relation to the world is *not* one of knowing it."[8] For we are in the midst of things that affect us even before we get to know them, and things continue to affect us even though they remain unknown to us. We are, in other words, affected by the world and through this affection we acknowledge the world before, or independently of, our knowing it.

To gain access to a world that remains unknowable to us by means of our epistemological procedures is, according to Cavell, Thoreau's major concern. That is to say that Thoreau radically reverses Kant's project: "epistemologically [*Walden's*] motive is the recovery of the object, in the form in which Kant left that problem . . . viz., a recovery of the thing-in-itself."[9] Thus, Kant's "untruth," his thing in itself that is "nothing at all" to us, becomes Thoreau's everything, because what concerns him is how to settle into a world that doesn't answer us and is instead lost to us. Thoreau approaches what "doesn't answer to our conditions of knowledge" without imposing on it any conditions at all, and therefore, works against the categories of his self-consciousness: "Here are the elements of *Walden*'s solution to the problem of self-consciousness. . . . What we know as self-consciousness is only our opinion of ourselves . . . it is hearsay, our contribution to public opinion. We must become disobedient to it, resist it, no longer listen to it."[10] No longer listening to our self-consciousness, though, requires us to completely undo and reposition the one who wants to know. Whereas in Kant it is the subject who, by always listening to his self-consciousness—by following protocols of "transcendental . . . preparation"—imposes its categories on the object, in Thoreau the subject abandons such mental projections, because they, and not the senses, block access to things. Thus, as Cavell understands it, Thoreau requires that "we . . . become disobedient to [our self-consciousness] . . . by keeping our senses still, listening another way, for something indescribably and unmistakably pleasant to all our senses."[11] When the mind suspends its obedience to self-consciousness and stills the senses, a feeling of the external world nearing us appears, a feeling so intense—so "indescribably pleasant"—that we accept where we are, letting where we are access us, and in that way find the world: "This is why the writer's readings of nature do not feel like moralizations of it, but as though he is letting himself be read by it, confessed in it, listening to it, not talking about it."[12] When the mind is suspended so radically that all representation of reality is cancelled, things come so close to us that we are disclosed by or "confessed in" them, settling among them.

Thinking with Geological Velocity

The invention of a mind that could access materiality distanced from it by mental categories is already a project of *A Week* even though it becomes more explicit in *Walden*. As the "Monday" chapter of the earlier work reveals, Thoreau's means of exiting the protocols of knowing imposed on thought by modern Western philosophy is derived from Asian sources. More technically, and following Charles Wilkins's 1785 translation of *The Bhagvat-Geeta*, both in *A Week* and in *Walden* Thoreau identifies this thinking, which "listens to [reality]" as contemplation.[1] Contemplation, for Thoreau, doesn't have the status of a metaphor for thinking in general. Instead, he understands it as a specific practice of thinking that is fundamentally opposed to the Western metaphysical tradition: "Western philosophers have not conceived of the significance of Contemplation in their [Indian philosophers'] sense" (W, 137), which is why "the comparatively recent literature of Europe often appears partial and clannish, and . . . the European writer who presumes that he is speaking for the world, is perceived to speak only for that corner of it which he inhabits" (W, 142).

A crucial passage from *A Week* summarizes the difference between two traditions of thought as a difference in how thought relates to speed, action, and appropriation:

> Behold the difference between the oriental and the occidental. The former has nothing to do in this world; the latter is full of activity. The one looks in the sun till his eyes are put out; the other follows him prone in his westward course. There is such a thing as cast, even in the West; but it is comparatively faint . . .

> The former class says to the latter, When you have reached the sunset, you will be no nearer to the sun. To which the latter replies, But we so prolong the day. The former "walketh but in that night, when all things go to rest, the night of *time*. The contemplative Moonee sleepeth but in the day of *time*, when all things wake." (W, 141)

When Thoreau suggests that the contemplative Moonee thinks during the night, not wanting to look at things in the sunlight, he isn't denying that for each of the traditions he investigates, truth is in fact imagined as some type of illumination and so emblematized by light. Rather, he wants to suggest that the two traditions differ in respect to how—with what kind of mind, what kind of logic and methodology—such an illuminative revelation of truth might be achieved. As Thoreau's remarks rightly signal, the Western tradition intuitively supposes that walking in the dark can only leave one in the dark, that is, that the light of truth can be achieved only if the path to it (method) is itself enlightened (by clear and distinct perceptions), organized (by categories and taxonomies) and controlled (by logic). Nocturnal thoughts are famously excluded from what counts as thinking at the very beginning of Descartes's "Meditations," on the pretext that a sleepy or melancholic mind's natural light *(lumen naturale)* has less capacity to illuminate the darkness of beings.[2] The "Meditations" thus epitomize Thoreau's claim that in the West the search for truth—being an enlightened search—always happens within the truth; hence the paradox Thoreau points to when he says that Western philosophers looks only in the sun in their search for truth; what he means is that Westerners look at what already reveals itself to them, that they are searching only for what they already know, leaving, just as Kant did, whatever is not visible in the realm of the un-knowable and so deeming it irrelevant. But by leaving what hides in the shade, they leave out a lot of truth, which is why Thoreau can say that Westerners are no nearer the truth once they think they have reached it ("when you have reached the sunset, you will be no nearer to the sun").

In contrast, Thoreau suggests that the night is the time that the Eastern mind judges favorable for thinking: the oriental "walketh but in that night, when all things go to rest, the night of time." Thoreau is here quoting from the *Gita*'s second lecture, on the soul, in which Krishna explains what wisdom is, how can it be obtained by contemplation, and who can be called wise (who can be called the "contemplative Moonee"). Contemplation, as Krishna details, can't tolerate movement or the rush of activities and so occurs only when everything that "hurries us up," is slowed down to the point of being abandoned. It emerges when the thought of causality and consequence is abandoned (for the idea that one thing leads to another moves us to something else and makes us expect reward, fruit, or gratification). It comes about when the inclination of the senses is halted and desires are stilled. Krishna's lecture is in fact an insistence that the condition of stillness substitute for the mobility of thinking, allowing thought to see the world as restful too. Because the world at rest is reminiscent of the repose of bodies in the night, Krishna calls this contemplative mind "nocturnal": "Such a one walketh but in that night when all things go to rest, the night of *time*."[3]

However, this "night of time," as Thoreau rightly interprets it, is not quite extra-temporal, but instead, "infinitely stagnant" (W, 136). In other words, it is less immobile than extremely slow, and it appears as immobile and a-temporal only to senses that rely on the perceptible speed of beings moving in the rush of day. Because it is too slow for human senses to register its movements, which can indeed be longer than the span of a human life, Thoreau understands this night of time not as subjective but as an impersonal temporality coinciding with the time of the elements and vegetation. The mind that enters time in repose wouldn't quite exit temporality but would instead enter a time zone—a "day of Brahma" as Thoreau calls it—in which mental coincides with the geological and historical with geographical:

> Menu understood this matter best, when he said, "Those best know the divisions of days and nights who understand that the day of Brahma, which endures to the end of a thousand such ages, [infinite ages, nevertheless, according to moral reckoning,] gives rise to virtuous exertions and that his night endures as long as his day . . ." And do we live but in the present?

> How broad a line is that? I sit now on a stump whose rings number centuries of growth. If I look around I see that the soil is composed of the remains of just such stumps, ancestors to this. The earth is covered with mould. I thrust this stick many aeons deep into its surface, and with my heel make a deeper furrow than the elements have ploughed here for a thousand years. If I listen, I hear the peep of frogs which is older than the slime of Egypt. (W, 153)

Thoreau imagines the contemplative mind to operate as a type of earth, whose remains, far from signifying the absence of what has been, compose a "living" now of the soil. Just as the soil he stands on keeps the past in the present, so the contemplative mind is slowed down to such an extent that it "endures" many ages. In other words, its time is imagined as capable of stretching the line of its present over its past, disturbing the distinction between what is and what isn't. This marvelous deceleration, by which the contemplative mind refuses the rhythm of a human clock so that it may coincide with the pace of a swamp, maintains temporality while defying the categorial division of historical time into what has passed and what is to come. For if, from the historical point of view, the slowness of geological time makes its past contemporaneous with its present—in terms of historical time one can say that a stone now is what it was "aeons ago"—then a stone's past catches up with its present. In the night of time all manners of time are simultaneous; the future and past glitter in the present, and one who contemplates sees them. This seeing is wisdom.

How to Greet a Tree?

The *Gita* specifies that the condition favorable to contemplation can be attained only by means of difficult willed relinquishments. For instance, Krishna requires the mind to disinvest from its environment, and in willfully extracting itself from the motion of external phenomena to "keep the outward accidents from entering it."[1] Similarly, the mind is required to disinvest itself from the "inclination of senses," feelings (Krishna specifically mentions anger, pride, grief, fear, and care) and thoughts. This is not to say that the contemplative mind will stop perceiving or sensing or thinking. Rather, its relation to sensations and thoughts will change. Disinvestment from a perception or thought doesn't make it vanish but instead alters it from something subjective into something external and inconsequential to the mind. In a sentence Thoreau quotes in the *Journal*, such disinvestment is described as the mind's renunciation of its boundaries, understood to be defined by the "the city" in which it lived: "The Brahman Saradwata says the Dharma Sacontala, was at first confounded on entering the city- 'but now,' says he, 'I look on it, as the freeman on the captive, as a man just bathed in pure water, on a man smeared with oil and dust'" (J, 3, 217–118). By divesting the mind of its content, mental states are transformed from something that is of the mind, something that fashions it by affecting it, to something not constitutive of the mind, which the mind can regard (in which it can "bathe" as Saradwata says) as if it had nothing to do with it. The contemplative man regards the renounced content of his mind merely as if it were something objective, something not his at all, a sort of natural phenomenon. Or, as Thoreau puts it, interpreting Krishna: "The philosopher contemplates human affairs as calmly & from as great a remoteness as he does natural phenomena" (J, 3, 217).[2]

Those *Journal* remarks from 1851 testify to Thoreau's continuing interest in Hinduism, which culminated with *Walden.* The culmination can be registered not only in the fact that, as Robert Kuhn McGregor reconstructs, in revising *Walden* Thoreau kept adding "series of quotations from Asian literature," or in the fact that, as Paul Friedrich detects, many of the symbols Krishna employs in the *Gita* (ax/axis, the upside-down tree, the field) found their place in it. More importantly, as Cavell attests, it is also evident in that some of the teachings central to *Walden*—such as relinquishment, divestment, and renunciation—took their direction from the *Gita:*[3]

> The *Bhagavad Gita* is present in *Walden*—in name, and in moments of doctrine and structure. Its doctrine of "unattachment," so far as I am able to make that out, is recorded in *Walden's* concept of interestedness. (This is, to my mind, one of Thoreau's best strokes. It suggests why "disinterestedness" has never really stabilized itself as a word meaning a state of impartial or unselfish interest, but keeps veering toward meaning the divestment of interest altogether, uninterestedness, ennui. Interestedness is already a state—perhaps the basic state—of relatedness to something beyond the self).[4]

Cavell here suggests that Thoreau in fact radicalized the *Gita's* understanding of disinvestment. The problem with that understanding—a condition in which the mind impartially observes its content as if it were something objective—is that it allows the mind still to maintain an albeit minimal relation with the object of its observation, and through that relation to maintain itself. To cancel even this relic of a relation, Thoreau, on Cavell's account, destabilizes it, having it veer toward a complete relinquishment of the self. Instead of Krishna's "I" detached from sensations or thoughts, there is now, in Thoreau, an absolute renunciation of the "I." That is why all the actions Thoreau undertakes at Walden would be actions of renunciation: "[Thoreau's] actions specifically declare the sense of leaving or relinquishing as our present business."[5] Moreover, on

Cavell's understanding, it is the idea of relinquishing that motivates Thoreau's central preoccupation at Walden: the process of building and acquiring the house. For, as he argues, Thoreau builds his house not so that he can have it and settle in it, but quite the contrary, to learn how to renounce it once the experience of settling in it has made him attached to it: "Of course, the central action of building his house is the general prophecy . . . that the day is at hand for [him] to depart from [his] present constructions. . . . Walden was always gone, from the beginning of the words of *Walden*. (Our nostalgia is as dull as our confidence and anticipation.)"[6] The future that motivates Thoreau, then, is one in which he finds himself completely impoverished, having let everything go, letting it be as it is when he doesn't relate to it.

Keeping in mind that Thoreau's actions are deliberate—always thought through—I suggest that the actions of renunciation that Cavell identifies were predicated on a certain type of thinking that Thoreau carefully cultivated, which most acutely registers the *Gita's* influence on him. How such thinking came to be is detailed in "Where I Lived, and What I Lived For," a chapter of *Walden* that might be called, before Heidegger, Thoreau's "What Is Called Thinking," an unfolding of his 1851 *Journal* remark that to think is to transform what we are attached to into something we are not attached to, into something that is as remote to us as any objective natural phenomenon could be. Its opening famously describes how Thoreau "surveyed the country" in and around Concord to find a spot where he could live, to find a house he would buy. The search starts from an identification and inspection of a farm and a house but immediately leads to his desire to appropriate it, as if the mere encountering and identifying of an object can't be differentiated from knower's irresistible desire to possess it:

> I have thus surveyed the country on every side within a dozen miles of where I live. In imagination I have bought all the farms in succession, for all were to be bought, and I knew their price. I walked over each farmer's premises . . . took his farm at his price, at any price, mortgaging it to him in my mind; even put a higher price on it . . . and withdrew when I had enjoyed it long

> enough leaving him to carry it on. . . . Wherever I sat, there I might live. . . . Well, there I might live, I said; and there I did live, for an hour, a summer and a winter life . . . and then I let it lie, fallow perchance, for a man is rich in proportion to the number of things which he can afford to let alone. (Wa, 81–82)

Imagination is what seems to fuse the act of seeing—coming to know predicated on seeing—with the desire to possess the seen. This is because its fabrication of representations not only subjectivizes what is represented but, in so doing, invests it with desire, turning it into a fantasized possession (it is thanks to imagination that Thoreau desires "*all* the farms" and all the houses he sees, for once they are mediated by imagination, they all become beautiful). There is thus no knowing without desiring and without appropriating.

However, the appropriative thinking Thoreau here describes is not peculiar to a certain historical moment. For instance, this intimate alliance of cognition and imagination in the service of possession was not generated by American modernity, which Bill Brown explains in terms of the culture's obsession with things—their "tyranny" over the mind that unleashes a will to possess. It is, of course, true that in 1850s Thoreau would, as Brown puts it, "everywhere . . . have found that man he had denounced," and that such a cultural environment would facilitate his "conviction that Americans should expropriate themselves of their possessions."[7] But it is no less true that Thoreau sees such a cultural turn only as the intensification of how we generally think. For we are all—his antebellum contemporaries, but also those of us reading him from distant historical and cultural futures—anticipated by him as thinking appropriatively: "the future inhabitants of this region, wherever they may place their houses, may be sure that they have been anticipated" (Wa, 81). We are anticipated in our thinking by appropriative identifications: in naming and knowing a thing so as to let it appear only as what it is for us, in the manner by which it alters our desires and inhabits our fantasies, as our own. Thus, when Thoreau insists on relinquishment and urges the "expropriation of our possessions," he is urging us to overcome the style of thinking that is universal to the Western mind, predicated on the belief that knowing a thing must lead through naming, identifying, and appropriating. If the violence of appropriations is to be suspended, a new form of thinking will be required, one that will expropriate us not from the desire to possess this or that particular thing but from the desire to possess as such (a direction signaled by Thoreau's remark that "man is rich in proportion to the number of things which he can afford to let alone").

Leaving beings as what they are, letting them lie, is not about abandoning them after we have used them up and need them no longer. Instead, it is about our allowing them to be without our thinking interfering with what they are. It is about respectfully freeing them from the demands of our desires or the judgments of our identifications, which is why it is those desires and judgments—in other words, our own selves—that must first be suspended and relinquished. Only self-relinquishment when it comes to accessing things promises that what we appreciate or love about them will not be what we alter about them through possessive force of imagination; it promises that we won't love ourselves in them, that our relationship to them won't be narcissistic.

To a Western philosophical ear such "unimaginative" thinking, in favor of objectivity, striving to receive what it encounters without identifying or having it, might seem dubious if not impossible, suggestive of a barely understandable ethics. But a hundred years after Thoreau proposed it, Heidegger identified it as the only practice of thinking that should be counted as thinking. To describe how such nonpossessive thinking works, Heidegger describes the encounter with a tree:

> We stand before a tree in bloom, for example—and the tree stands before us. The tree faces us. The tree and we meet one another, as the tree stands there and we stand face to face with it. As we are in this relation of one to the other and before the other, the tree and we are. This face-to-face meeting is not, then, one of these "ideas" buzzing about in our heads. . . . What happens here, that the tree stands there to face us, and we come to stand face-to-face with the tree? . . . When ideas are formed in this way, a variety of things happen presumably also in what is described as the sphere of consciousness and regarded as per-

> taining to the soul. But does the tree stand "in our consciousness," or does it stand on the meadow? Does the meadow lie in the soul, as experience, or is it spread out there on earth? Is the earth in our head? Or do we stand on the earth? . . . When we think through what this is, that a tree in bloom presents itself to us so that we can come and stand face to face with it, the thing that matters first and foremost, and finally is not to drop the tree in bloom, but for once let it stand where it stands. Why do we say "finally"? Because to this day, thought has never let the tree stand where it stands.[8]

Except, I would argue, Thoreau's thought. Unlike the encounters generated by "our traditional nature of thinking"—which Heidegger identifies as "the forming of representational ideas"—Thoreau's encounters are never in himself, but outside himself.[9] Thoreau doesn't encounter a tree in his head, as a mental picture, but leaving his head, he encounters the tree where it stands, on the earth. Such an encounter is what Heidegger understands as thinking "underway" to the world, and what Thoreau understood as an effort to recover things as they are.

Contemplating Matter

Walden AND "Walking" offer a detailed discussion of the mind's self-relinquishment practices that Thoreau found in the *Gita*. However, as I am suggesting, the practices described there are far from a simple application of Hindu teachings. As Thoreau correctly interprets it, the *Gita* imagines self-relinquishment as the mind's "eternal absorption in Brahma" (W, 136), as its "transportation" into the "divine essence . . . the eternal abode of Vichnou" (J, 3, 216). But although the *Gita* always—and specifically in chapter two, from which Thoreau draws the quote that contemplation occurs only in the "night of time"—insists that the divine being into which personhood dissolves is incorporeal (the contemplative mind is said to "mix with the incorporeal nature of *Brahm*"), Thoreau will come to understand it, on the contrary, as material.[1] However, his materialistic understanding of a divine nature that absorbs contemplative persons is not a simple misunderstanding, for it is in fact proposed in Henry Colebrooke's commentary on *The Sānkhya Kárika* and in Horace Wilson's discussion of *The Bhashya, or Commentary of Gaudapada*. As Ellen M. Raghavan and Barry Wood document, Thoreau read those commentaries in the late 1840s—and quoted them in *A Week*—but took lengthy notes from them only in 1851.[2]

Both Colebrooke's and Wilson's commentaries register their own difficulties in reconciling the purely spiritual nature of *Brahm* with the creation of the material world. Following the *Gita*, the Sānkhya school understands *Brahm* to be a purely spiritual being, without form or personality; preceding space, time, and causation; and being always one and formless and utterly passive. That passivity of the *Brahm* means, however, that it creates nothing, neither powers nor qualities, neither causes nor effects, for such a creation would generate the discreteness of qualities and beings and so destabilize *Brahm*'s oneness. As the *Gita* has it, quoted by Thoreau, "the Almighty creates neither the powers nor the needs" (J, 2, 256). Accordingly, Sānkhya teachings ascribe all the power of generation not to *Brahm* but to a self-caused "primordial matter" *(Prakriti)*. This matter is also the first cause of all creation, whether of bodies, energy or particular minds *(Parusha)*, yet despite its creativity it is not elevated to the status of the supreme being of *Brahm*. Instead, as the *Sānkhya Kárika* has it, *Prakriti* only "intimates, that which precedes, or is prior to, making; that which is not made from any thing else."[3] As Colebrooke explains, *Prakriti*—which he renders synonymously as either "nature" or "matter"—is "at once the origin and substance of all things."[4]

The problem Colebrooke and Wilson register in their commentary on *The Sānkhya Kárika* concerns precisely the utterly passive nature of supreme being. Obviously guided by the Western philosophical sensibility, which assimilates being to acting and identifies God as the cause of all creation, they raise the following question: if *Brahm* (or as they render it, "*Brahme*") can't act, why would it be given any priority over creative matter? Colebrooke's commentary implicitly answers this question by proposing that *Brahm*—because of its unrelatedness to anything created—be understood as sheer "abstract existence," and he defines abstraction as what can't be conceptually accessed by the mind, hence a fantasy of no concern to "philosophical investigation." What, on his understanding, does concern "philosophical investigation . . . and must be known," is the power that generates real existences, nature or matter *(Prakriti)*.[5] In the edition of *Sānkhya Kárika* that Thoreau reads, Colebrooke's thesis is reinforced by Wilson's commentary. Wilson suggests that creative matter (which he brings even closer to the Western philosophical tradition by rendering it "*materia*") or "nature" (for him, too, in the "*Sānkhya* system . . . nature is not distinct from matter itself")[6] is neither inert nor formed but is instead without parts;[7] it is not crude—palpable or visible—but instead subtle *(saukshmya)*, the "parent and progeny" of beings, intrinsic but not reducible to its products.[8] As an inexhaustible flow of generation, this subtle corporeality is thus to be understood as "plastic nature that acts," "undergo[ing] modification through its own" activity, while generating and animating all existences; it is the unending and imperceptible, yet "substantial" or material life of all that is.[9]

The ontological adjustment made by Colebrooke and Wilson, their turning material life into the highest existence, rendering it the privileged object of thought—a gesture that crucially influences Thoreau—complicates the final goal of Hindu contemplation. For Colebrooke proposes that it involves two phases. Contradicting his own claim that *Brahm* is an abstraction inaccessible to thought, later in the commentary he refrains from denying that the final aim of contemplation is indeed the absorption of a personal mind into the completely spiritual *Brahm.* But he suggests along the way that such absorption can be achieved only after the mind's lengthy contemplation of nature: "The union of soul with nature is for its contemplation. . . . As nature . . . can neither enjoy nor observe . . . its existence would be without an object, unless there were some other one capable both of observation and fruition. This other one is soul."[10] When contemplation starts eroding the boundaries of the personal mind, the sensations and thoughts contained in it are released into the objectivity of nature, rendering it, as it were, contemplative. That is why Colebrooke can assert that the contemplative fusion of mental and natural means at the same time the becoming reflexive of sensitive but nonreflective natural life: "the object of the union of soul and nature, or the final liberation of the former by its knowledge of the latter, is here explained."[11] But immediately after claiming that the union of soul and nature is the soul's final liberation, Colebrooke goes on to identify liberation as simply a stage in the contemplative practice. The contemplative union of the soul with nature "slowly matures," as Colebrooke phrases it, into a fruit *(bhoga).* In fusing with nature the soul comes to know it and subsequently achieves the mature feeling that it can also withdraw from nature. In other words, the soul comes to feel itself capable of fulfilling all of the *Gita's* requirements for successful contemplation: the mind extracts itself not from its sensations and thoughts only, but from the natural world also, thereby "keep[ing] the outward accidents from entering it" as Krishna puts it. This divestment of everything that is invested with natural life—which amounts to a divestment of everything there is—is the final fruit. The fruit of the contemplative union of the mind with nature is the mind's relinquishment of the natural, or, as Colebrooke terms it, its "final liberation" into the purity of uncreated spirit: " 'Contemplation,' *dar'sana*, is considered to comprise 'fruition,' *bhoga*," leading to "separation [of the] purified soul, which cannot be attained without previous union with nature."[12]

The notes that Thoreau takes from Colebrooke's commentaries in 1851 register his almost exclusive interest in the first phase of contemplation, which allows for mental fusion with the natural. Following Colebrooke, he understands that in Sānkhya philosophy persons are a sort of compartmentalization of infinite spirit into bundles of sensations, generated by the mobility of corporeal life. Each of these "compartments" is identified as the individualized "spiritual substance" called *"Purusha,"* which through contemplation can be released into infinite and eternal embodied life *(Prakriti)*: "The Harivansa describes a 'substance called *Poroucha*, a spiritual substance known also under the name of Mahat, spirit united to the five elements, soul of beings, now enclosing itself in a body like ours, now returning to the eternal body" (J, 3, 215).[13] And the release of the spiritual into something embodied is, on Thoreau's understanding, enacted precisely by contemplation:

> The Richis mingle with nature. . . . Luminous & brilliant they cover themselves with a humid vapor, under which they seem no more to exist, although existing always, like the thread which is lost and confounded in the woof. . . . Thus the Yogin, absorbed in contemplation, contributes for his part to creation: he breathes a divine perfume, he hears wonderful things. Divine forms traverse him without tearing him, and united to the nature which is proper to him, he goes, he acts, as animating original matter. . . . A commentary on the Sankhya karika says "By external knowledge worldly distinction is acquired; by internal knowledge, liberation." (J, 3, 216)

In Thoreau's interpretation the dissolution of personal identity signifies its "return" not to a purely spiritual entity but to "nature," which he in turn understands, precisely as Colebrooke did, as "animating original matter," as infinite generative vitality. Far from being merely a technique for abstracting the personality into pure spiritual being, contemplation, in Thoreau's understanding of the *Sānkhya*, becomes above all the process of merging of matter and sensing, emancipated from the partiality of the personal senses, into impersonal yet corporeal life. However bizarre it might sound, contemplation is thus understood as a technique for gaining entrance to a vital materiality that "does not know . . . of death" (J, 3, 215).[14]

The Noontime of the Mind

Although Thoreau did use Hindu philosophy to empower, or to make less eccentric, his own ideas of vital matter articulated as early as 1840—"Like some other preachers-I have added my texts-(derived) from the Chineses & Hindoo scriptures-long after my discourse was written" (J, 3, 216)—he also disregarded central aspects that didn't fit into his larger epistemological project of recovering materiality. For Thoreau wants the mind to detach itself from its mental content without keeping "outward accidents" away from itself, which is what Krishna also demands. If Thoreau's thought could be called ecological, it is above all because he wants to adhere only to Henry Colebrooke's first phase of contemplation, in which the mind fuses with vital materiality and is exposed to its environment, tuned in to "outward accidents." Only through such exposure can the mind access what is true and become passive to the point of self-cancellation—it doesn't mediate the sensations it receives. Instead, it receives them "raw," welcoming the marvelous presence of things in themselves in the mind. They arrive from the outside, and because there is nothing in the mind of its own, the outside therefore determines what will be in the mind, converting it this way or that. That is already how Thoreau describes the process in *A Week*, intimating the later, more complex epistemology of *Walden*.

In *A Week* it is as if the outside world directly fashions the content of the brothers' minds, through an infusion of the sensual that is—as understood there—to be characteristic of Eastern contemplation:

> There are moments when all anxiety and stated toil are becalmed in the infinite leisure and repose of nature. All laborers

> must have their nooning, and at this season of the day, we are all, more or less, Asiatics, and give over all work and reform. While lying thus on our oars by the side of the stream, in the heat of the day, our boat held by an osier put through the staple in its prow, and slicing the melons, which are a fruit of the east, our thoughts reverted to Arabia, Persia, and Hindostan, the lands of contemplation and dwelling places of the ruminant nations. (W, 125–126)

The "noontime" of the mind is the moment when it relaxes in coincidence with its surroundings. "Nooning" is the activity, or rather passivity, in which the mind is withdrawn so as to let what is external to it—landscape, trees, fruit—not only navigate it toward a certain way of thinking, but also in fact determine it. It is not the mind's intention that determines what it will think. Instead, an external object generates a thought: the mental condition of "anxiety" is becalmed by the repose of the external world, and sliced melons, "the fruit of the East," revert the mind to "Hindostan." This reversion is not to be understood as a reminiscence or association generated by an active mind (for instance, Thoreau doesn't say "we were reminded of Hindostan"), but rather as something the mind unwillingly suffers, being passively transported into a different condition. As if melons were endowed with a "magical" power to alter the mind concretely, moving it from anxiety to calmness and changing the visible field from Concord to a Persian landscape. Thoreau in fact reinforces the interpretation that melons can alter the mind without the mind's participation, when, continuing the passage I've just quoted, he compares the melon's effects on the mind to a thought generated by consumption of the kat-tree fruit specifically, and opium more generally. The sight (and texture, aroma, or taste) of Hindu melons, he asserts, has the same power over the mind as does a drug that determines thinking by materially (chemically) affecting the brain. In both cases, it is suggested, the mind suffers what is outside it to determine its condition:

> Slicing the melons, which are a fruit of the east, our thoughts reverted to Arabia, Persia, and Hindostan, the lands of contemplation. . . . In the experience of this noontide we could

> find some apology even for the instinct of the opium, betel, and tobacco chewers. Mount Sabér, according to the French traveler and naturalist, Botta, is celebrated for producing the Kát tree, of which "the soft tops of the twigs and tender leaves are eaten," says his reviewer, "and produce an agreeable soothing excitement, restoring from fatigue, banishing sleep, and disposing to the enjoyment of conversation." (W, 126)[1]

In *The Pharmaceutical Journal* of September 10, 1837, which Thoreau here refers to, Botta testified to a contradictory effect of the kat-tree, for its "leaves when chewed had an agreeable exciting action, which imparted the desire to spend the night rather in quiet conversation than sleeping," adding that kat "provoked lovely dreams," in the awake state of the mind.[2] What attracts Thoreau in this description of pharmaceutically induced contemplation is thus precisely its contradictory nature: its passivity is not inert but excited, yet its excitement, far from exhausting the mind, restores it; similarly, it banishes sleep, but instead of ending in insomniac tiredness, it sooths and "disposes" to conversation, which is claimed, paradoxically again, not to be verbal but quiet. In contrast to the Western understanding of passivity as mental stupor, in Thoreau's version the nonlaboring mind experiences the clarity of an enthused perception. In a *Journal* entry from May 12, 1851—and in the midst of discussion of Hindu "nooning"—Thoreau precisely describes how he achieved that condition through his own experience of drugs:

> By taking the ether the other day I was convinced how far asunder a man could be separated from his senses. You are told that it will make you unconscious-but no one can imagine what it is to be unconscious-how far removed from the state of consciousness . . . until he has experienced it. The value of the experiment is that it does give you experience of an interval as between one life and another—A greater space than you ever travelled. . . . You expand like a seed in the ground. (J, 3, 218)

On Thoreau's account, ether allowed him access to the "interval between" two conscious positions. It thus testified to him—and that is precisely what so astonishes him—that against what both common sense and philosophy render credible, so-called consciousness (the presence of a gathered personality) doesn't appear necessary for intense mental and sensorial experience. Instead, it is precisely the absence of personality, an "unconscious . . . far removed from the state of consciousness," that intensifies Thoreau's sensorial experience, precious in its incredulity. Thoreau describes the event of this incredulity as an experience of de-creation leading to a new creation ("an interval as between one life and another"). What is so incredible about this experience of a passage is that it afforded Thoreau an intense feeling of living while transitioning from one personality to another, that is to say, while being nobody. When Thoreau says that he was "given" or was induced to the experience of an interval from which his personhood was absent, he is claiming something fabulous: that during the interval when he was nobody, he (somehow) knew himself to be nobody. What is negative—the absence of personhood—is experienced affirmatively, as existence. And what made this experience additionally remarkable, as if adding to its intensity, was its slowness. Voyaging through an interval from one personhood to another, Thoreau-nobody is so slowed down that he feels as if he will never regain a personality. The instants of time are slowed down to the point of becoming "great spaces" too large to travel, so that the mind feels abandoned to a repose in which its life, much like a plant's, impersonally "noons," languid like a seed.[3]

Thinking with the Body 1: Walking

THOREAU NEVER GAVE up the belief that it should be possible for the mind to experience objectivity and hence, truth. But the practices he designed to that end while at Walden Pond show that his answer to the question of how exactly such access of thought to the "thing in itself" is to be achieved gained in rigor and complexity. In *A Week* such access was imagined as an immediate event: the sensation of a material object—melon or tree—imprints itself on the senses and directly affects the mind.[1] In contrast, *Walden* testifies to Thoreau's realization that mind can't be immediately affected by the external world, because senses are not its neutral transmitters but are themselves mediated by the condition of the body. This is less obvious than it might seem. For it helped Thoreau realize that perception of an object is neither immediate nor conditioned by the mind, as Kant thought, but is instead fashioned by the body. How we perceive, and how those perceptions affect the mind, depends on how our body is doing. Our body determines our thoughts, or, as Thoreau explicitly puts it "the states of mind answer to the states of body" (J, 3, 4). If externality were to inhabit the mind, Thoreau realized, work on the body would be needed; the body needs to be in a condition that enables such inhabitation. Hence the circular strategy that Thoreau develops in *Walden*, as if adopting his personal yoga: by an intentional effort of the mind, the body is brought into a posture that assists in the process of weakening the mind's intentions, thus enabling the suspension of self-reflexivity.

Walden describes two such practices. One is famously and somewhat counterintuitively identified in the chapter "Village" as walking. I say counterintuitively, because various Western traditions had typically understood walking as enabling what Thoreau wants to disable. Major Western walkers—from Aristotelian peripatetics to English romantics—walked not to weaken the self but to strengthen it. It is said of Aristotle that he "lectured and taught while walking up and down," because such monotonous motions enabled one to forget the body and focus on the mind, whereas Hazlitt testifies that Coleridge "could go on in the most delightful explanatory way over hill and dale . . . and convert a landscape into a didactic poem or a Pindaric ode."[2] Coleridge's walk is thus opposite to Thoreau's in that it doesn't make the mind susceptible to the outside world, but instead "converts" the outside into the mind, adjusting it to its own (lyrical) rhythm. Thoreau's walks are sometimes compared to Wordsworth's, but they differ from those also, since Wordsworth, much like Coleridge, used them to focus on his mind and its poetic perspective (an interpretation authorized by Wordsworth's own titles, such as "Lines Composed a Few Miles above Tintern Abbey on Revisiting the Banks of the Wye during a Tour"). According to his own testimony, Hazlitt used his walks to revive his childhood memories and, by reappropriating the past that was no longer immediately available to him, to gain access to his self: "Then long forgotten things . . . burst upon my eager sight, and I begin to feel, think, and be myself again."[3] But it was Kant who gave to walking as self-gathering its most radical formulation, arguing in "What Is Orientation in Thinking?" that no walk is possible unless guided by self-centered subjectivity. To walk, Kant argues, is to traverse space in an oriented way; it is to "use given directions" by "means of certain guidelines." However, such usage of directions is possible only if, within itself, the subject knows and feels its differences in a directional way; that is, only if the subject is able to distinguish within itself "between movement from left to right and movement from right to left without reference" to any external space. He will then apply, a priori, that internal and purely subjective difference to the space through which he walks, "orient[ing] [him]self *geographically* purely by means of a *subjective* distinction."[4] Or, as Derrida puts it, walking for Kant "depends in the last analysis on no objective or objectifiable datum but merely on a principle of subjective differentiation." If subjectivity were not a priori gathered, and did not remain so during the walk, it wouldn't know how to define the positions of objects in space—"I would not know whether to locate west to the right or to the left of the southernmost point of the horizon"—and so would "inevitably become *disoriented*."[5] And getting lost, ambulating in a disoriented way, is not walking at all, but represents a disaster of subjectivity. Thus, for philosophy walking is performed by the mind; the body can't walk, it is instead always taken for a walk.

In contrast to the philosophically minded walks, we might consider the more modern walks and excursions of explorers, scientists, and tourists, intended not to focus the self but rather to enable the walker to encompass epistemologically the environment. In the first instance such practices are closer to Thoreau's in that they distract the observer from the self. But, opposite to what he advocates, they repeat the gesture of appropriation. Thus, Susan Scott Parrish depicts eighteenth-century scientists as tireless walkers collecting specimens (of plants, "quadrupeds, amphibians, fish, seeds, stones") to be analyzed, and so made readable and controllable.[6] Alternatively, as de Certeau famously argued, the drive to know and control typical of the naturalist walk also characterizes modern urban walking.[7] Describing Manhattan walkers, he suggested that they are driven either by a scopic drive, in which case the "all-seeing power" of walker's "totalizing eye" desires to make "the complexity of the city readable" and predictable; or else by tactile behavior that—despite the momentous joy a walker can draw from the motion of the body in a space too vast to represent—also aims at the "appropriation of the topographical system," finally rendering the opaqueness of the city's space controllable.[8] To be sure, Thoreau often walked to come to know his surroundings. But he describes the walks motivated by such interest as expeditions or tours ("Our expeditions are but tours" (Wk, 162), suggesting that they are not what he has in mind when he talks about walking.

We might, in fact, better describe Thoreau's idea of walking by outlining what it isn't. Unlike Aristotle or Hazlitt, Thoreau does not treat walking as an occasion for the mind to remember or to reflect on personal or communal aspects of daily life. The cases in which the mind remembers are so failed as to become alarming: "I am alarmed when it happens that I have walked a mile into the woods bodily, without getting there in spirit. In my afternoon walk I would fain forget all my morning occupations and my obligations to society" (Wk, 165). Nor is what Thoreau calls a walk moved by a scopic drive. In fact, *Walden* identifies the most successful walk as one that preempted the scopic drive altogether, for it took place in darkness so intense that the "eyes could not see the path" (Wa, 170). The idea that darkness is a favorable condition for a walk proper also indicates that walking is designed to cancel both aesthetic and scientific impulses. Darkness would, for example, disable Coleridge's aesthetic "conversion of a landscape into a didactic poem," and block the scientific ambition to render readable the complexity of a forest space. However, it encourages a radical neglect of the self and the mind's activity: "Sometimes, after coming home thus late in a dark and muggy night, when my feet felt the path which my eyes could not see, dreaming and absent-minded all the way . . . I have not been able to recall a single step of my walk" (Wa, 170). Instead of seeing, not-seeing; instead of remembering, forgetting; instead of focusing (on poems, like Wordsworth; on a phenomenon, like a scientist; on a problem, like a philosopher; or on beauty, like an aesthete), absent-mindedness, so that the path the walker "has travelled . . . a thousand times, he cannot recognize a feature in it, but it is as strange to him as if it were a road in Siberia" (Wa, 171).

And, finally, Thoreau's walk is radically anti-Kantian. A *Journal* entry of August 21, 1851, describes the process of weakening the mind's desire to apprehend either itself or what is external to it as the "perfect freedom" generated by a walk: "Granted that you are out of door-but what if the outer door is open, if the inner door is shut. You must walk sometimes perfectly free-not prying nor inquisitive-not bent upon seeing things" (J, 4, 6). The walk is designed to open what Thoreau calls the "inner door" of the mind, a door into the body as the mind's most immediate exteriority, a process described in "Walking" as settling "where my body is," into its senses (Wk, 165). *Walden* famously describes this walk of the self into the body—the mind's becoming pure sensations, so many things in themselves—as the self's complete self-abandonment: "It is a surprising and memorable, as well as valuable experience, to be lost in the woods any time . . . and not till we are completely lost, or turned round,-for a man needs only to be turned round once with his eyes shut in this world to be lost,-do we appreciate the vastness and strangeness of Nature" (Wa, 170–171). For Thoreau, as for Kant, getting lost generates a disaster of the self. But in opposition to Kant, Thoreau claims that only such a disaster, the moment in which the self is "completely lost," constitutes a walk. The perfect freedom granted by a walk, then, is not freedom *for* thinking, but *from* it. Walking for Thoreau takes place only when the mind is walked by the body. To avoid any ambiguity, "Walking" explicitly claims that a walk proper generates a disaster of the self so radical that it equals death: "We should go forth on the shortest walk . . . never to return,—prepared to send back our embalmed heart only as relics to our desolate kingdoms. If you are ready to leave father and mother, and brother and sister, and wife and child and friends, and never see them again,—if you have paid your debts, and made your will, and settled all your affairs, and are a free man, then you are ready for a walk" (Wk, 162). In a walk, the self dies; the relic of its embalmed heart is sent to the mourners to whom the self that took a walk never returns.

But death is a strange thing in Thoreau. For he also inserts a poem in "Walking," "The Old Marlborough Road," that depicts how a walker—one who gets "completely lost" and so dies while on a walk—in fact loses himself into life. We are told that the one who walks on that road "liv'st all alone, / Close to the bone, / And where life in sweetest / Constantly eatest." Such a walker can't "get enough gravel / On the Old Marlborough Road. / Nobody repairs it / For nobody wears it; / It is a living way, / As the Christians say. / Not many there be / Who enter therein, / . . . What is it, what is it, / But a direction out there, / And the bare possibility / Of going somewhere?" (Wk, 168–169). If, as Thoreau has already suggested, the walker's self dies while on a walk, here it appears that the road of that death is to be understood as an incessant process, interminable self-generation: "nobody repairs it; / It is a living way, / As the Christians say" (Wk, 168). The Christian saying to which the poem refers seems to be John's famous understanding of life as a way ("I am the Way and the Truth and the Life" [John 14:6]), which, as philosophers of Christianity explain, is less an allegory of human existence as a journey from birth to death into resurrection, than an ontological proposition regarding life itself. As Michel Henry puts it, John's formulation says that thinking about life must rule out the concept of being and substitute becoming for being For "life 'is' not. Rather, it occurs and does not cease occurring. This incessant coming of life is its eternal coming forth in itself, a process without end, a constant movement. In the eternal fulfillment of this process, life plunges into itself, crushes against itself, experiences itself, enjoys itself, constantly producing its own essence . . . thus life continuously engenders itself."[9] If the walker that Thoreau describes in "Walking" has both to settle into the senses "where [his] body is," and die on the Marlborough road into a life that "continuously engenders itself," that is because the walker's sensuous body is precisely a living way, a path "close to the bone / And where life in sweetest / Constantly eatest" (Wk, 167). *Walden* repeats the same idea, calling the most "vital experience" afforded by the body "life near the bone where it is sweetest" (Wa, 329). At the price of losing the self, the walker is resurrected into the glory of corporeal life, and this immersion into life is precisely Thoreau's version of the mind's mingling with animated matter as he understood from the *Sānkhya Kárika.*

Because one would like to "constantly eat" from this bone, never leave the truth, the process of selfless becomings called walking finds its emblem not in a circle, as in Emerson, but instead in an infinite curve: "The outline which would bound my walks would be, not a circle, but a parabola, or rather like one of those cometary orbits which have been thought to be non-returning curves" (Wk, 169). However, Thoreau allows that walks end, and that a self appears after them. Hence the central question raised by a walk involves how to reconcile the insistence on a walk of "no-return" for the self—its disaster and death—with an emergence of the self subsequent to the walk. A series of *Journal* entries in which Thoreau meticulously analyzed his experience of walking offer an answer to that question. Thoreau's analyses focus on the interval of coming "back" from the walk, during which a self gathers itself. In the *Journal* and, famously, in *Walden,* Thoreau calls this interval an "awakening," because the first stirring of the self after a walk is structurally coincident with its gathering after a dream. A *Journal* entry from September 12, 1851, presents walking and dreaming as reciprocal experiences:

> Let no man be afraid of sleep. . . . He who has travelled to fairy-land in the night-sleeps by day more innocently. . . . That kind of life which sleeping we dream that we live awake-in our walks by night, we, waking, dream that we live, while our daily life appears as a dream. (J, 4, 74–75)

Because, as this vertiginous reasoning asserts, walking and dreaming mind are interchangeable ("in our walks . . . we, waking, dream"), in that they are both states of selfless perception, awakening from a dream can provide the answer to Thoreau's question concerning what an awakening into the self after a walk looks like. Six months after the comparison between waking from a walk and from a dream, Thoreau offered this precise description of awakening:

> I catch myself philosophizing most abstractly-when first returning to consciousness in the night or morning. I make the truest observations & distinctions then-when the will is yet wholly asleep.—& the mind works like a machine without friction. I am conscious of having in my sleep transcended the limits of the individual-and made observations & carried on conversations which in my waking hours I can neither recall nor appreciate. As if in sleep our individual fell into the infinite mind-& at the moment of awakening we found ourselves on the confines of the latter- On awakening we resume our enterprise take up our bodies & become limited mind again. We meet & converse with those bodies which we have previously animated. There is a moment in the dawn-when the darkness of the night is dissipated & before the exhalations of the day commence to rise-when we see things more truly than at any other time. (J, 4, 392–393)

Awakening is a condition between what and who. Because the mind's will, hence its self-determination, is not yet operative but is instead "wholly asleep," Thoreau can say that in awakening the mind is still not gathered, that it remains undetermined (or, as Thoreau counterintuitively phrases it, the mind finds itself still "confined" by what has no confines, the infinite and impersonal life). But because in awakening the mind is not quite asleep either, being on its way to the confines of a separate self, it also starts perceiving (as Thoreau puts it, it starts working "on its own," "like a machine without friction," without the intervening power of reflection, habit, or memory). Awakening is thus a transition from formlessness into a form, from impersonality to personhood (which is why Thoreau compares it to dawn, the moment in the morning when light appears but shapes are not yet distinct, when one can see how everything is before figures are determined). In awakening, then, the mind is absent enough not to reflect and interpret but is awake enough to perceive. That is to say, against all normal logic, that in awakening the mind is still captured by the impersonal life while simultaneously perceiving it. Awakening thus affords a glimpse of what "no waking hour" knows: things as they are during the walk, phenomena as they are in the mind's absence. In awakening the mind gets the feel of its own outside; it catches the unfathomable and astonishing sight of the world not seen through its perspective (in awakening, "I am conscious of having in my sleep transcended the limits of the individual").

That changes everything for the awakening mind. For as *Walden* notoriously detailed, the feel of life dislodged from the personal mind, the feel of the body animated, moved, in walking, as opposed to a body only "conversed with," causes in an awakening "I"—"when there is a dawn in me" (Wa, 90)—a transformation so radical that Thoreau understands a new self to have been generated, an appearance of "whole new continents and worlds within you, opening new channels . . . of thought" (Wa, 321). The sight of what is unimaginable to the mind once awake—the shimmer of things as they are when the mind is not there to see them—is so stunning that it unsettles everything in the "I"; its world dies, as Thoreau puts it, and without nostalgia or regret. With no "patriotism . . . for the soil that makes [its] grave," reconstituted otherwise by the death of its world, the awakening "I" migrates to a "new continent" as someone else. The walk glimpsed at in awakening, the feel of animation registered, transforms the "I" from what it was before the walk. The "I" that goes for Thoreau's walk is, thus, never the "I" that comes back from it, which is why Thoreau can say that a walk is always one of no return, even if a self comes back from it. Whenever an "I" in Thoreau says "I am," it in fact is saying: "Look, *I am changed.*" The central paradox of Thoreau's walk is thus the emergence of an impersonal interval connecting two different "Is," generating continuity between what is different.

"Walking" provides a strange insight into how this uprooting of self-identity and crossing of space "between continents" reworks the walker's past on the model of the migratory movements of certain animals. A walk is "akin to the migratory instinct in birds and quadrupeds,—which, in some instances, is known to have affected the squirrel tribe, impelling them to a general and mysterious movement, in which they were seen . . . crossing the broadest rivers, each on its particular chip, with its tail raised for a sail, and bridging narrower streams with their dead" (Wk, 170). The walker might be required to leave mother and father and friends before the walk but is not asked to leave the dead. Instead, those who depart for a walk travel, like the squirrels mentioned here, with their dead. And even though Thoreau is not explicit about the position of the dead during this mysterious movement, it is safe to presume that he doesn't imagine them as existences separate from the walker's who remind him of the past and call him back to the point of departure. It is safe to presume that because, as Thoreau explicitly states in "Walking," a walk is a passage over Lethe.[10] In other words, the "I" has no memory of itself before the walk, for otherwise the change wouldn't be "wholly complete." Still, its dead are not quite lost. For as the walker's mind changes during the walk—as its sensations, images, and memories are reconfigured—the dead become part of a different ontological alloy that will eventually wake up as a new "I." The self that awakes from its walk, while being new and unknown to itself, is nevertheless an admixture of others. In Thoreau, then, just as in ancient tales of metamorphosis, one never metamorphoses alone; we owe ourselves to others, we are composed with our dead, for they are mixed and born again in the birth of the awakening "I."[11] One could thus say that just like nature in mourning, which Thoreau mentions to Emerson after John's death, a walk too finds what was lost "under new forms" (C, 64).

Thoreau had an obsessional dream. It came to haunt him many times over a number of years. In 1857 alone he dreamed it at least twenty times. On October 29 the dream visited him again, but that day he decided to record it:

> There are some things of which I cannot at once tell whether I have dreamed them or they are real. . . . This is especially the case in the early morning hours, when there is a gradual transition from dreams to waking thoughts . . . as from darkens . . . to sunlight. Dreams are real, as is the light of the stars and moon. . . . Such early morning thoughts as I speak of occupy a debatable ground between dreams and waking thoughts. . . . This morning, for instance, for the twentieth time at least, I thought of that mountain in the easterly part of our town (where no high hill actually is) which once or twice I had ascended. . . . I now contemplate it in my mind as a familiar thought which I have surely had for many years from time to time . . .
>
> My way up used to lie through a dark and unfrequented wood at its base,—I cannot now tell exactly, it was so long ago, under what circumstances I first ascended, only that I shuddered as I went along (I have an indistinct remembrance of having been out overnight alone),—and then I steadily ascended along a rocky ridge half clad with stinted trees, where wild beasts haunted, till I lost myself quite in the upper air and clouds, seeming to pass an imaginary line which separates a hill, mere earth heaped up, from a mountain, into a superterranean grandeur and sublimity. What distinguishes that summit above the earthy line, is that it is unhandselled, awful, grand. It can never become familiar; you are lost the moment you set foot there. You know no path, but wander, thrilled, over the bare and pathless rock. . . . It is as if you trod with awe the face of a god turned up, unwittingly but helplessly, yielding to the laws of gravity. . . . In dreams I am shown this height from time to time, and I seem

> to have asked my fellow once to climb there with me, and yet I am constrained to believe that I never actually ascended it. It chances, now I think of it, that it rises in my mind where lies the Burying-Hill. You might go through its gate to enter that dark wood [interlineation: perchance that was the grave], but that hill and its graves are so concealed and obliterated by the awful mountain that I never thought of them as underlying it. Might not the graveyards of the just always be hills, ways by which we ascend and overlook the plain? But my old way down was different, and, indeed, this was another way up, though I never so ascended. I came out, as I descended, breathing the thicker air. . . . There are ever two ways up: one is through the dark wood, the other through the sunny pasture. That is, I reach and discover the mountain only through the dark wood. (Jn, 10, 141–143)

Because the dream recalled here concerns a walk, awakening from it—as if in a mise-en-abyme of Thoreau's epistemology—coincides with the self's return from a walk. In this dream-walk the walker sets out for the woods and reaches a pathless site—alternatively remembered as a rocky ridge, a bare rock, heaped up earth or simply a hill—where, disoriented, wandering among wild beasts, he gets lost. Thoreau doesn't clearly remember whether the site that so stunned him was a single grave ("perchance that was the grave") or a burying site, a place of many graves as heaped up earth most likely would be (Thoreau's question "might not the graveyards always be hills?" can also be heard in reverse as "might not all hills be graveyards, isn't all nature of the dead?"). But whether one or the other, it is at that site that the walker's transformation occurs; as if in a religious conversion, where one is addressed by God, here the dead appear to the walker as the face of an awful god to whom the walker finally yields, "helplessly falling" out of himself, letting himself go ("It is as if you trod with awe the face of a god turned up, unwittingly but helplessly, yielding to the laws of gravity"). And whereas the walker leaves his dead self with the dead, the new self that emerges from the walk carries those dead within itself. Thoreau allows that the new self, the one for whom everything old is different and new (so that even the "old way down was different"), is made of the dead too when he claims that the metamorphosis of the self must take place by passing through the world of the dead ("I reach and discover the mountain only through the dark wood"). No transformation, then, without those who were before us, no new self that doesn't bear the dead. Thus, even if it happens that the walker is "out overnight alone" and suffers the awe that the walk generates all by himself—for Thoreau's walker is not a man of trivial joy who walks through sunny pastures, but one who knows real terrors—his metamorphosis is communal, embracing those who were, carrying them out of the woods, into the "thicker air" of the living. Because the self doesn't remember his previous forms, his future avoids nostalgia and yet the dead are not, for all that, left behind. The forgetting necessary for auspicious beginnings is thus reconciled with care for those who have perished.

Thinking with the Body 2: Sitting

Sitting is the second posture befitting Thoreau's version of contemplation. It is more obviously related to Eastern practices of contemplating than walking is. Thoreau himself refers to sitting, both in the *Journal* and in *Walden*, as a bodily posture required for contemplation by Eastern schools of thought. He finds such a requirement in Persian sources ("The poet Mir Camar uddin Mast of Delhi . . . says 'Being seated, to run through the region of the spiritual world'" [J, 3, 21]);[1] in the Hindu tradition ("The Hindoo thinks so vividly & intensely that he can think sitting or on his back" [J, 3, 22]); and, of course in Buddhism where the selfless condition of the mind known as "Zen" is accomplished by the practice of "zazen," which means precisely "to be sitting." Despite differences in mobility and speed that at first glance distinguish walking from sitting, sitting is not simply opposed to walking, as passivity would be to activity. Like walking, sitting is an activity in its own right. As Roland Barthes explains in his analysis of sitting in Buddhism, which he differentiates from the Christian posture of kneeling and the Fascist posture of standing: "despite its 'strong' negativity, the gesture (posture) shouldn't be flattened: to be sitting is active = an act, antonymic to 'letting oneself fall down.'"[2]

Despite detailed descriptions of the walks Thoreau undertook while living at the Pond, sitting is the unquestionably privileged posture in *Walden*. Moreover, *Walden* is nothing but a detailed description of building a seat and the process of sitting as settling into that seat. Thoreau explicitly states that to build a house is in fact to build a seat ("what is a house but a *sedes*, a seat?" [Wa, 81]) It is thus the possibility of settling into sitting that motivates the art of living that Thoreau practices at Walden Pond, the possibility of being at home in the posture necessary for contemplation. But the coincidence in *Walden* is not just between a seat and the house, or sitting and living, but more importantly, between dwelling and thinking. In Thoreau, the house is never simply an existential site without also being an epistemological one; a house always embodies an image of thought. Thoreau's motto is: "tell me what your house is like, I'll tell you what you think."

However, if not just any ambulation constitutes a walk, so not just any seat is good for dwelling; one can't settle into any frame of mind. That is why, in Thoreau's famous description of his deliberations, the seat of house cannot be chosen without first considering "every spot as the possible site of a house" (Wa, 81). No less significantly, the making of a seat, indeed building—and by extension thinking and living—starts in *medias res;* the foundation of the house is an effect of metamorphosis. As always in Thoreau, there are no pure origins; to begin is to begin reforming, to start is to start living from the first day of change. That is a fundamental message of his experiment, for his cabin comes into being as the transformation of another house. Thoreau built his cabin with materials he obtained after pulling down the "the shanty of James Collins, an Irishman who worked on the Fitchburg Railroad," which he "bought . . . for boards" (Wa, 42–43). Some scholars have read Thoreau's negative description of Collins's shanty and his gesture of pulling it down as the expression of an anti-Irish and anti-labor stance. According to those interpretations, Thoreau's gesture signals an elitist contempt for the poor, as well as his xenophobia. As one critic puts it, when Thoreau finally builds his own cabin, he comes to "regard [it] in every way as infinitely superior to those of the Irish."[3] However, it seems to me difficult to support the argument that Thoreau's negative description of Collins's shanty represents his derision of the dwellings of the poor, because it is precisely the houses of the poor that, in the same chapter, he identifies as the most architecturally interesting: "The most interesting dwellings in this country are the . . . humble log huts and cottages of the poor commonly" (Wa, 47). Similarly, his negative description of the Irish shanty is only one among many negative descriptions of "much of the architecture of man" (Wa, 48) that proliferate in the same chapter. Thus, the dwellings of rich Yankees are dismissed as no better than "common dilettantism," while the houses of the middle class are repudiated as "uninteresting . . . citizen's suburban box" (Wa, 47). All of them, houses of the rich and poor, of Irish and Yankee, share the features of darkness and enclosure that Thoreau criticizes in Collins's shanty; all of them amount to nothing but a "few sticks . . . slanted over [man] or under him, and . . . colors . . . daubed upon his box"; all of them, as Thoreau rephrases it in the same chapter, are "of a piece with . . . a coffin,- the architecture of the grave" (Wa, 48). And when Thoreau finally builds his cabin, he will come to regard it not "in every way as infinitely superior to those of the Irish," but in every way infinitely superior to any house: "I intend to build me a house which will surpass any on the main street in Concord" (Wa, 49).

Thus, although I do not want to discount the possibility that Thoreau harbored vexed sentiments toward the Irish, I prefer to pursue a reading of Thoreau's metamorphosis of Collins's house that is preoccupied less with political economy than with epistemology. Thoreau himself authorized such an approach by explaining that his experiment in house building and, by extension, his critique of contemporary architecture, was motivated by issues of philosophy: "even the *poor* student studies . . . only *political* economy, while that economy of living which is synonymous with philosophy is not . . . professed" (Wa, 52). If the art of living is synonymous with philosophy, then the question of the house as the site of such an art must, for Thoreau, be a properly philosophical one. I thus come back to my previous claim: in Thoreau, a house is the embodiment of a certain style of thinking and of a subjectivity contrived by it. I therefore read Thoreau's description of Collins's shanty as his critique of an image of thought.

What kind of a house, then, is Collins's shanty and what kind of subjectivity is it suggestive of?

1. The house is *isolationist*—a feature that can be surmised because Thoreau "walked about the outside, at first unobserved from within, the window was so deep and high" (Wa, 43). A highly positioned window makes the inside invisible from and secret to the outsider,

thus securing the dweller's privacy, as if fencing off and so protecting the inhabitant's interiority.

2. The house is *perspectival*—that the house has only one window (instead of two, which would be suggestive of two eyes, or many, suggestive of many senses) signals a focused perspective generated by the mind's unification of different perceptions; it thus embodies a subjectivized point of view.

3. The house is *ideational*—because the window is too high and deep, the house doesn't receive light from the outside, through the window, but needs to be illuminated by its own internal light: "Mrs. C. came to the door and asked me to view [the house] from the inside . . . It was dark. . . . She lighted a lamp to show me the inside" (Wa, 43). Epistemologically rephrased, the house is suggestive of a mind that does not receive the world through the senses and the body, but instead thinks it through its internal light. It is much like the mind that Augustine contrived, asking the meditating Christian to leave behind the physical world enlightened by the light of the "sun, moon, and stars," and move "step by step" into "internal reflection . . . and enter into our own minds,"[4] residing there by the lamp of the mind's inner light alone. That would also be much like Descartes's *cogito*, conceiving itself without hands to touch or eyes to see, stubbornly believing only in the internal lamp of the mind as source of truth: "I shall think that the sky, the air, the earth, colours, shapes, sounds and all external things are merely the delusions of dreams . . . I shall consider myself as not having hands, eyes, or flesh, or blood, or senses; . . . I shall stubbornly and firmly persist in this meditation."[5]

4. The house is *arborescent*—in Collins's house, there is a cellar under the bed, "a sort of dust hole two feet deep" (Wa, 43). That there is a cellar even in the most humble house—belonging to a household so impoverished that it wouldn't have anything to store—just as in "the most splendid" houses; the fact, in other words, that there is a cellar in all our houses, as if we were unable to think of a house that wasn't founded on it, signals for Thoreau, even if counterintuitively, that cellars are in fact conceived of less for stacking goods than for storing the past: "Under the most splendid house in the city is still to be found the cellar where they store their roots as of old, and long after the superstructure has disappeared posterity remarks its dent

in the earth. The house is still but a sort of porch at the entrance of a burrow" (Wa, 44–45). The cellar in which roots are stored roots the house. It makes it immobile—as Minerva, Greek goddess of wisdom made it, and as Momus, with whom Thoreau sides, ridiculed it—bounded to a genealogy out of which it grows like a tree. Such a house is territorial, planted in the land, and lodged in its ancestrality; as Robert Pogue Harrison remarks, commenting on Collins's shanty, its cellar is like a "fold, a crypt, a wrinkle" of the dweller's past.[6] But Thoreau does not romanticize this past by invoking a reactionary nostalgia for it. Instead, the past will, catastrophically, come to be the dweller's only future: long after the house is gone, the cellar marks its "dent in the earth," thus assuming the function of a tombstone to commemorate the dweller's life. The genealogical house, Thoreau suggests, grows out of and into its past, as if the dweller's future life had been fatalistically determined by the identity of his ancestors, as if promised to it. Far from being a house of life and process, the ancestral house—indeed the house of the Western mind—is so identitarian that it functions only as a porch over its long-dug grave. In this respect too, Collins's shanty is but a humble version of most Western houses, all of which are like a "museum, an almshouse, a prison, or a splendid mausoleum" (Wa, 28), all of them versions of the grave.

To the architecture of the grave Thoreau opposes the architecture of life, to what is closed what is open, to what is heavy what is light, to what is vaulted what disturbs borders. He intends to transform our thinking about architecture—spaces for dwelling and thinking—and in the process the mind that generates it. To do so he again starts from the middle. For if Thoreau pulled down Collins's shanty, he didn't abandon it as ruins; taking it down wasn't a gesture of total destruction, as if to suggest that he could do without the tradition he sought to overcome. Instead, after dismantling Collins's house, after annulling its form, Thoreau attends to its material. Just as, in the operation of natural mourning, what is gone is "found without loss under a new form," the matter of Collins's house is preserved by migrating into a new form: "I took down this [Collins's] dwelling the same morning, drawing the nails, and removed it to the pond side by small cartloads, spreading the boards on the grass there to bleach and warp back again in the sun" (Wa, 44). Having been "turned from [its] natural or true course"—which is how the dictionary defines the verb "to warp"—by means of the form Collins gave it, the matter of his house is now allowed, with the help of the elements, to warp back; it comes back to its "natural course," back to a life that is not formed—perspectival or biased—but instead refrains from inclination or tendencies ("to be biased," "to have a particular inclination" are further dictionary definitions of the verb "to warp," hence to "warp back" is to obviate such tendencies). Matter thus warps back into a neutral life, and it is from this life—the life given by another—that Thoreau begins to build his own house. But not only does he start from the life of another; he also starts with others. He constructs the frame of the new house using a borrowed axe, for "it is difficult to begin without borrowing," and "with the help of some of [his] acquaintances," whom he has invited less out of necessity than because "it is the most generous course thus to permit your fellow-men to have an interest in your enterprise" (Wa, 41). The life of another, the tools of others, together with others: in opposition to interpretations of Thoreau's experiment that view him as isolationist, he keeps reminding us that we are never alone in our transformations and never change simply for ourselves but always with and for others.

The new house that Thoreau builds represents a complete metamorphosis of the old one; everything about it is different.

1. Unlike Collins's genealogical house, Thoreau's is rhizomatic. It is not rooted in a past stored in its cellar, for the cellar is not in the house at all: "I dug my cellar in the side of a hill sloping to the south, where a woodchuck had formerly dug his burrow, down through sumach and blackberry roots, and the lowest stain of vegetation" (Wa, 44). And because death—the cellar, the past—has been expelled from it, Thoreau's house is not a porch at the entrance to the grave, but instead a dwelling for life. This is to say that the life Thoreau imagines to have settled in his house is less conditioned by what doesn't change (ancestral identities, for instance), than unfounded, free to travel through whatever identities it desires, entrusted to its own metamorphoses. He understands his unfounded seat to be so airy and light that it gives the feeling of floating and moving (much like the house he made for Emerson with Alcott, for it too was de-

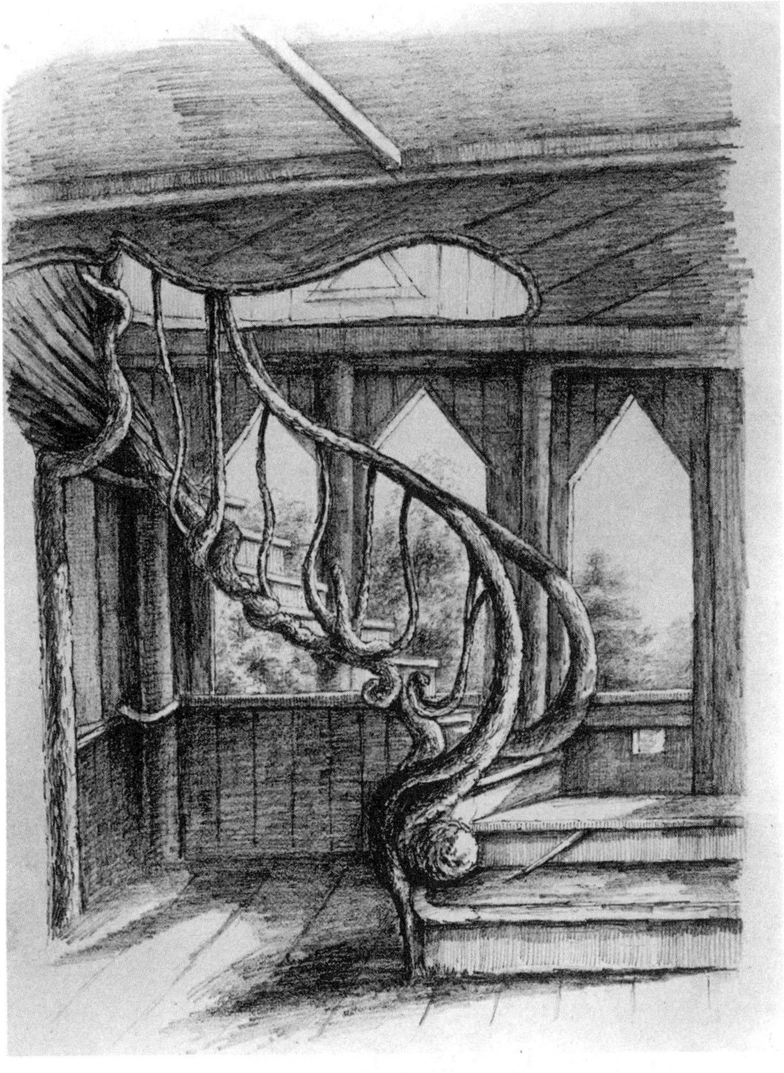

Stairs, sketch. Lousia May Alcott, *Concord Sketches, Consisting of Twelve Photographs from Original Drawings by May Alcott* (Boston: Fields, Osgood & Co., 1869).

signed to be drifting, reduced in that experiment to nothing more than meandering stairs).

His seat is one that only nomads could desire: "This was an airy and unplastered cabin, fit to entertain a travelling god" (Wa, 85). Nomadic rather than genealogical, Thoreau's house is alive,

Walden Cabin, adjacencies. Courtesy Joshua Bartlett.

Walden Cabin, chimney. Courtesy Joshua Bartlett.

proliferating laterally as he keeps building its annexes. For not only is a small shed adjoined to it ("I have also a small wood-shed adjoining" [Wa, 49]), but its chimney stands on its own, adjacent to the house ("I laid the foundation of a chimney at one end" [Wa, 45]).

The chimney, that, in a genealogical house, is rooted to the roof, growing out of the house that grows out of foundations rooted in

Walden Cabin, windows. Courtesy Joshua Bartlett.

the cellar, is here juxtaposed to the dwelling, touching it yet self-standing.

Opposing the logic of a tree, Thoreau's house grows not in a linear fashion, in time; it is rather dispersed horizontally in space, like grass, demonstrating that different modes of time—the past of the cellar, the now of the kitchen—can be contemporaneous without causally determining one another.

2. Unlike Collins's perspectival house, Thoreau's is nonperspectival: instead of having one window positioned high above the visual field—instead of a focused perspective—it has two large windows positioned low on each side, so that one's focus is always disabled by the simultaneity of perceptions coming in from two opposite sides, or from other sources, such as the large door he keeps open: "I have thus a . . . house . . . with a large window on each side, two trap doors, one door at the end, and a brick fireplace opposite" (Wa, 48).

Additionally, Thoreau is careful to explain that the absence of a unified perspective is not generated by the multiplicity of simultaneous perceptions alone but, more importantly, by the lack of a power that could mediate them into a focused unity. That is what he suggests in an extraordinary passage where he explains that if there are no walls or floors dividing his house, it is not for economic reasons (not because his small budget didn't permit him to build more, for instance), for even a much bigger and more luxurious house would have to be a one-story loft without interior divides:

> I sometimes dream of a larger and more populous house, standing in a golden age, of enduring materials . . . which shall still consist of only one room, a vast, rude, substantial, primitive hall, without ceiling or plastering, with bare rafters and purlins supporting a sort of lower heaven over one's head; . . . where some may live in the fire-place, some in the recess of a window, and some on settles, some at one end of the hall, some at another, and some aloft on rafters with the spiders; . . . a house which you have got into when you have opened the outside door . . . where you can see all the treasures of the house at one view . . . ; at once kitchen, pantry, parlor, chamber, store-house and garret. . . . A house whose inside is as open and manifest as a bird's nest. (Wa, 243–244)

Thoreau's insistence that a "dream house" should be a loft thus signals epistemological rather than economic motivations. Epistemologically rephrased, his house is suggestive of a mind whose perceptions—whatever enters the mind from the opening of the senses, represented by its large windows and doors—are not amalgamated into a homogeneity by the act of reflection, which would think them from the aloofness of its perspectival position (from an upper floor or from high windows). Instead, in this mind, perceptions, sensations, feelings, imaginations, and concepts occur simultaneously and at the same level. Yet, as Thoreau is careful to make clear, his loft-mind avoids freezing its content into "oneness" by remaining constantly open: moreover, it is structurally so—being made without a ceiling—such that something is always moving into it, which means pressing something else out. The loft-mind is thus agitated and processual. Thoreau explicitly imagines it as the simultaneity of processes (cooking, reading, conversing, sleeping, or, epistemologically put, perceiving, sensing, thinking), all of which, because they occur on the same plane, affect one another. It would thus be possible to say that Thoreau's loft-mind constitutes sheer affect, not in the sense that it is made exclusively of sentiments or that it privileges them, but more literally in the sense that everything it contains is affected and so reshaped by whatever else encounters it. Each of these "affective" bondings are little entities in and of themselves, as it were, for they will never be integrated into a single subjectivity whose mind requires, but lacks the sophistication of, self-reflection (and precisely because the mind lacks such a refinement Thoreau can call it "rude, substantial, primitive" [Wa, 243]). In other words, this mind will always be a multiplicity of variable affects, an understanding Thoreau conveys by the image of his ideal house as not only open but also communal. For when he depicts this one-person house as "populous"—offering the strange image of its being inhabited by an anonymous crowd, "some" living in the window, some with spiders, some in the fireplace—he is suggesting precisely that the mind he dreams of is not mediated into the unity of a single personhood. Instead, some of us, within us, are receiving perceptions (from the window), while others in us are reading, some are engaged in visiting dark spots (where something inhuman is), while others think fine thoughts. Or, as he explicitly describes it in a sentence I have already referred to, "every part of the body has its thought—What do the hands think? What the feet? The loins?—the back? They should all be thoughtful" (J, 3, 4). Each part of the body, then, constitutes a little affective bond between sensations and thoughts, perceptions and feelings, feelings and thoughts: a little soul perhaps. An "I" is always a crowd of such little souls, it is always a "we." As Thoreau keeps insisting: every personhood is a community.

3. Unlike Collins's crypt-house, Thoreau's house is open. The changed structure of its interior—the fact that, in its ideal version, it would not only "consist of . . . one room" but also wouldn't have a ceiling—alters its relation with what is external to it. Anticipating in an astonishing way Philip Johnson's Glass House, the inside of Thoreau's house becomes "as open and manifest as a bird's nest," allowing, as I've already suggested, what is external to it to settle in, whereas things internal might find themselves outside.

The description that Thoreau offers of his Walden cabin conveys his effort to build an open settlement very much in accordance with the ceilingless "ideal." Thus, the frame of his house was traversed by a "window on each side" and "door at the end," all of which were so large that when they were open, the frame of the house would be light and fragile enough to stop functioning as something determining an interior, being instead transformed into an "outline" merely, too weak to establish firm borderlines: "This frame, so slightly clad, was a sort of crystallization around me . . . it was suggestive somewhat as a picture in outlines" (Wa, 85). As a result of the absent or unstable divide between inside and outside, processes that typically take place inside, in the privacy of one's home or mind, move outside. Thus, Thoreau's bathtub was the pond; he cooked his food "out of doors on the ground, early in the morning," in a primitive cooking apparatus, which as Harding explains, consisted of a "hole made in the earth and inlaid with stones, upon which the fire was made, as at a clambake."[7] When Thoreau cleaned his house, rather than move his furniture around it, he would instead "set all [his] furniture out of doors on the grass," letting books, paper, and household objects mix with trees and flowers, again collapsing the threshold separating "inner space" from outside wilderness: "It was pleasant to see my whole household effects out on the grass . . . and my three-legged table, from which I did not remove the books and pen and ink, standing amid the pines and hickories. They seem glad to get out themselves, and as if unwilling to be brought in" (Wa, 113). And while the objects were outside—glad, as suggested by Thoreau's undecided pronoun, not only to get out of the house but also out (of) themselves, as if "willing" to become otherwise—what was outside passed through the inside, such as birds "[singing] around or flitt[ing] noiselessly through the house" (Wa, 111), as if they were themselves confused about where their own outside was.[8] As a result of the collapse of the interior, Thoreau felt as though he were living in the open: "I did not need to go out doors to take the air, for the atmosphere within had lost none of its freshness. It is not so much within doors as behind doors where I sat" (Wa, 85).

Thoreau's practice of contemplation takes place right there: at the threshold where what is inside is glad to get outside itself. For between inside and outside his house, there is a stone step, less a threshold separating two spaces than a link fusing them, making transitions and exits gradual.

Walden Cabin, seat/threshold. Courtesy Joshua Bartlett.

And on that step, as he details in "Sounds," Thoreau sits to contemplate: "Sometimes, in a summer morning . . . I sat in my sunny doorway from sunrise till noon . . . in undisturbed solitude and stillness . . . while the birds sang around or flitted noiseless through the house. . . . I grew in those seasons like corn in the night. . . . I realized what the Orientals mean by contemplation" (Wa, 111–112). But in sitting to contemplate he is also altering the "oriental" practice of contemplating, generating a specifically American version of it. For unlike Eastern contemplative sitting *(zazen)*, which eventually leads the mind away from the body and the outside world, Thoreau, as I've been suggesting all along, makes his mind behave like his things, glad to be outside itself, by disciplining it into "looking always at what is to be seen" (Wa, 111) and listening to what is to be heard. There is thus nothing strange in the fact that the chapter of *Walden* in which Thoreau describes his technique of contemplation is nothing more than a detailed list of sounds he hears while seated in his doorway. For contemplation, as Thoreau practices it, is above all about being exposed to what is there to be heard.

The sounds he hears occur regularly and are repetitive. "The whistle of the locomotive penetrates [his] woods" and the "train rattles past [him]"; and its cars also ("I watch the passage of the morning cars with the same feeling that I do the rising of the sun, which is hardly more regular" [Wa, 116]); "now that the cars are gone" the "faint rattle of a carriage" comes into his meditation; sometimes, on Sundays, he hears church bells and listens to their "vibratory hum" (Wa, 123). As he reports, the automated sounds of machines are no less repetitive than those released by birds. In fact, sounds released by whippoorwills are as regular as the tick-tock of a clock: "They would begin to sing almost with as much precision as a clock, within five minutes of a particular time, referred to the setting of the sun, every evening" (Wa, 124); "regularly at half past seven, in one part of the summer . . . the whippoorwills chanted" (Wa, 123). What Thoreau hears and sees is monotonous, repetitive, habitual: "I had a rare opportunity to become acquainted with [whippoorwills's] habits" (Wa, 124). Sitting thus for hours and days on end, exposed to the repetitive behavior of beings, not letting his mind to return to itself but instead keeping it in the habit of a bird, or corn, Thoreau grows into that habit, he "grows like corn" (Wa, 111), or he causes a bird to confuse him with another bird. As Emerson testified: "[Thoreau] knew how to sit immoveable, a part of the rock he rested on, until the bird, the reptile, the fish, which had retired from him, should come back, and resume its habits, nay, moved by curiosity should come to him and watch him."[9] That is the main lesson of Thoreau's practice; sitting is for him not a matter of being sedate but of enabling a type of errancy or vagrancy, and the stillness of the mind activates fabulous transformations. Hence he can say that "he who sits still . . . may be the greatest vagrant of all" (Wk, 161). With his body stilled to the point of becoming "a part of the rock" it rested on, and his mind stilled into that immobile body to the point of having nothing in it except the habits of birds and animals it listens to, Thoreau contracts those habits and trespasses into another.

But the point of this contemplative metamorphosis is not to substitute one habit for another, the old habits of one's own mind for a habit of a bird. Instead, assuming the habits of others teaches Thoreau how others move and change and so leads him into ever more fabulous migrations. For instance, the owl he listens to for days on end releases a repetitive sound that is heard as "gl, gl, gl," which Thoreau strives to repeat accurately even if part of him, resistant to becomings, hears the owl's sound as "expressive of a mind which has reached the gelatinous mildewy stage in the mortification of . . . thought" (Wa, 125). The symbol of Western wisdom howls like an idiot until suddenly everything changes, until another owl's response alters the repetitive "gl" sounds, disturbing the habitual pattern: "But now one answers from far woods in a strain made really melodious by distance,- *Hoo hoo hoo, hoorer hoo*" (Wa, 125). The series of "gls" leads to the series of "hoos" as repetition engenders difference, and boring inanity is transformed into an event.

That instantaneous shift of one into another—another's sudden exit from its habit, a journey Thoreau also made—changes everything in the soul that contemplates it. The different sound registered by the contemplator suddenly inscribes in his soul a new quality, which Thoreau, at the beginning of "Sounds," calls an "event . . . without metaphor" (Wa, 111). If that event is not linguistic—not metaphorical—it is because it is the immediate imprint of what arrives from the outside onto the contemplator, and which, because of its immediacy, instantaneously contracts its own quality. In other words, if such an event is literal, it is because it is a unity of perception and the affect or thought it generates. It is thus a unity of the corporeal and mental, a perception-thought that, Thoreau suggests, would be possible in an ideal loft-mind. The contemplation he has in mind is precisely that: a contractile power that contracts vibrations, agitations, or elements into a new sensation-thought compound. That is moreover how Thoreau explains it when in the same chapter he describes his listening to the repetitive sound of bells: "There came to me in this case a melody which the air had strained, and which had conversed with every leaf and needle of the wood, that portion of the sound which the elements had taken up and modulated and echoed from vale to vale. The echo is, to some extent, an original sound, and therein is the magic . . . of it. It is not merely a repetition" (Wa, 123). The sound generated by the bells might be automatic and thus always the same, but because each time it reaches the contemplative soul, it has traveled through a different modulating element, the soul contracts it as a novel intensity. Thoreau's contemplation thus attains what Kant's reasoning couldn't: access to the world in process, where there are no repetitions but only events, and no categories but only qualities, or, as Thoreau puts it, modulations. Strange as it might seem to a sophisticated mind like Kant's, for which the acategorial is not only untrue but inconsequential; strange as it might seem even to common sense (which is used to relying on categorial divides), it is modulations, these small literal contractions manufacturing an instantaneous unity of mind and body, and escaping categories and concepts, that Thoreau calls facts: "My facts shall all be falsehoods to the common sense. . . . Facts which the mind perceived-thoughts which the body thought with these I deal-I too cherish vague & misty forms-vaguest when the cloud at which I gaze is dissipated quite & nought but the skyey depths are seen" (J, 4, 170–171). Facts in Thoreau are not the fixed and long-lasting entities of philosophers and scientists. Rather, they are an acutely lived unity of perception and thought, a unity that Thoreau understands as a perception clear and intense enough to perceive right through the perceived and thus to equal the most profound thought; and a thought intense enough to traverse the body in the form of its strongest sensing, thus becoming a corporeal occurrence par excellence.

At this point, where only intermittent corporeal thoughts remain, where we are past self-reflection and the bounded personhood it generates, what subsists—what is contracted as the intense identity of perception and thought, agitation, and insight—is the feeling of pure life. It could thus be said that the facticity that Thoreau's contemplation treats of is that of life as such. Such a claim is to be taken literally: it doesn't mean that Thoreau wants to contemplate life as something different from contemplation, but as what contemplation is. That would be the final, ontological consequence of Thoreau's epistemology: if contemplation requires the abandonment of personhood to the extent that a body can contemplate, then there is no reason any body wouldn't contemplate. The sound of bells that reaches Thoreau "converse[s] with every leaf and needle of the wood." Through that conversation the trees modulate the sound into a difference Thoreau will hear, but, because a conversation is by definition something that goes both ways, the sound also modulates the leaves, making them contract vibrations differently and taking them out of their vegetal boredom. The contraction of intense qualities belongs to all life; life itself is contemplation. Hence, Thoreau can not only say that he is interested in thoughts that the body thinks but he can also risk a more startling claim: "All matter indeed is capable of entertaining thought" (J, 2, 146). Contemplation, then, is not something brought to matter by the mind; rather, in Thoreau's account, all matter is treated as contemplative, alive, and thoughtful.

Personhood: Who or What?

The idea that all matter contemplates is foreign enough to Eastern traditions to distance Thoreau from them. But for Western thinkers too, it was so strange that few risked it, Plotinus being perhaps the most important exemption. That Thoreau read Plotinus and commented on *Enneads* in his 1840 *Journal* thus raises the possibility that in his travels from West to East to formulate his own philosophy of contemplation, Thoreau was influenced by Plotinus. That all matter contemplates is in fact the central hypothesis that Plotinus articulates in the *Enneads*. Thus, Thoreau could have read in Plotinus's third book the astonishing—albeit on Plotinus's own account, unendurably odd idea ("could anyone endure the oddity of this line of thought?")[1]—that all matter contemplates: "but now let us talk about the earth itself, and trees, and plants in general, and ask what their contemplation is, and how we can relate what the earth makes and produces to this activity of contemplation, and how nature . . . has contemplation in itself and makes what it makes by contemplation . . . it is contemplation and object of contemplation."[2] Everything that exists, then, produces and relates by contemplation. Contemplation, however, is not something different from material beings, reaching them from an exterior and so mobilizing them (it is not, for instance, a force that God bestows on inert matter to animate it, while maintaining it as substantially different from matter). Instead, the very motion of nature that makes beings is contemplation. Contemplation is what, in moving itself, creates itself into a being; it is thus one with the created (according to Plotinus's phrasing, nature is "contemplation and object of contemplation"). Or as Pierre Hadot states, in Plotinus "life is immediate self-contemplation."[3] All creatures contemplate; all bodies, through the relations they establish—by touching the earth or other beings, by breeding, flying or crawling, eating or sleeping—contract new qualities. These qualities are not concepts or ideas but sensations though which all matter feels itself without ever being self-conscious. As Deleuze and Guattari sum it up in their discussion of contemplation in Plotinus: "These are not Ideas that we contemplate through concepts, but the elements of matter that we contemplate through sensation. The plant contemplates by contracting the elements from which it originates—light, carbon, and the salts—and it fills itself with colors and odors that in each case qualify its variety, its composition: it is sensation itself. It is as if flowers smell themselves by smelling what composes them."[4]

As is indicated by Thoreau's *Journal*, the idea that all matter contemplates by contracting sensations is precisely what attracts Thoreau to Plotinus. In an entry from 1840 he quotes the following in Greek from the *Enneads* VI: "a kind of tactual union, and a certain presence better than knowledge, and the joining of our own centre, as it were, with the centre of the universe" (J, 1, 127). Contemplation is here understood to be a "tactile" or sensuous fusion of the mind with what transcends it; it is understood as the moment in which a mind contracts itself and what is external to it into a new quality. Elsewhere in the *Enneads*, Plotinus is explicit about this transformation of the mind into what it is not, constituted by contemplation: "We should remember that, even in this world, when we contemplate . . . we do not turn towards ourselves intellectually . . . rather . . . our activity is directed towards the object, and we become the object."[5] In the sentence that Thoreau quotes, this becoming a new entity of the mind and the object Plotinus identifies as a new "presence" ("tactual union, and a certain presence"). Presence is the "suchness" or objectivity of a new creation and its sensing itself. It is a sensing existence that, by being accessible to itself in its immediacy, is claimed by Plotinus to be "better than knowledge." As Deleuze and Guattari explain again, what is better than knowledge is this sensing existence's proximity to itself, its being "filled with itself by filling itself with what it contemplates: it is 'enjoyment' and 'self-enjoyment.' "[6] Just as in Plotinus all life contemplates, so in Thoreau all matter entertains the contemplative way of life.

But there is a cardinal difference between Thoreau's and Plotinus's understanding of contemplation. For while Thoreau's understanding of it as what fuses the mind with matter differs from the Hindu understanding, the value he affords contemplative matter distances him in turn from Plotinus. It is true that for Plotinus everything contemplates. A natural object can contemplate a natural object, because nature is nothing other than contemplation ("all things come from contemplation").[7] But there are more possibilities still: a mind can contemplate another body; or a mind can contemplate another mind, in which case the corporeal is excluded from contemplation; and finally, a mind can contemplate its "eternal model [God], as the latter exists within divine Thought."[8] However, not all of those contemplations are equally valued by Plotinus. In Hadot's account: "[Plotinus] situates himself . . . within a hierarchy of realities which extends from the supreme level—God—to the opposite extreme: the level of matter."[9] Thus, although the human mind can indeed contemplate the material, such contemplation means its debasement, for it fuses the superior with something inferior to it. Conversely, the mind contemplating the eternal and purely spiritual divine self is elevated, because through fusion with God it becomes incomparably superior to other beings that partake in matter.

When Thoreau says that the "thoughts the body thought" is what chiefly interests him, he is therefore radically reversing Plotinus's values, just as he had with Hinduism. Always democratic in his ontology and epistemology, Thoreau insists that instead of trying to think absolute being, such as God, which is only fantasized and nowhere to be seen, we should "*always* [look] at what is to be seen" in front of us. This injunction to dedicate our thought to what is corporeally in front of us advocates as an epistemological principle that only earthly creatures are, strictly speaking, knowable to us. Ethically, it suggests, that creaturely life must be of highest value for us, because it is with creatures—human, animal, and vegetal—that we build communities. In fact, precisely because creatures partake of our daily life, which is by definition social, Thoreau can assert that "socialness" is to be experienced in the contemplative encounter with the natural: "I experienced sometimes that the most sweet and tender, the most innocent and encouraging society may be found in any natural object. . . . I was suddenly sensible of such sweet and beneficent society in Nature" (Wa, 131–132). This also suggests that in contrast to Plotinus's contemplation leading to an ecstatic union with God, Thoreau's experience of the "sweetest" sensation through contemplation of the corporeal ("any natural object") radically severs his understanding of contemplation from mysticism. Or, as Cavell cautions us, if one is still to call Thoreau's sensing of sweetness in the presence of a natural object "mystical"—and Thoreau himself did it at least once—such mysticism should be understood as radically different from anything we recognize as such: "you may call this mysticism, but it is a very particular view of the subject."[10] For in mysticism a human being and its world are dependent on God's sovereignty and so are founded on something different from them. In Thoreau, there are only beings related to one another through relations that found the world. Thoreau's treatment of the world in terms of relations, which incessantly remake it, additionally marks how radically his redirection of contemplation from God to earth alters traditional metaphysics. To remove God from contemplation is to cancel the force that bestows on the world a stable ontological reality. Instead of owning an essence generated in advance, Thoreau's nonessentialist world makes itself as it proceeds.[11] In this world in the making, the only essence there is consists of what is fleeting. In accord with the biology that allows for anomalous becomings, Thoreau's epistemology also converts the world from fixed metaphysical being into transitory becomings: "Transiting all day . . . it is only a *transjectus*, a transitory voyage, like life itself" (W, 118).

Finally, existentially, Thoreau's treatment of contemplation—as what enables thought to access this new and flexible ontological reality—revises the way in which metaphysics traditionally thinks personhood. Indeed, such a revision of personhood is mandatory, because the epistemology of modern philosophy, from Descartes to Kant, involves responding to the ontology of the eternal being of the world by construing a corresponding unchangeable mind to think it. For both Descartes and Kant, the "I" is determined by its a priori posited unity. This "I" precedes what is experiential and is not affected by it. As Descartes programmatically states in the Synopsis of his *Meditations:* "the human mind . . . is a pure substance. For even if all the accidents of the mind change, so that it has different objects of the understanding and different desires and sensations, it does not on that account become a different mind."[12] Just imagine: a world completely changed, all objects in it new and unforeseen, manufacturing different desires, changing the subject's senses and fantasies and yet, despite all those marvelous transubstantiations, the Cartesian subject changes not an iota.

However, for Thoreau, because the mind doesn't just think the world but is of it, something like a radical dismantling of the modern transcendental mind—this pure, clean substance—has to happen. That is suggested in a *Journal* passage, which is a longer and clearer version of a famous moment in *Walden:*

> I know . . . that Sadi entertained once identically the same thought that I do—and thereafter I can find no essential difference between Sadi and myself. . . . Sadi possessed no greater privacy or individuality than is thrown open to me. He had no more interior & essential & sacred self than can come naked into my thought this moment. Truth and a true man is something essentially public not private. If Sadi were to come back to claim a personal identity with the historical Sadi he would find there were too many of us-he could not get a skin that would contain us all. . . . In his thoughts I have a sample of him a slice from his core . . . I only know myself as a human entity-the scene, so to speak, of thoughts & affections-and am sensible of a certain doubleness by which I can stand as remote from myself as from another. However intense my experience-I am conscious of the presence & criticism of a part of me which as it were is not a part of me-but spectator sharing no experience, but taking note of it-and that is no more I than it is you.- When the play-it may be the tragedy-is over, the spectator goes his way. (J, 5, 289–290)

In Thoreau's description of thinking there is an "I," but it provides no constancy or continuity for thoughts and affections. This "I" neither synthesizes nor essentializes those thoughts and affections but instead stands "remote from them." Thus nothing affects this "I," leaving it cool and distant, a mere observer "sharing no experience, but taking note of it." Even if something tragic were to happen, this "I" still wouldn't change: "when the play—it may be tragedy—is over, [the "I"] goes away"). Just as in Descartes or Kant, it seems. But unlike modern epistemologists who entrust true thinking to this "I" because it is withdrawn from the experiential, Thoreau, for the same reason, renders the "I" inconsequential; because it stands remote from "affects & thoughts" that make me, it is not me at all. In fact, precisely because there is nothing of "me"—of my thoughts and affects—in this "I," Thoreau can risk the most radical epistemological proposal, claiming that what I call my "I" is so abstract that it can equally be yours: "and it is no more I than it is you." There are thus "thoughts & affections"—there are contemplative contractions—but because there is nothing "interior & essential & sacred" about them that would affix them to an interiority of an "essential" "I," they don't belong to the "I." Instead of Descartes's epistemological mantra "*Ego cogito, ergo sum*," Thoreau proposes a thinking that occurs outside an "I," an "I" that does not think. His formula would thus be: *cogitations sunt, ergo Ego ne est*—where there are thoughts there is no "I."

The sensations and affects that are contracted without being mediated by an essential "I" possess nothing that would gather them into unity and confer on them continuity (like Thoreau's house, they are without roots and genealogy). They are Thoreau's pure, literal events, attributed to nobody but instead executing themselves as they come to be (sensations executing sensing, thoughts thinking). In so doing, they affect one another and thus enter into relations. And since there is no essence to which they can be attributed, those specific constellations are all that a personality consists of, all that can be called a personhood. It is such a constellation of "naked" affects and thoughts, as Thoreau specifies, that is called "Sadi," and not his "interior and sacred" self. And because no such interiority intervenes to claim it, it becomes possible that "identically same thought," as Thoreau puts it, or the same constellation of affects and thoughts to be contracted across vast spatial and temporal distances. When that happens, a multiplication of the same person occurs ("there were too many of us"); Thoreau becomes Sadi. Thoreau's language insists on the literality of such a becoming: he doesn't simply say that he entertained thoughts similar to Sadi's, while otherwise inhabiting his own self. Still less does he say that on reading Sadi he remembered his thoughts, thus retaining his self while commemorating the past. More radically, he says that while thinking those thoughts—in their lived and living present—he was nothing but them, nothing else of him remaining. He has become Sadi, without any remainders: "and thereafter I can find no essential difference between Sadi and myself." Sadi doesn't die in Shiraz six hundred years before Thoreau reads him in Concord, Massachusetts, but still survives, not *in* Thoreau but *as* Thoreau: "by the identity of his thought and mine his still survives." This, then, is what contemplation does: neither memory nor understanding, it contracts thoughts and affects once lived into instants of the now living present. It revives, it enacts survivals, it turns what was into what is.[13]

Coda: Benjamin's Seagulls

Giving what is fleeting and ephemeral the relevance of what is essential and irreplaceable; suspending the self to let what is external to it, whether human or nonhuman, inhabit it; recording (in *Walden*) how such strange inhabitings produce an incapacity to differentiate between himself and "Mill Brook, or a weathercock, or the north-star, or the south wind, or an April shower, or a January thaw, or the first spider in a new house" (Wa, 137); exalting over his experience of the world as a being in which everything is alive, in which not just animals but also "gentle rain," rainbow, landscapes, "even scenes which we are accustomed to call wild and dreary," look at him, are with him as if "kindred" or "friend[ly]" to him (Wa, 132)—all that might appear, as Thoreau himself suspects, to betray a "slight insanity in [his] mood" (Wa, 131). But the lived intimacy with animate or inanimate objects, and even transacting one's nature for theirs, can be considered less eccentric than Thoreau self-diagnoses, or than it might seem to us today. Walter Benjamin, for example, famously described it as an experience of the world that he considered to be characteristic of all premodern Western humanity. For Thoreau's description of a mind that fuses with an object in the same "atmosphere" is precisely what Benjamin famously called the "experience of the aura": "[aura] arises from the fact that a response characteristic of human relationships is transposed to the relationship between humans and inanimate or natural objects. The person we look at, or who feels he is being looked at, looks at us in turn. To experience the aura of an object we look at means to invest it with the ability to look back at us."[1] On Benjamin's understanding, rapid stimuli and images distributed suddenly as if in a flash—generated by new technologies from photography to cinema, telephone to gas lamps—constitute the modern world in which the auratic unity of humans and natural or artificial objects is shattered into an experience of shock, danger, and isolation. If we Moderns have a lived experience of unity with a natural object at all, it is, as Benjamin claims (quoting Valéry), only in a dream. And modern dreams, unlike Thoreau's, don't render what is real but only the imaginary: "Valéry's characterization of perception in dreams as an auratic perception: To say 'Here I see such-and-such an object' does not establish an equation between me and the objects. . . . In dreams, however, there *is* an equation. The things I look at see me just as much as I see them."[2] Thoreau's auratic reality is thus banished to the outskirts of the modern imaginary, turned into something fantastic if not downright occult. That would seem to suggest Thoreau's datedness. But was Thoreau really so unfashionable, so much out of tune with his times?

It is true that most of the figures that Benjamin lists as allegories of modern life—conspirator, Apache, detective, journalist, whore, rag-picker, aesthete, gambler, fencer, lesbian—have come to exemplify various ways in which the mind, reacting to an "overabundance of information," takes its distance from things while simultaneously trying to gain possession of them. Yet, the flâneur, the figure that, more than any other, came for Benjamin to emblematize modernity, is a figure neither of distance nor of speed. Instead, he is the figure of an auratic fusion with objects generated through a walk in a way that is structurally identical to Thoreau's walker. It is this coincidence of the flâneur and Thoreau walker that positions Thoreau's experiences of the material and sensuous as genuinely modern.

Like Thoreau's walker, Benjamin's hypermodern flâneur walks differently from other Moderns, for instance, tourists or historians: "The great reminiscences, the historical shudder—these are a trumpery which [the flâneur] leaves to tourists."[3] He also walks differently from the aesthete or philosopher—"completely distanc[ing] himself from [a] . . . philosophical promenader"[4]—the walker, who, as in Kant, separates himself from the stimuli the city emits by translating them into "abstract knowledge" supposed to help him decipher what he sees. The flâneur neither reminisces nor conceptualizes, then; but still less does he follow traces, for he doesn't walk as a scientist or detective either, progressing evidentially toward the truth of an event. In fact, he doesn't conduct his steps at all but "speechlessly, mindlessly, aimlessly" strolls until, as Benjamin interprets, the "street [starts to] conduct the flâneur." Rather than imagining the flâneur as a figure who takes a walk, it would be more precise to see him as a creature walked by the streets—as Thoreau's walker is walked by the woods—until he gets completely lost at a point where the recognizable contours of the city turn into a maze, leading the flâneur still more "downward . . . into a past that can be all the more spellbinding because it is not his own, not private."[5] Spellbound by impersonal exteriority and lost in it, his memories and thoughts cancelled, the flâneur is taken out of himself while the things and beings he passes by move in to inhabit and "instruct" him. This condition, in which flâneur's self is possessed by external objects, is identified by Benjamin as one of "anamnestic intoxication" or, alternatively, as aura, for aura is precisely the experience of the world in which "[the thing] takes the possession of us."[6]

If the flâneur is annulled into the things he encounters, he does not for all that objectivize himself in the world of fetishes. The flâneur doesn't walk through the city as a living being moving among inert objects. Instead, like Thoreau, the flâneur is a vitalist. His world is all alive and animated, its matter imbued with sensations, and even "abstract knowledge" and "dead facts" are transformed into something "lived": "That anamnestic intoxication in which the flâneur goes about the city not only feeds on the sensory data taking shape before his eyes but often possesses itself of abstract knowledge—indeed, of dead facts—as something experienced and lived through."[7] In the auratic world of the flâneur (or Thoreau) there are no dead facts but instead the whole city—stones, trees, iron constructions, churches, candelabras, clothes—becomes "the landscape built of sheer life,"[8] of which he is a part, traversed by sensations and the temporality of others. In fact, if the flâneur can become a passage of sheer life woven of sensations that are not privately his—if his body can turn himself into a vital arcade, as it were—it is precisely because, as Sylviane Agacinski remarks, the flâneur has through self-abandonment "given up" his time and "made himself strangely available to [the] time" of others. The flâneur "must not just 'make time pass'; he must also 'invite it home' *(einladen)*. [He must] 'take on time' *(laden)*. But what can it mean to 'take on time'? It could be this: to accept not having time oneself, not to become involved in the temporality of an action, thus making it possible to accommodate the event . . . and to grant things their own temporality, their own particular rhythm."[9] The flâneur grants things their own time by taking on their rhythms, becoming them, rather than being absorbed by his own progress. And so, when Benjamin says that the "flâneur is the virtuoso of empathy with commodity," which is "fundamentally empathy with the exchange value itself,"[10] he doesn't suggest, as a Marxist reading might have it, that the flâneur reifies himself into a commodity. For the flâneur is not a Marxist. To him, things and beings can be exchanged not because they are commodities but because they are not. It is precisely because nothing can be commodified but everything is unique in its living preciousness, that, in Benjamins's precise formulation, the flâneur is not in love with this or that thing but with the "exchange value itself"; he is in love with the process of exchanging natures, his for those of others. Much like Thoreau's walker, the flâneur is a creature of becomings. He is a creature of transformations, such as Flaubert describes himself to be in a passage that Benjamin comments on as the paradigmatic experience of flânerie: "Regarding the intoxication of empathy by the flâneur, a great passage from Flaubert may be adduced. It could well date from the period of the composition of *Madame Bovary:* 'Today, for instance, as man and woman, both lover and mistress, I rode in a forest on an autumn afternoon under the yellow leaves, and I was also the horses, the leaves, the wind, the words my people uttered, even the red sun that made them almost close their love-drowned eyes.' "[11]

As this example assures us, the flâneur is a creature auratically possessed by a tree, a horse, or a bird, becoming them. That is why, I am proposing, a privileged example of the flâneur is less Baudelaire walking the streets of Paris, than Benjamin himself on his trip to Sweden in 1930, on "board of a ship, moving through the sea at nightfall," looking at seagulls:

> What now took place with the birds—or was it with me?—happened by virtue of the place that I, so commanding, so

> solitary, had chosen out of melancholy in the middle of the afterdeck. All of a sudden, with a single stroke, there were two seagull populations, one the Eastern, the other the Western, left and right, so totally different that the name 'seagulls' dropped off them. Against the backdrop of the moribund sky the birds on the left retained something of its brightness, glittering with every swoop up and down, got along or avoided each other and seemed not to stop weaving an uninterrupted, unpredictable series of signs, an entire, unspeakably changing, fleeting winged web—but one that was legible. Except that I slipped, finding myself obliged to start all over again with the other. Here nothing stood before me, nothing spoke to me. Scarcely had I followed those in the East, a pair of deep black beating wings flying toward a final glimmer, losing themselves in the distance and returning, when I was no longer able to describe their movement. So entirely did it seize me that I myself came back from afar, black with what I had been through, a soundless flurry of wings.[12]

As Agacinski suggests in her interpretation of this experience, "the one whose eyes follow the flight of a gull over the sea adopts the temporality of that flight; his time becomes the gull's time. . . . [He] yields to the rhythm of a movement that is not [his] own. In forgetting his own movement, and thus his own time, [he] embraces the time of things."[13] Whether it is Thoreau's walker suspending his time to become corn, fish, or bird; whether it is Baudelaire becoming a passerby on the sidewalk of a big city; whether it is a passerby "standing before Notre Dame de Lorette," his soles becoming it;[14] whether it is Flaubert, becoming leaves or horses; or whether it is Benjamin on the Nordic Sea, seagulls flying through him, seizing him into becoming gulls, and returning from such a visitation blackened by it, the same modern experience is always in question. It is an experience that Benjamin identifies as anamnestic and intoxicated; the experience of self-surrender, exchange, and versatility; the experience that Thoreau's walker—the artist of becoming—generates and embodies no less than a Parisian flâneur.

Part IV

Ossossané Village, Ontario, 1636: Acts of Recollecting

Introduction: The Ethics of Communal Mourning and the Flight of the Turtle Dove

MANY IMPORTANT readers of Thoreau have argued that his understanding of community is predicated not on abundance and fullness but, to the contrary, on death, loss, and lack. It is impossible to determine—for Thoreau made no explicit references to it—whether in understanding community thus he had in mind its premodern, Pauline, understanding, as voiced in Romans and Philippians, where, as Esposito explains, Paul insisted that "taking part" in the life of a community "means everything except 'to take'; on the contrary, it means losing something, to be weakened."[1] Unlike *Res publica*, predicated on appropriation and participation in what is called the "public sphere," community was understood to be "in close contact with death, 'a society from and with the dead.'"[2] Yet, even if we cannot tell whether or how much Thoreau was influenced by the Pauline understanding of the communal, his own conception of it is very much in accord with what was developed in those Epistles.

In a remarkable essay on *Cape Cod* Mitchell Breitweiser argues that Thoreau doesn't experience community as something that safeguards and protects but instead as a precarious network of relations impossible to root or stabilize, always exposed to fragmentation, loss, and wreckage. Moreover, in Breitweiser's reading, the citizens of Cape Cod are emblematic of Thoreau's understanding of what community is in general terms, precisely because they "do not seek 'dry land,' that is, an impossible solidity safe from wreck. Rather, they live in the wreck, in the midst of it, marked by it, themselves such pieces of affection as the items they collect" from the ships wrecked on the beach.[3]

Similarly, in Barbara Johnson's and Cavell's readings, *Walden* expands the experience of loss from something private into a phenomenon that centrally pervades all communal reality. In the *Walden* passage that Cavell identifies as "perhaps [Thoreau's] most famously cryptic,"[4] the history of an individual life is said to be entangled in communal losses whose nature is described, in the preceding paragraph, as arcane:

> You will pardon some obscurities, for there are more secrets in my trade than in most men's, and yet not voluntarily kept, but inseparable from its very nature. I would gladly tell all that I know about it, and never paint "No Admittance" on my gate.
>
> I long ago lost a hound, a bay horse, and a turtle-dove, and am still on their trail. Many are the travellers I have spoken concerning them, describing their tracks and what calls they answered to. I have met one or two who had heard the hound, and the tramp of the horse, and even seen the dove disappear behind a cloud, and they seemed as anxious to recover them as if they had lost them themselves. (Wa, 17)

The paragraphs are so famous that it is almost trivial to quote them. Following Emerson, most readers have understood Thoreau to have been formulating a riddle. The three animals referred to are thus typically taken to be symbols, emblematizing loss while obscuring its content.[5] Johnson and Cavell, in contrast, intervene in this long tradition by proposing a change of strategy. Instead of trying to determine what the animals symbolize, they suggest we try to understand how the passages in question not only talk about but also generate obscurity. This is important, because, as both Cavell's and Johnson's otherwise different readings agree, Thoreau's obscurities are indeed involuntary, in that they are effects not of Thoreau's mind but of loss itself, of what eludes clear cognition—for otherwise it wouldn't be lost—and withdraws where neither knowledge nor desire can access it.

But while agreeing that what is at stake in Thoreau's paragraphs "is not any set of lost things but rather the very fact of loss," Cavell and Johnson differ concerning what Thoreau understands losing to be and how loss is related to the communal. Johnson argues that for Thoreau something is lost, properly speaking, only when it is lost to our cognition, that is, only when one who has lost something doesn't *know* what he has lost and, therefore, doesn't know that he has lost anything. According to Johnson, Thoreau suggests this understanding of loss when he proclaims that he will "admit us" only to all he knows about his losses, thus signaling that all he knows about it "is not all there *is* about it."[6] What there is about Thoreau's loss is precisely what he doesn't know.

For Johnson, loss becomes communal in Thoreau, because only others—represented by the travelers Thoreau meets on the road—can know that we have suffered losses. They see on our faces that we have lost what we can't identify, just as we do for them. The experience of loss is never personally accessed but instead becomes the vehicle of our "transference" with others: "To follow the trail of what is lost is possible only, it seems, if the loss is maintained in a state of transference from traveler to traveler, so that each takes up the pursuit as if the loss were his own. Loss, then, ultimately belongs to an other; the losses we treat as our own are perhaps losses of which we never had conscious knowledge ourselves."[7]

Whereas for Johnson, one would desire to recover one's losses if only one could come to know them, in Cavell's reading one knows one's losses ("Everything he can list he is putting in his book; it is a record of losses") but doesn't desire to recover them: "the myth ["the hound, a bay horse, and a turtledove" passage] does not contain more than symbols because it is no set of desired things he has lost, but a connection with things, the track of desire itself."[8] And because life for Cavell is always a life of losses ("to grow . . . humanly . . . is always a loss of something"),[9] to lose the desire to connect with losses translates into losing the desire to connect with one's life. The loss of "the track of desire itself" is thus the loss of access to life, which for Cavell means the loss of "everything there is to lose."

However, the desire to recover the connection with what is gone as a desire to recover one's life is not, in Cavell's understanding of

Thoreau, motivated by others with whom we travel. For him, the travelers Thoreau meets on the road are not, as in Johnson, in possession of some knowledge about us that is inaccessible to us; on the contrary, they are just like Thoreau himself, knowing their losses but having lost the connection with them, spectrally journeying, severed from their lives. Although in Johnson community saves us from our losses by diagnosing them for us, in Cavell community is, like we are, lost: "the writer fully identifies his audience as those who realize that they have lost the world, i.e., are lost to it."[10] Writing *Walden* is, in Cavell's reading, Thoreau's effort to restore this colossal loss ("completing the crisis by writing his way out of it"). In writing it, Thoreau carves his way back into relations with the living and dead who are lost to him; by rediscovering the track of desire, he recovers his self. For "the fate of having a self—of being human—is one in which the self is always to be found; fated to be sought, or not; recognized, or not. . . . In the passage in question, *Walden*'s phenomenological description of finding the self, or the faith of it, is one of trailing and recovery; elsewhere it is voyaging and discovery."[11] According to Cavell's understanding then, there is no recovery unless it is a self-recovery, no loss unless lived as one's own.

In contrast to Johnson's reading, I argue that Thoreau in this passage does know his losses, for he records them and registers when they occurred (long ago). But in contrast to Cavell's findings, I consider that the one thing Thoreau didn't lose was the desire to recover losses: "I long ago lost a hound, a bay horse, and a turtle-dove, *and am still on their trail*." Time doesn't therefore weaken Thoreau's desire to recover what has been lost long ago, a desire that should properly be called "grieving," because to spend one's life endlessly calling the names of the lost, to evoke them as Thoreau does (even explaining to travelers he meets on the road what to call his losses), is precisely to turn one's life into an endless lament. This desiring lament is, then, still active by tracking what is desired. In fact, if anything, the passage depicts a grief intensified by time to the point of obsession, as the mourner identifies his whole life with it. His life consists of journeying on his interminable search for the lost ("long ago lost . . . still on their trail"). Yet, despite the unending nature of the search, the traveler never seems defeated or tired. Instead, he keeps on watching, always ready to talk about his losses, to make inquiries about them, to instruct others where to look and what to listen for. It is as if mourning, far from suspending life into a melancholic stupor, in fact refreshes it, energizing the mourner with strength to keep on traveling.

The same holds for the community. In Thoreau's paragraph the others he meets on the road not only know of their own losses and, like Thoreau, are mobilized by the desire to find them. They are additionally mobilized by the losses of others, for they search also for what the others have lost. Among the strangers Thoreau meets on his journey are some who have indeed found Thoreau's losses, lived with them, perhaps briefly, but nevertheless long enough to see their beauty ("I have met one or two who had heard the hound, and the tramp of the horse, and even seen the dove disappear") and are now "as anxious to recover them as if they had lost them themselves." What we have lost others have found, become attached to, even loved, and then lost again. Conversely, because in journeying after our losses we come across the losses of others—which remains possible, because in Thoreau's passage the losses are animated, they are all still living animals, leaving tracks, answering calls, releasing sounds, tramping, or flying, as if the losses migrated into interminable lives of their own—we love them as they did, or as they love ours. In Thoreau's democratic view, then, the losses of an individual are shared not because such sharing can take someone's grief away, but because to actually participate in the grief of others—to mourn their losses—is the way to reach them. Thoreau's wager, supposed to rebuke hopelessness without cheering us into hope, is this: as long as everybody is on the trail, something will be recovered by somebody.

Deathways of Things

Throughout his work Thoreau addresses manifold ways in which a community is mobilized by losses. In *Walden* such mobilization is discussed through the way the community relates to things. *Walden* is as obsessed with things as it is with water and oak trees: from Concord houses to pyramids and coffins; from Native American baskets to oversized English suitcases; from the pots and pans Thoreau uses for cooking to chairs, lamps, and pencils; from trains to the Kouroo artist's equipment, objects are always either juxtaposed on or intertwined with natural things (landscapes, apple trees, leaves, flowers, birds, or flows of wet sand).

If things assume a central role in Thoreau's understanding of communal life and its relation to its losses, it is because a thing for Thoreau is precisely inherently communal, embodying what it means for a community to remember and recover. For Thoreau, a thing is thus not an isolated, inert entity that we can confront, use, and possess. A thing does not reduce to the obsessive focus of a collector's desires, imbued with qualities it doesn't have but which the obsessed mind of its owner infuses into it, turning a commodity into a fetish, as Marx would put it. Moreover, a thing is not a commodity at all; it is not something that can be reified and exchanged for money, as capitalists desire. No flows of capital circulate through it. But neither is a thing a pile of formed matter that the human invests with sentiments, nor is it matter from which affects emanate in some remarkable way; a thing is neither souvenir nor archaic totem. Things in Thoreau are not inert relays relinking a present perception to a past experience, as they become in the modernist experience formulated by Proust. But nor are they bodies hosting images that render them corporeal, as they become in the modernist experience formulated otherwise than in Proust, according to William Carlos Williams's famous dictum "no ideas but in things." Above all, what counts as a thing in Thoreau is not a thing in the sense that Heidegger famously gave it: it is not a shrine full of meaning transcending human existence, which even when withdrawn from humans, lets them inhabit the world by assuring them that the world is here to stay. In Heidegger assurance of "staying" is the thing's principal ontological operation; made of the earth and the air, of clay and heavens and thus, by extension, of mortals and immortals—at the very intersection of gods, humans, and animals—the "thing" combines in mutual belonging the corruptible and the eternal, the material and the divine. The thing is what preserves this mixture through its own immobility or staying, which assures mortals of their own "staying" in immortality.[1]

For Thoreau, in contrast, the thing is fashioned less by its form or usage, by its solidity or impenetrability, than by the intense relations in which it dwells and through which it moves, relations capable not just of reshaping past into future but also of enacting marvelous exchanges between what is and what isn't human. Thoreau's things are thus never stable objects, but exist in their thinghood only when caught in a process that makes them toss and turn, rendering them volatile. That is what Thoreau suggests when he describes how things in his house, as well as the house itself, relate to phenomena that are external to the house. For instance, while we are told that there is a frying pan and a jug of oil in the cabin, we also understand that there is no stove to confirm that the food is indeed cooked inside the house. That is because Thoreau's stove is not inside the house at all but is located outdoors, made of stones in front of the cabin. In Thoreau's own testimony, he was "doing [his] cooking . . . out of doors on the ground, early in the morning," just as he would bake his bread on the fire he would light in front of the house, so that "when it stormed before [his] bread was baked [he] fixed a few boards over the fire, and sat under them to watch [his] loaf" (Wa, 45). From table to stones and the grass-ground, the frying pan and jug of oil located inside the cabin shift places, and unlike Heidegger's jug that stays and perseveres, Thoreau's things journey among objects and elements (earth, fire, rain); their shifting positions register how their meanings—or their purposes—alter as their bodies are materially affected by the weather or heat changes, by their encounters with the earth.

Hence, the central question that *Walden* raises regarding circulation of things, communal life, and its experience of loss is the following: what happens to meanings and relations embodied and generated by unstable things when a thing has fallen out of the flow of communal life? *Walden* discusses two different communal responses to the phenomenon of things that have exhausted their concrete engagement in communal exchange, that of the Mucclasse Indians and Mexicans, and that of the New Englanders, each of which embodies a radically different response to loss, forgetting, and memory. Although Thoreau prefers the former, he does not side with either.

Thoreau predicated his theory of the Mucclasse Indian understanding of a thing on William Bartram's *Travels through North and South Carolina*, quoting approvingly and at length from Bartram's description of the their "busk" ritual dedicated to the cremation of "exhausted" things:

> When a town celebrates the busk . . . having previously provided themselves with new clothes, new pots, pans, and other household utensils and furniture, they collect all their worn out clothes and other despicable things, sweep and cleanse their houses, squares, and the whole town, of their filth, which with all the remaining grain and other old provisions they cast together into one common heap, and consume it with fire. After having taken medicine, and fasted for three days, all the fire in the town is extinguished. . . . The Mexicans also practiced a similar purification at the end of every fifty-two years. (Wa, 68)

What is so appealing to Thoreau about the understanding of things described by Bartram is that it appears to be guided by the Mucclasse Indians' nondualistic understanding of the world. According to them, a meaning of an object or a body cannot be abstracted into a pure idea separated from its materiality. Meaning must be embedded in the material, but not as something simply hosted by matter while preserving its substantial difference from that matter, rather as something that is itself concrete and indistinguishable from what is material. And conversely, what is material is never simply mute but is always considered as the concretization of meaning in a body. When the thing falls out of or withdraws from the relations that incorporate it into tribal life, it loses its meaning and, concomitantly, its body and becomes, for the tribe, what is lost. In suggesting that the past meanings of the thing can't be salvaged in some ideated way, even though it doesn't participate in present life, the Mucclasse Indian holds that something can be rendered historical only if it still participates in the current circulation of communal life in a concrete and material way. The past can't preserve its status as history if it is something archived or remembered as previously generative but no longer operative. And because past meanings function only to the extent that they are also vitally present, the thing in Mucclasse Indian ontology can't be relegated to the status of a souvenir, monument, or archive. Instead, the withdrawal of the thing's meaning from the current senses of tribal life also amounts to the loss of the archive, the extinction of monuments. So things burn.

Walden patiently delineates how things die in an opposite way for New Englanders, closer to the Christian denigration of matter and belief in purely spiritual resurrections. In New England a thing exhausts its meaning and "dies" into the world on the basis of a dualistic ontology, where matter serves spirit while never being confused with it. In New England a thing that has long stopped affecting other things or persons (whether because the assemblage in which it was participating has dissolved, or the persons it was affecting are dead)—a thing whose body has long been severed from its meaning—is given an afterlife. It continues to function as a repository of the memory of the dead and gone long after those alive remember them. In New England, then, a thing is afforded a strange capacity to gather worldless and impersonal affects and memories—worldless and impersonal, because they no longer belong to anybody—thus constituting something like a surplus of the present. Assuming the function of the shrine, the body of a thing hosts the obscure spirit of the past world, whose obscurity (that is at least what New Englanders hoped for, in Thoreau's understanding) could be rendered transparent by some future archivist or archaeologist.

Walden is full of stories about people who can't leave dead things, but not because of their love for possessions. The New Englanders Thoreau talks about are not the stylish collectors Walter Benjamin identifies as rescuers of things, bestowing on them a love expressed through appropriation and belonging. In fact, the things that New Englanders can't leave are removed from the economy of belonging, for as Thoreau remarks, even the mere idea of storing them burdens the owners. They would even go as far as to move out of their houses just so they can leave their things behind: "pray, for what do we *move* ever but to get rid of our furniture, our *exuviae*" (Wa, 66). If New Englanders have difficulty casting off the dead skin their things have become, it is because they trust them to be memorials, repositories of times past, something like the chattering clutter of bygone ordinary lives. Not "hav[ing] the courage to burn them," they pile them up into "a great deal of baggage, trumpery which has accumulated from long housekeeping," which they then store. On Thoreau's estimate, everybody in New England has things and furniture stored somewhere—a garret, a shanty, a barn or "other dust holes" (Wa, 67)—and "even those who seem for a long while not to have any, if you inquire more narrowly you will find have some stored in somebody's barn" (Wa, 66). If things are still there, persisting in storage—even though they are withdrawn from all relations—it is precisely because they are imagined as an archive whose meaning still lurks in the darkness of a garret, waiting to be deciphered, should anybody want to excavate it. This, then, is the premise of the Concordian ontology of things: garrets and barns, like contemporary garages, are imagined as archives of the quotidian, an ordinary-life version of annals and museum basements hosting mute objects.

Thoreau believed there was something decidedly deeply erroneous about this ordinary archival imagination predicated on the belief that pure meanings can be salvaged so long as the coffins harboring them—the matter of things—remain. That is his impression after visiting the "deacon's effects" estate sale, described in *Walden:*

> As usual, a great proportion was trumpery which had begun to accumulate in his father's day. Among the rest was a dried tape-

> worm. And now, after lying half a century in his garret and other dust holes, these things were not burned; instead of a *bonfire*, or purifying destruction of them, there was an *auction*, or increasing of them. The neighbors eagerly collected to view them, bought them all, and carefully transported them to their garrets and dust holes, to lie there till their estates are settled, when they will start again. When a man dies he kicks the dust. (Wa, 67–68)

What Thoreau understood from the deacon's sale was much closer to the ontology of Mucclasse Indians. What his neighbors imagined as an archival site was for him simply an accumulation of things from one generation to another; far from turning the deacon's garret into an archaeological site, such an accumulation had in fact turned it into a geological one, where glasses and plates, armchairs, and clothes started to host dried tapeworms, as if things were on the verge of becoming limestone, hosts for fossils or herbaria for vegetal corpses. In enabling Thoreau's neighbors to move the deacon's things into their own garrets, the estate sale reversed the process of geologization by turning the geological back into the archaeological: the things the neighbors carefully transported to their garrets were temporarily rescued from becoming fossils, their worms removed, and the dust cleaned. But despite being redeposited into yet another "dust hole," they were not for all that restored to their lively thingness. Instead, they were merely restored as mute archeological objects, for rather than revealing their meaning, the estate sale simply enabled them to prolong their archival silence. As Thoreau moves among the deacon's dead things, he therefore realizes that the garret-archive contains only junk covered in dust, and to move it around from one garret to another is only to "kick the dust"; it is a matter of relocating things from one landscape of oblivion to another. Objects in garrets or in archives, exhibits in museums, are therefore mute stuff with no lingering meanings, which is what Thoreau had already remarked in a short discourse on museums recorded first in his *Journal* on September 24, 1843, and subsequently revised on several occasions throughout 1844. In this text, which anticipates in many ways *Walden*'s thing ontology, he explicitly writes that archived things are not like dormant creatures waiting to be awakened. For unlike dormant creatures "who have such a superfluity of life . . . enveloped in thick folds of life," stored and archived things have fallen out of their vital folds, so that no meaning attaches them to life; they are mute "preserved death" (J, 1, 465), testifying nothing: "I hate museums-there is nothing so weighs upon my spirits. They are the catacombs of nature. . . . They are dead nature collected by dead men. I know not whether I muse most-at the bodies stuffed with cotton and saw-dust-or those stuffed with bowels and fleshy fibre outside the cases" (J, 2, 77).[2] Later he suggests that such "dead nature collected by dead men" is lethal even for those who observe it: "men have a strange taste for death who prefer to go to museums to behold the cast off garments of life-rather than handle the life itself" (J, 2, 472). What Thoreau realizes, observing the objects at the estate sale, is what Carolyn Steedman, in her remarkable study, identified as the void and forgetting essential to the archive:

> It is a common desire—it has been so since at least the end of the nineteenth century—to use the Archive as metaphor or analogy, when memory is discussed. . . . But in actual Archives, though the bundles may be mountainous, there isn't in fact, very much there. The Archive is not potentially made up of *everything*, as is human memory; and it is not the fathomless and timeless place in which nothing goes away that is the unconscious. . . . The Archive is made from . . . mad fragmentations . . . that just ended up there. . . . And *nothing happens to this stuff, in the Archive.*[3]

Rather than being the site of power, the archive reveals itself as a site of weakness, for it hosts not memory but forgetting and contains only unrelated fragments. The story or meaning that would relate them into a whole is precisely missing; the archive, then, is a storyless void hollowed out by what is no more.

But Thoreau's critique of New England ways with things amounts to more than a critique of archives and museums. For what Thoreau realizes by looking in his neighbors' garrets is that there are no pure meanings, for no recollected image or idea can endure unless it is concretized in what is material; he realizes the fallacy of idealism. When Thoreau criticizes his neighbors for endeavoring to keep the meaning of a thing alive in a disembodied way by accumulating things in storage, he isn't accusing them of materialism from some idealistic point of view, as is too often presumed. Rather, I argue, he is accusing them of idealism while himself speaking as a radical materialist. For he is saying that his neighbors accumulate things not because they are preoccupied with material possessions, but because they are obsessed with ideal essences. And he is telling them that their garrets are just like museums, or in Miguel Tamen's phrasing "places where bits of matter are reduced . . . to meaning,"[4] where the corporeal is imagined to be miraculously transubstantiated into the persevering mental. Thoreau formulates for them the central paradox of museums and archives: in storing material objects, they nevertheless function as sites not of material but of immaterial culture. That is also what the Mucclasse Indians knew when they burned their dead things; they cremated them not because they were idealists careless about the destruction of the material, but, conversely, because of their commitment to embodied meanings—because they realized that just as there are no pure meanings, so matter severed from its meanings is already dematerialized, already gone and dead.

Ragged Fragments and Vital Relics

But the choice Thoreau makes in *Walden* between Concordian and Mucclasse Native American ontologies of things is conditional. That is to say, he would prefer the Indian ontology to the Concordian if those were the only two options he had. However, his description of furnishing the Walden cabin shows that what counts as a thing might include yet another possibility, because he says there that whereas he built some of the furniture in the cabin, other things were taken precisely from village garrets that hide "plenty of such chairs as I like best" (Wa, 65). In fact, Thoreau often frequented estate sales and visited his neighbors' storage, long before and after his experiment at Walden Pond. What are we then to make of this habit that distinctly contradicts both the customs of the Mucclasse Native Americans (for Thoreau doesn't burn the old things he gets from garrets) and those of the Concordians (for the things salvaged from their garrets are not relocated in another storage place)? Although neither *Walden* nor the *Journal* offers a theoretical statement explaining Thoreau's habit of frequenting storage places and acquiring archived things, his detailed description of things recovered, not just through sales but also in various (often disconcerting) ways, shows that he was experimenting with a third type of thing-ontology, one whose vitalist orientation—which I have reconstructed throughout this book—redefines what counts as a thing. In defying the idea that meanings of things need be exhausted at all, it also offers insight into how lost things and—because things receive their meaning from community—how the lost history of communal life can be recovered by what I am calling his practice of "communal mourning." In the context of the community's lost history, Thoreau's vitalist stance is articulated in a

Journal entry in which he declares his hatred for the "preserved death" of museums: "the life that is in a single green weed is of more worth than all this death. They are very much like the written history of the world" (J, 1, 465). In what way is a weed more historical than the archive, and what kind of traces does a leaf chart so that its surface can become much like the written history of the world? A single green weed writes history, because the shapes preceding and generating its leaves—the seed, the stem—are not distant or severed from it but instead generate the life that the leaves in turn vitalize. This is not an organicist metaphor. Thoreau doesn't want to say that one form smoothly becomes another, following the neat narrative order of organic development. Rather, the metaphor of the weed functions counter to organicism to suggest a radical disturbance of the narrative ordering of phenomena, for what it accentuates is that a transition whereby one organic form vanishes in the arrival of another is not in fact what happens to what lives. The weed tells a story not of linearity but of simultaneity: through the living link of the stem the seed nourishes the life of the leaves that nourish it; the leaves feed the life of what preceded them. This is less trivial that it might seem, for it says that causes neither simply generate consequences nor continue to inhabit them. Instead, less intuitively, they are regenerated by what they produce. The weed thus tells a different ontological story—one that reveals the logic of life—according to which consequences come to cause what caused them. If only life—as opposed to archives and museums—has the capacity to commemorate, as Thoreau says, it is because, in this nonlinear way, it can make what comes after keep causing, gathering anew, and recollecting what preceded it. Therefore when I suggest that what counts as a thing in Thoreau is related to his "vitalist ontology," I am arguing that Thoreau desires to recover the things that died for New Englanders by turning them into animated relics or living monuments: neither purely antiquarian nor geological sites, things in Thoreau appear as the corporeal assent of living memory.

What Thoreau comes to call a "thing" or a "relic," however, violates everything that common sense so identifies. For his new understanding of a thing confuses things with animated bodies, with water, earth, and words, while disturbingly calling living human beings relics or memorials. To outline the ontology and ethics that motivated his practice of recovery of things from others—the losses of others—I discuss three examples from his early to his later years: the obituary he published in 1837; his *Journal* description of objects recovered after *Elizabeth* was wrecked off Fire Island, killing Margaret Fuller and her family; and his description of the things he obtained from the deacon's sale. Together these examples testify to Thoreau's commitment to materialized meaning, his commitment to the past and his effort to embody it in a vital present. But above all else they connect Thoreau's understanding of the losses of others (as what everybody must work to recover), to his understanding of lost things (as what belongs to everybody, what is everybody's to find and restore).

Living relics. On November 25, 1837, a twenty-year-old Thoreau published his first text. It was an obituary to Anna Jones, an eighty-six-year-old woman, who died in the Concord poorhouse. Thoreau didn't know her but nevertheless went to interview her there in the first week of November, hoping, as Harding suggests, that she would "still be able to recall vividly incidents in the Revolutionary War." But "she was already on her deathbed" when he came and "too ill to tell him much."[1] So, even if he saw her, in the end Thoreau didn't come to know her. Yet, he decided to write an obituary for her, which appeared anonymously in the *Yeoman's Gazette*, on November 25, 1837, and which I quote here only partially:

> DIED:
>
> In this town, on the 12th inst. Miss Anna Jones, aged 86.
>
> When a fellow being departs for the land of spirits, whether that spirit take its flight from a hovel or a palace, we would fain know what was its demeanor in life—what of beautiful it lived.
>
> We are happy to state, upon the testimony of those who knew her best, that the subject of this notice was an upright and exemplary woman, that her amiableness and benevolence were such as to win all hearts, and, to her praise be it spoken, that during a long life, she was never known to speak ill of any one.

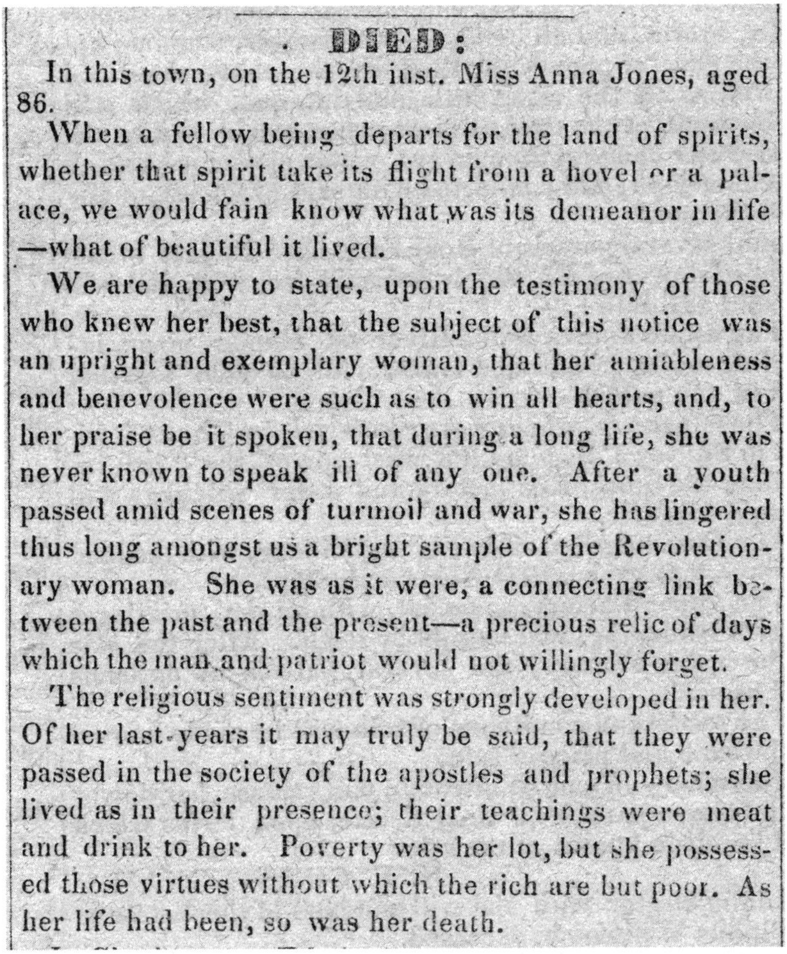

DIED:

In this town, on the 12th inst. Miss Anna Jones, aged 86.

When a fellow being departs for the land of spirits, whether that spirit take its flight from a hovel or a palace, we would fain know what was its demeanor in life —what of beautiful it lived.

We are happy to state, upon the testimony of those who knew her best, that the subject of this notice was an upright and exemplary woman, that her amiableness and benevolence were such as to win all hearts, and, to her praise be it spoken, that during a long life, she was never known to speak ill of any one. After a youth passed amid scenes of turmoil and war, she has lingered thus long amongst us a bright sample of the Revolutionary woman. She was as it were, a connecting link between the past and the present—a precious relic of days which the man and patriot would not willingly forget.

The religious sentiment was strongly developed in her. Of her last years it may truly be said, that they were passed in the society of the apostles and prophets; she lived as in their presence; their teachings were meat and drink to her. Poverty was her lot, but she possessed those virtues without which the rich are but poor. As her life had been, so was her death.

Anna Jones obituary, *The Yeoman's Gazette*. Courtesy Concord Free Public Library.

> After a youth passed amid scenes of turmoil and war, she has lingered thus long amongst us a bright sample of the Revolutionary woman. She was as it were, a connecting link between the past and the present—a precious relic of days which the man and patriot would not willingly forget.[2]

The obituary departs from the typical notices published in the *Yeoman's Gazette* or *Concord Freeman* between 1835 and 1845, which

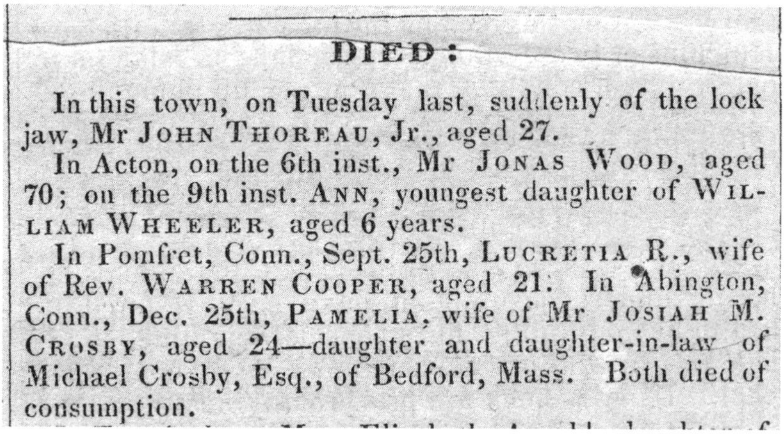

DIED:

In this town, on Tuesday last, suddenly of the lock jaw, Mr JOHN THOREAU, Jr., aged 27.

In Acton, on the 6th inst., Mr JONAS WOOD, aged 70; on the 9th inst. ANN, youngest daughter of WILLIAM WHEELER, aged 6 years.

In Pomfret, Conn., Sept. 25th, LUCRETIA R., wife of Rev. WARREN COOPER, aged 21. In Abington, Conn., Dec. 25th, PAMELIA, wife of Mr JOSIAH M. CROSBY, aged 24—daughter and daughter-in-law of Michael Crosby, Esq., of Bedford, Mass. Both died of consumption.

John Thoreau death announcement, *The Concord Freeman*. Courtesy Concord Free Public Library, William Munroe Special Collections.

Thoreau read and which must have served as the paradigm of the genre. Sometimes simple death notices ("Died: In this town," such as that announcing the death of Thoreau's brother John in the *Concord Freeman*, obituaries were usually composed by relatives of the deceased and increasingly registered the middle class sentimentalization of grief as well as, wherever possible, the social status of the deceased.

Very few were as long as Thoreau's, and they differ from his in other fundamental ways. For even though some were unsigned or simply initialed—as in the case of the obituary for Hannah Leighton, published in the *Concord Freeman* on January 7, 1842, initialed by J. T. W., whose logic and length comes closest to Thoreau's—they were typically written by a community pastor who would find a way to disclose his identity by referencing his role as the community's spiritual leader. Or else—as in the case of an obituary for Martha E. Hunt, a nineteen-year-old woman who "in great depression of spirits, and a temporary insanity, threw herself into Concord river and was drowned," from the *Concord Freeman* on July 11, 1845[3]—a long-anonymous obituary would be written by someone who knew the deceased well (in Hunt's case it was somebody who was familiar with the history of her mental instability and who had access to her now-lost diary). But regardless of the differences among them, long obituaries were written in the tradition of the Puritan typology, and for the same reasons: they turned the personal life of the deceased into an exemplar of a Christian value to be preached to the community.

None of that characterizes Thoreau's obituary for Anna Jones. It is unequivocally anonymous. In fact, Thoreau covered his traces so well that scholars discovered the obituary and attributed it to him only in the mid-twentieth century. Additionally, the obituary is written not by a relative or a friend and so is not marked by personal grief. Thus, in its anomalousness, it raises the twofold question of *what* and *why* Thoreau here commemorates.

It is not quite clear what constitutes the loss, because the writer admits that all he knows about the deceased is based "upon the testimony of others," anonymous travelers he meets as he journeys on the trail of losses. Indeed, what these travelers tell him, and what is ultimately related in the obituary, is common to the point of being impersonal, as if Thoreau were at a loss to know what to say about Miss Jones. The only thing that seems to be memorialized by this obituary is the ontological possibility of a living relic. For nothing about Miss Jones is remembered except the fact that her living yet ancient body fused the archive with the present. That such a fusion was worthy of Thoreau's memorialization suggests that the idea of a "living link" was, as early as 1837, constitutive of his understanding of the past. It signals that for him past events and meanings can be connected to the present only corporeally, as if, much like for the Mucclasse Native Americans, the past couldn't be transmitted in idealized ways—through images, dreams, or legends—but only by bodies rendering it materially contemporaneous with the present.

Yet, to say that Jones's body was relevant as a "living link" beween past and present still doesn't answer the question of what exactly is it that it links and so archives? For the body doesn't seem to be connected to any immediately relevant particularity. The obituary almost redundantly claims that Miss Jones was young during the Revolutionary War, without mentioning anything specific about it that will presumably be in need of commemoration. After that, nothing very specific is known about her, except that for more than sixty years she was our "fellow being." Like many, perhaps

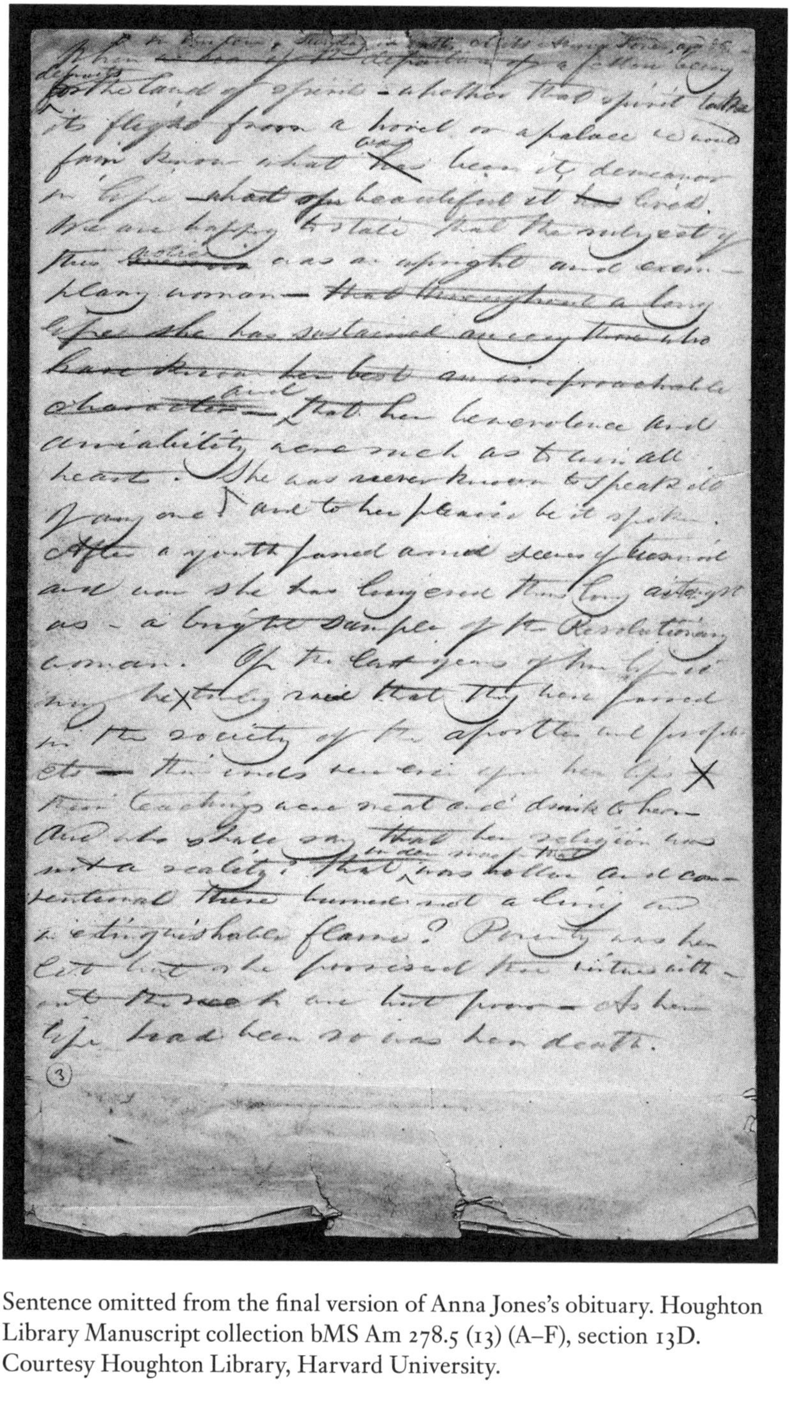

When [illegible] the departure of a fellow being to the land of spirits — whether that spirit take its flight from a hovel or a palace we would fain know what was its demeanor in life [illegible]. We are happy to state that the subject of this notice was an upright and exemplary woman — [illegible] and that her benevolence and amiability were such as to win all hearts. She was never known to speak ill of any one [illegible]. After a youth passed amid scenes of [illegible] and war she has lingered thus long as a bright sample of the Revolutionary woman. Of the last years of her life it may be truly said that they were passed in the society of the apostles and prophets — their words were ever upon her lips — Their teachings were meat and drink to her. And who shall say that her religion was not a reality? that [illegible] was hollow and conventional? there burned not a living and inextinguishable flame? Poverty was her lot but she possessed those virtues [illegible]. As her life had been so was her death.

Sentence omitted from the final version of Anna Jones's obituary. Houghton Library Manuscript collection bMS Am 278.5 (13) (A–F), section 13D. Courtesy Houghton Library, Harvard University.

like everybody, she was simply a frail "lingering" existence among us. What then does this "living link" link?

The answer to that question is implied in the second draft of the obituary—two drafts survive, the printer's manuscript being lost[4]—which contains the following variant of the last paragraph: "And who shall say that under much that was conventional there burned not a living and inextinguishable flame?"[5]

The editors of Thoreau's *Early Essays* speculate that the sentence was dropped, whether by the printer, the editor, or Thoreau himself, on the grounds of the sentiment's unorthodoxy. To the editor of the *Yeoman's Gazette* it might have appeared strange, they opine, that Thoreau would talk with passion about the conventional life of a person he didn't know.[6] While that might be correct, it is precisely the unorthodoxy of Thoreau's passion about a life that passes for insignificant, that, I argue, answers the question of what it is that Thoreau here commemorates. For when in the deleted sentence he suggests that each conventional life—a life presumably worthy of commemorating only by those who were personally related to it, but forgettable from the point of view of history and community—constitutes a unique passion generative "of the beautiful," and "inextinguishably" living under the conventional, he is in fact claiming the following. Each conventional life on its way to utter anonymity in unmourned death is composed of experiences of unparalleled beauty that makes it uniquely excellent and worthy of commemorating. To lose this uniqueness is thus a loss for all, a communal loss, which is why any member of the community—any traveler—should record, salvage, and so restore it as if it were the member's own loss.

A fragment Thoreau wrote for *A Week*, but which never made it into his book, further develops this insight, suggesting not only that all lives are ordinary yet unforgettable, but also that all extraordinary events—historical, political, religious, or scientific—are of a collective nature, brought about by anonymous communities of forgettable lives: "Fame itself is but an epitaph, as late, as false, in time. . . . See how mankind remembers Newton . . . but forgets the large circle of Newton's neighbors and acquaintances, as if indeed those were only transient appendages to him."[7] It is our ideology of fame and exceptionality that renders so many lives conventional, allowing us to pay no attention to them and thus to miss what unique beauty there was in them. Thus, lives such as that of Miss Jones are doubly lost, lost not only to death but to our memory also, and "we would fain know what was its demeanor in life—what of beautiful it lived." The office of the communal mourner works to record and commemorate them, to relink them, and thus to make their passion "a living and inextinguishable flame." In that way the communal mourner recovers the losses of others as if they were his.

Wreckages. In July 1850 the ship *Elizabeth*, "bearing Margaret Fuller . . . her husband and young son as well as Charles Sumner's brother Horace, had been wrecked on Fire Island just off Long Island."[8] As he explained in a letter to Marcus Spring of July 23, 1850, Emerson decided, without explaining why, that Thoreau was best suited to go the Island himself, to try to recover Fuller's manuscripts, if there were any: "At first I thought I would go myself and see if I could help in the inquiries at the wrecking ground, and act for the friends. But I have prevailed on my friend, Mr Henry D. Thoreau, to go for me and all the friends. Mr Thoreau is the most competent person that could be selected; and in the dispersion of the Fuller family, and our uncertainty how to communicate with them, he is authorized by Mr Ellery Channing [Margaret Fuller's brother-in-law] to act for them all. . . . Mr Thoreau is prepared to spend a number of days in this object, if necessary."[9] Charles Sumner—the abolitionist and later Massachusetts senator—also asked Thoreau, if possible, to recover the body and belongings of his brother Horace. Thus, because he is, as Emerson says, somehow "more competent" to take care of the dead, Thoreau travels to Fire Island on behalf of a whole community of mourners: Fuller's dispersed family, Emerson himself, "all [Margaret's] friends," and the Sumners.[10]

What remains of the *Journal* Thoreau kept while on the Island and immediately on returning to Concord—the surviving *Journal* entries he wrote in the Fall of 1850,[11] and a fragment of manuscript probably written while on the Island, which Steve Grice has recently recovered—suggest that Thoreau's experience as a communal mourner, "recovering property" of the dead who were not his own, as well as their bodies, allowed him a glimpse into a vitalist ontology of things.

In a manuscript leaf that Grice has recovered Thoreau lists the belongings of the dead he found on Fire Island beach:

> I found the engravings at
> Oakes'. They said that they
> were left out of the trunk. The
> gown and one article of the
> child's dress at Daniel Jones',
> Patchogue—and the other arti
> cle of the child's dress-at John Hein-
> ners in the same village. They
> said that they picked them up 1 ½ or 2 miles east of the wreck. There
> were more things there and else-
> where which were either not worth taking–or not worth waiting to see.
> I saw a calico dress like the
> pattern which I bought at Skinners
> It had silk fringes & was much torn
> also some drawers and a night gown all torn & without mark.
> At Carman Rowlands Patchogue
> I saw a gentleman's shirt.
> At Wm Gregory's in the same village
> a cart load of rags & remains of a
> childs petticoat. He said that
> his brother had much more.
> At Wm Smiths, near Patchogue
> a childs striped apron & a lady's
> skirt fringed.
> Geo Curtis of Sayville–the skirt of
> a silk dress–"lilac ground, middling
> dark stripe" which I could not wait to
> get.[12]

Thoreau was never certain about what ontological status to attribute to the clothes of the dead. Were they part of the living persons who wore them and, if so, could it be said that a person dies fully only when his clothes decay completely? "My friend died long ago-why follow a body to the grave yard? . . . There still remain his clothes shall we have a third service when they are decayed?" (J, 3, 48). But the Fire Island experience recorded on this leaf signals more confusing taxonomical dislocations, since Thoreau's language, as if in a tropological mise-en-abyme, represents a concentrated attempt to connect the ragged remnants—incoherent fragmentation each archive is composed of—to things that he can only imagine or remember. For example, the materiality of the fringed skirt is mixed in his report with the imagined silk dress with "lilac ground," and the much torn "calico" dress merges with the memory of patterns on clothes that Thoreau bought at Skinners. Thoreau's list crosses boundaries between what is formed and disseminated, destroyed and restored, actual and imagined, reassembling ragged reality into new compounds that are now inseparably attached to his life by virtue of being recorded on his list of losses. Back home, after July 29, 1850, Thoreau writes a long entry regarding how this encounter with bones and ragged belongings of the dead affected him:

> Whatever actually happens to a man is wonderfully trivial & insignificant—even to death itself I imagine. He complains of the fates who drown him that they do not touch *him.* They do not deal directly with him. I have in my pocket a button which I ripped off the coat of the marquis of Ossoli on the sea shore the other day—held up it intercepts the light & casts a shadow, an actual button so called- This stream of events which we consent to call actual & that other mightier stream which alone carries us with it-what makes the difference. . . . We are ever dying to one world & being born into another-and probably no man knows whether he is dead in the sense in which he affirms that phenomenon of another-or not. . . . There was nothing remarkable about them they were simply some bones lying on the beach. . . . I should think that the fates would not

> take the trouble to show me any bones again. I so slightly appreciate the favor. (J, 3, 95)

The material and ideal entities are in this experience so mixed that it is difficult for Thoreau to determine any differences among them.[13] Even the dead matter of Ossoli's button is endowed with the light and translucence one typically relates to what is spiritual and animated, as if to suggest that all matter is living. What we call "actual" and what we call "ideal," or else what we call "material" and what we call "mental," is differentiated as such only because of our "consent" to do so. Without our classification, things called material or ideal, dead or living are in fact intermingled so that the difference between them can't be stated. Moreover, the mixtures of life and death are incessant, making it hard for any one to determine with confidence in which existential condition he finds himself ("and probably no man knows whether he is dead in the sense in which he affirms that phenomenon of another-or not").

A long *Journal* entry written about three months later—after October 31, 1850—revisits the Fire Island experience and formulates more explicitly the bizarre idea of mental and corporeal mixtures that enable a movement from life to death and back again:

> I once went in search of the relics of a human body—a week after a wreck- -which had been cast up the day before onto the beach–though the sharks had stript off the flesh. I got the direction from a light house— . . . Pursuing the direction pointed out-I expected that I should have to look very narrowly at the sand to find so small an object-but so completely smooth & bare was the beach-half a mile wide of sand-& so magnifying the mirage toward the sea-that when I was half a mile distant the insignificant stick or sliver for which marked the spot–looked like a broken mast in the sand
>
> —As if where there was no other object, this trifling sliver–had puffed itself up to the vision to fill the void–& there lay the relics in a certain state—rendered perfectly inoffensive to both bodily & spiritual eye by the surrounding scenery—A

> slight inequality in the sweep of the shore Alone with the sea & the beach-attending to the sea–whose hollow roar seemed addressed to the ears of the departed–articulate speech to them. (J, 3, 127)

The inoffensive combination of things of different orders—flesh, bones, things, shells, stones, sand, beach grass, and water—affords the spectator the "vision" of an unexpectedly diverse ontological order. For the perceiver comes to understand how decaying relics merge into other elements as if they have been picked up by the flow of life. Thus, Thoreau's earlier statement that it was difficult to differentiate between corporeal and mental, dead and living, is now translated into the insight that what is called death is less the cessation of being than its transformation into a different manner of existence. For when he comes to the beach to attend to the dead, he finds *them* attending to the Ocean, developing different senses, assembling new ears, learning a new language, relics still living.

Obituaries. Walter Harding has suggested that there are numerous hints that following the Anna Jones obituary Thoreau wrote "other [similar] anonymous pieces . . . in the Concord newspapers," even though "no one has succeeded as yet in identifying them."[14] However, the failure to identify other obituaries in local Concord newspapers shouldn't mean, I suggest, that Thoreau in fact wrote no more, but simply that he changed his mind concerning where they should be placed. For Thoreau didn't stop writing obituaries for people he didn't know, but after 1837, started recording them in his *Journal.* The *Journal* is packed with obituaries of people most of whom Thoreau didn't know personally, but which closely correspond to the tone of that written for Jones. For instance: "On the 11th (instant,?) [May '49] on Rappogenes Falls, Mr John Delantee of Orono, Me., was drowned while running logs. He was a citizen of Orono, and was 26 years of age. His companions found his body, inclosed it in bark, and buried it, in the solemn woods" (J, 3, 13). Or: "Col. Elisha Jones was the owner & inhabitant of an estate in Weston Mass before the Revolution. He was a man of standing & influence among his neighbors. He was a tory. He had 14 sons & 1 daughter"

(J, 3, 15); similarly, "Mr Thomas Russel-who cannot be 70-at whose house on Leyden st. I took tea & spent the evening-told me that he remembered to have seen Ebeneezer Cobb a nat. of Plymouth who died in Kingston in 1801 aged 107 who remembered to have had personal knowledge of peregrine White" (J, 3, 348).

On January 22, 1856, Thoreau embeds in his *Journal* a long obituary that in many ways resembles the one he wrote for Anna Jones, but with one big difference: the "old citizen" that he buries and commemorates this time will in fact be a large elm tree:

> I have attended the felling and, so to speak, the funeral of this old citizen of the town . . . I was the chief if not the only mourner there. I have taken the measure of his grandeur; have spoken a few words of eulogy at his grave, remembering the maxim de mortius nil nisi bonum (in this case *magnum*). But there were only the choppers and the passers-by to hear me. Further the town was not represented; the fathers of the town, the selectmen, the clergy, were not there. But I have not known a fitter occasion for a sermon of late. Travellers whose journey was for a short time delayed by its prostrate body were forced to pay it some attention and respect. . . . How have the mighty fallen! Its history extends back over more than half the whole history of the town. Since its kindred could not conveniently attend, I attended. Methinks its fall marks an epoch in the history of the town. . . . Its virtue was that it steadily grew & expanded from year to year about to the very last. How much of old Concord falls with it! The town-clerk will not chronicle its fall—I will—for it is of greater moment to the town than that of many a human inhabitant would be . . . we take all possible pains . . . to obliterate its stump, the only monument of a tree which is commonly allowed to stand . . . link that bound us to the past is broken. . . . How much of Old Concord was cut away with it! A few such elms would along constitute a township. (Jn, 8, 130–131)

The obituary to the tree tells us explicitly that the historical and commemorative is embedded in a living body that connects, relays, and revives different temporal sites. The tree's longevity functions, as in the case of Jones, as the "living link" that corporeally connects temporally distant moments without ceasing to change, mature, and develop ("its virtue was that it steadily grew"). That nobody in the anthropocentric community registers its death as a loss in the annals of human history ("how much of old Concord falls with it") is read by Thoreau as a failure to attend to the archives of that community's time, thereby losing them to forgetting. And it is that loss that Thoreau tries to restore through his funeral speech and a mourning that relinks his present to the tree's past. It is he, not the town clerk, who chronicles its death; it is he, not the clergy, who sermonizes; and he who, as "chief mourner," represents the citizens who remain unaware of the loss. His performance of those various tasks turns him into a veritable community, and, attending to the community, he stops the travellers who meet him on their journey to instruct them about what they have lost by losing the tree, putting them also on the trail of recovery. In so doing, he hopes to turn the broken link into a "stump" that will be a commemorative text, one composed of rings of the past relinked by the tree's very corpse into a simultaneity with the present.

If Thoreau's *Journal* practices communal mourning by chronicling the deaths of members of the community, it is also affected by the experience of salvaging the wreckage on Fire Island. Thoreau's *Journal* is in fact a gigantic ontological admixture echoing what he witnessed on Fire Island. It constantly confuses what is with what isn't, and itself functions as a vital relic, a memorializing collage of things, moments and words recollected from diverse unnamed persons, some dead, some still living. The *Journal* often speaks about the dead and often describes abandoned things: "A lone Indian woman without children-accompanied by her dog" (J, 3, 93); "A woman who wished to go to Nashua was left behind at Groton Junction" (J, 5, 336); "An old man who used to frequent Walden 55 years ago when it was dark with surrounding forests tells me that in those days he sometimes saw it all alive" (J, 6, 209); "I can just remember an old brown-coated man who was the Walton of this stream . . . his old experienced coat hanging long and straight and brown as the yellow pine bark, glittering with so much smothered sunlight" (W, 24–25). Hundreds of commemorating sentences similar to those are mixed into the *Journal* with detailed descriptions of landscapes, plants, or animals, as well as human-made tools or objects that Thoreau recovers while searching through estate sales.

Estate sales. In 1854, for instance, while at the deacon's estate sale described in *Walden*, Thoreau looks closely at "a pair of old snow shoes . . . almost regularly sold on these occasions," but decides to acquire instead "Ephraim Jones's decaying 'Waste Book' from 1742, found in Brown's garret since his death" in a condition closely resembling fossilized remnants (J, 7, 251).

He understands the waste book as a record of obituaries to things, and commemorating the life of a community by listing things traded in the winter of 1742: "The articles most commonly bought were mohair . . . rum, molasses, shalloon, fish, calicoe, some sugar a castor hat, almanac, psalter, & sometimes primer & testament, paper, knee-buckles & shoe-buckles-garters & spurs by the pair Deer skins-a fan-a cartwhip-various kinds of cloth . . . gloves-a spring knife-an ink horn" (J, 7, 250).

By mixing phenomena of different orders—food and ink, deer skin and testament, sugar and primer—Jones's book chronicles things the community had and lost, following the trail of communal losses. But as Thoreau fills pages of his *Journal* with extracts from the waste book, he mixes obituaries to things with the living text he is composing, transforming the waste book into a vital relic. Thus the recording, and the recombining of what was recorded into new ensembles, is turned into a process of revitalizing.

Throughout his *Journal* Thoreau inserts short chronicles of people dead and annals of things lost into long descriptions of nature in bloom and decay, which turns it into a waste-saving book of a certain kind also, meant to revitalize through commemorating, by performing communal mourning as Thoreau imagines it: the gathering of ragged memories into compounds of new life. His *Journal* is like a totality of things and beings mixed in together, designed to recreate new sensations by memorializing the old. It produces a memorial mingling of beings and entities of different ontological natures—

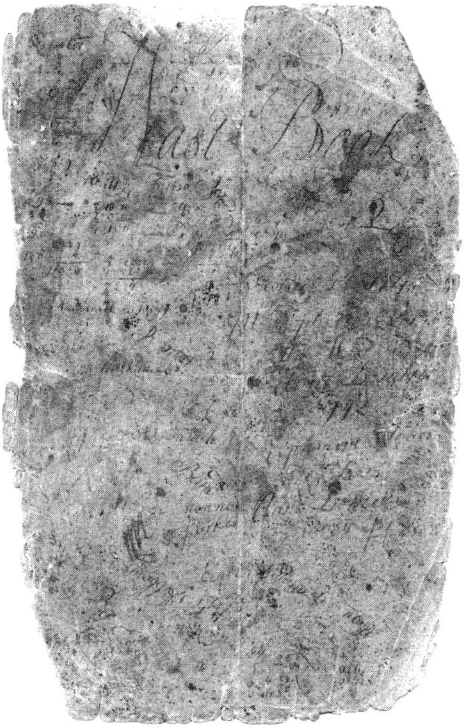

Ephraim Jones waste book—cover. Courtesy Concord Free Public Library, William Munroe Special Collections.

Ephraim Jones waste book—remnant. Courtesy Concord Free Public Library, William Munroe Special Collections.

oaks, huckleberries, owls, political events, crickets, pines, mad women's cries, the barking of dogs, dashes, arrow heads, frogs, the silence of a moose, smiles of a brother long dead, old snow boots, obituaries, flowers, sand, poems, decaying leaves, water, waste books, beans, bones, memories of an April shower, conversations, and buttons—all of that combines into a new flow we call the "*Journal.*" Because it records so many losses one could say about it what Breitwieser says about *Cape Cod*, that it is made of "wreckage from the start."[15] Yet that doesn't make it a pile of disconnected fragments

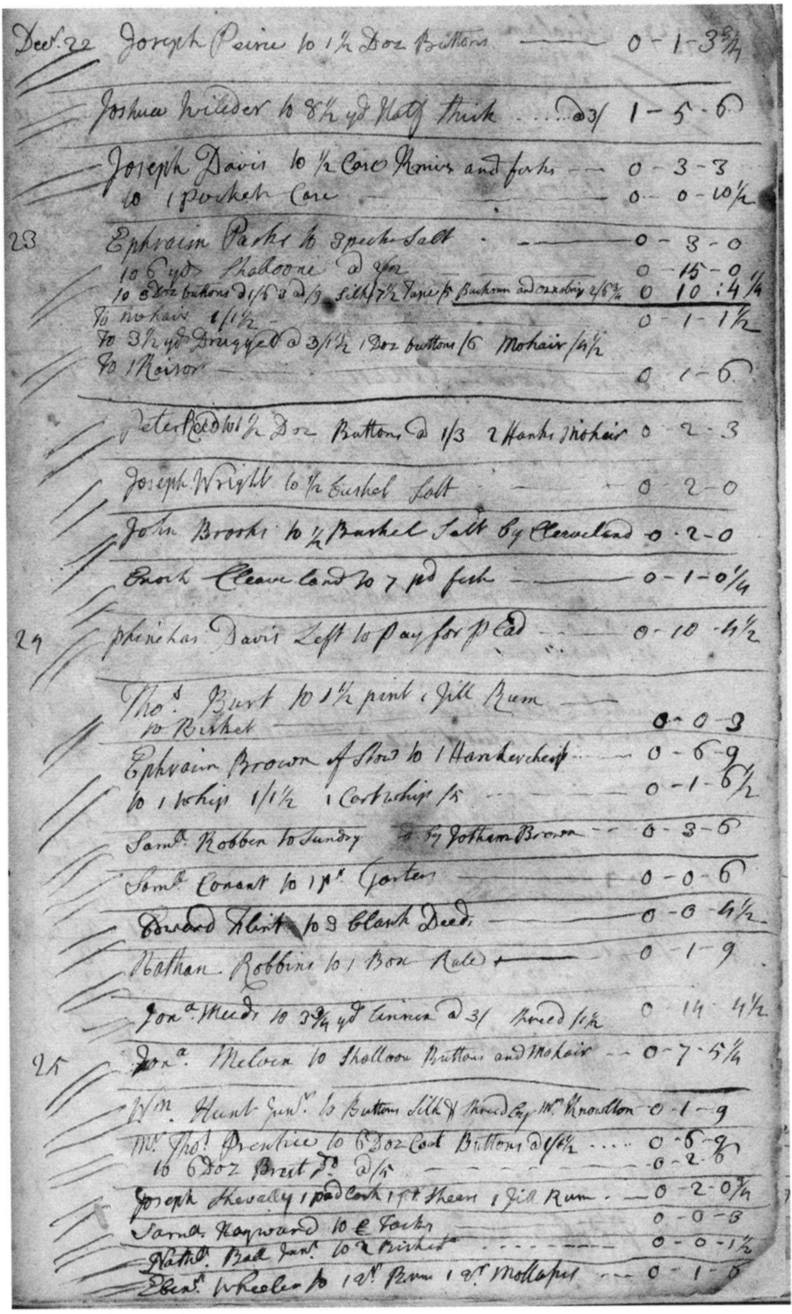

Decr. 22 Joseph Paine to 1/2 Doz Buttons — 0-1-3 3/4
Joshua Wilder to 8 1/2 yd Half thick .... @3/ 1-5-6
Joseph Davis to 1/2 Case Knives and forks — 0-3-3
to 1 Pocket Case — 0-0-10 1/2
23 Ephraim Parks to 3 peck Salt — 0-3-0
to 6 yds Shalloone @ 2/6 — 0-15-0
to 3 Doz buttons @ 1/6 3 @ /9 silk/7 1/2 Tape ... Buckram and oxenbrig 2/6 3/4 0-10-4 1/4
To mohair 1/1 1/2 — 0-1-1 1/2
to 3 1/2 yd Drugget @ 3/1 1/2 1 Doz buttons /6 Mohair /4 1/2
to 1 Raisor — 0-1-6
Peter Reed to 1/2 Doz Buttons @ 1/3 2 Hanks Mohair 0-2-3
Joseph Wright to 1/2 Bushel Salt — 0-2-0
John Brooks to 1/2 Bushel Salt by Cleveland 0-2-0
Enoch Cleaveland to 7 lb fish — 0-1-0 1/4
24 Phinehas Davis Left to pay for ... — 0-10-4 1/2
Thos. Burt to 1 1/2 pint 1 Jill Rum —
to Bisket 0-0-3
Ephraim Brown of Stow to 1 Hankercheif — 0-6-9
to 1 Whip 1/1 1/2 1 Cartwhip /5 — 0-1-6 1/2
Saml. Robbens to Sundrys by Jotham Brown — 0-3-6
Saml. Conant to 1 pr Garters — 0-0-6
Edward Flint to 3 blank Deeds — 0-0-4 1/2
Nathan Robbins to 1 Box Rale — 0-1-9
Jona. Meads to 3 3/4 yd Linnen @ 3/ thread /1 1/2 0-14-4 1/2
25 Jona. Melven to Shalloon Buttons and Mohair — 0-7-5 1/4
Wm. Hunt Junr. to Buttons Silk & thread by Mr. Knowlton 0-1-9
Mr. Thos. Prentice to 6 Doz Coat Buttons @ 1/4 1/2 0-6-9
to 6 Doz Brest Do. @ /5 — 0-2-6
Joseph Chevally 1 pad Lock 1 pr Sheers 1 Jill Rum — 0-2-0 3/4
Saml. Hayward to 2 Forks — 0-0-3
Nathl. Ball Junr. to 2 Bisket — 0-0-1 1/2
Ebenr. Wheeler to 1 qt Rum 1 qt Mollases — 0-1-6

Ephraim Jones waste book—page. Courtesy Concord Free Public Library, William Munroe Special Collections.

simply patched together. For when the wreckage of Jones's "Waste Book," for instance, is mingled with a discussion of turning leaves, frost near Worcester, or quotes from De Quincey's "Historical Essays," it stops being a ragged fragment decaying in Deacon Brown's garret, and, just like the dead and their flotsam on the beach at Fire Island, is rearticulated anew into a living flow. That commemorating flow constitutes a new life and turns Thoreau's *Journal* into memorial moments from the lives of many, human and nonhuman, living and dead, natural and artificial.

The Huron-Wendat Feast of the Dead

In 1850, on Robert F. Sayre's reconstruction, Thoreau began his Indian Notebooks, a vast collection of notes amounting to twelve volumes that he took from readings on Native American cultures. Sanborn would later call those notes "great heaps of matter."[1] The readings ranged from history, archaeology, and ethnography to more specific accounts of mythology and spirituality, and they dealt with Native American cultures from Canada to Mexico, from the East to the West coasts of the Americas.[2] Although most of the notes taken from a single source comprise no more than several pages of a notebook, those taken from *Jesuit Relations*—accounts of Jesuit missionary encounters with Huron Indians who lived in southern Ontario and today's Michigan—occupy almost a quarter of the recorded material. Thoreau started reading *Jesuit Relations* in late 1852. As Sayre remarks, he read *Relations* "as systematically as someone doing research." Moreover, taking into account that notes from *Relations* play such a central role in the Notebooks it is not impossible, as Sayre remarks, that research on *Relations* became for Thoreau a "project carried on for its own sake."[3] The end of Notebook 6 and a large part of Notebook 7 contain 141 pages of notes from *Relations*, whereas by June 21, 1858—when Thoreau wrote in Notebook 11 that "he was in Cambridge and had examined 'unbound added Relations' "—he took "nearly 200 pages more of notes . . . until he stopped with the ones for 1693–1694."[4] Most of the notes from the first cluster, however, are from accounts related by the earliest missionaries, Paul Le Jeune (1634), François Le Mercier, Barthèlemy Vimont, and Jean de

Brébeuf. When Sayre reports that from their accounts Thoreau took "pages of notes on [Huron] farming, hunting, marriage and sexual practices, burial customs, house-construction, medicine, games, magic"[5] he is not incorrect. Yet, to phrase it in that way gives the impression that Thoreau was equally interested in all those topics, whereas in fact most of the notes are on Huron's burial practices as reported by de Brébeuf, and above all de Brébeuf's description of the Huron Feast of the Dead, which Thoreau copied almost in its entirety, devoting to it more than twenty pages.

De Brébeuf's description of the Feast is a complex narrative relating Huron funerary customs interpreted through the lens of Huron spirituality. De Brébeuf describes, and Thoreau records, how a Huron family, each of which "has one who is taking care of the dead,"[6] would first flex a corpse into the fetal position—"put him on a peloton, almost in the same position in which infants are in the belly of the mother" (Notebook 6)—wrap it in a beaver robe and then take it to the local village cemetery for burial on a scaffold holding a bark tomb on the surface of the earth. That first funeral above the earth was, however, only temporary. As Thoreau records, the scaffolding "sepulchers are not perpetual, as their villages are not stabile" (Notebook 6). The second funeral, entailing inhumation, involved the depositing of all those who had been "temporarily" buried above the earth in a single grave. That second ceremony was followed by what is in fact called "the feast of the dead, which is made usually every dozen years" (Notebook 6). Such an event could happen after the "elders and notables of the country assemble to decide on a definite time to hold the feast that will be convenient for the whole country and foreign nations that may be invited."[7] Once the place and time were determined, all families from the nation, unknown to one another (being located in scattered villages) "transport all the bodies to the village where the common grave is located."

They "load" their dead "upon their own shoulders, and cover them with the finest robes which they have" (Notebook 6), turning the feast into a massive collective burial uniting the entire Huron nation, their living as well as their dead.

At the site of the common grave the communal preparation of the dead for the burial would begin. Thoreau quotes in detail from de Brébeuf's description of how the Huron would scrape and clean the

Ancestral Huron Feast of the Dead; Author: J.-F. Lafitau. Source: *Moeurs des sauvages amériquains*, II, 1724.

bones of all the corpses that had been brought to the village, then lay them together in the common grave. In the nondualistic Huron theogony preparation of the bones was esteemed particularly important, because bones were considered not just vehicles hosting souls but were indeed indistinguishable from souls. Contemporary historians and anthropologists recognize the semantic kinship between bones and souls in the Huron language, which maintains a single origin for mind and body while also allowing for their difference (*atisken*, bones; *esken*, souls).[8] But de Brébeuf heard it differently. In his account, as recorded by Thoreau, the Huron used the word bone (*atisken*) not as a similar word to that for "soul" but as meaning "soul," rendering souls and bones indiscernible:

> Returning from this feast with a captain . . . I asked him why they called the bones of the dead *Atisken*. He gave me the best explanation he could, and I gathered from his conversation that many of them think we have two souls, both of them being divisible and material, and yet both reasonable. One of them separates itself from the body at death yet remains in the cemetery until the Feast of the Dead, after which . . . according to a common belief, it goes away at once to the village of souls. The other is more attached to the body and, in a sense, provides information to the corpse. It remains in the grave after the feast and never leaves, unless someone bears it again as a child.[9]

Preparation of the bones was thus nothing less than preparation of the souls, and it was meant to liberate the two souls that each human possesses for new lives, one unearthly and the other still earthly. For at the moment of final burial, one of the souls was released—together with the souls of all those buried on the same occasion—to the village of eternally living souls located in the sky, which suggests that the Huron believed that a soul of the dead shouldn't start a new life alone, as it were, from scratch, but in the community with others immediately able to constitute themselves as a village. That is how in the Huron theogony those who were lost to the living would begin to be recovered into the community.

The other soul of the dead—the one that doesn't ascend to the skyscape of a new village—remains incorporated in the community also; it remains in or near the ossuary, close to the living (occasionally enacting, as de Brébeuf puts it, "metempsychosis," as for instance when a child bears it). Thus, for the earthly community of the living, the dead are not simply released into a transcendent eternity but, thanks to the proximity of their other soul to the earth and the village, remain in the neighborhood of the living and of their own material bones. It is as if the Huron imagined the dead as never quite lost, never quite departed from the life of the living, enabling the living to always stay on the trail of their own losses.

Finally, the Feast of the Dead restructures and reenacts communities of loss in yet another way. For after the bones have been buried in the common grave, there begins what Seeman calls a "great feasting." Those who have come from surrounding villages and who therefore don't know one another now gather to share a meal, served from the same kettle and prepared by the villagers hosting the burial ceremony. As Seeman puts it, "to feed one's friends and . . . strangers out of one's kettle was the characteristic gesture of Wendat [Huron] hospitality."[10] The ritual of communal eating, however, was only the closure of a week of highly structured social engagements, all of which were in preparation for the feast. Thoreau takes long notes from de Brébeuf's description of the rituals he witnessed in Ossossané, which I quote only partially, inserting in brackets the passages from de Brébeuf that Thoreau shortened:

> A day or two before setting on for the feast they carried all their dead to the largest cabin in the village where the captain made for them a magnificent feast—in the name of the dead capt. whose name he has. The relations made presents . . . of robes—porcelain & kettles. The Capt. sung the song of the dead Captain. . . . At the end they went out of the cabin imitating as they said the cry of the souls haéé, haé. . . . The seven or 8 days before the feast were passed in assembling with the souls & the strangers who were invited to it. . . . All the while the living were making presents to the young in consideration for the dead. The women contended with the bow for some prize . . . the young men contended for prizes. . . . They were continually arriving with the souls i.e. bones—packed on their back under fine robes [Every day more souls arrived. It is heartening to see these processions, sometimes of two or three hundred persons, everyone carrying their souls—in other words, their bones—done up in parcels on their backs]. Sometimes

> they arranged their packet in the form of a man adorned with collars of porcelain with a beautiful garland of . . . red fur.—uttering the cry of the souls by the way—This cry appeased the carriers greatly for their burden though of souls was heavy [On setting out from the village, the whole band cried out *haéé, haé*, and repeated this cry of the souls along the way. According to them, this cry greatly consoles them. Without it, this burden of souls would weigh very heavily on their backs and cause them a pain in the sides for the rest of their lives].[11]

The Huron lament is ambiguous in that the lamenting cry is not of the living but of the dead; *haéé, haé* is said to be the cry of the soul-bones themselves, hosted and channeled by the living and addressed to those who are encountering one another for the first time and who are in this way introduced to one another through the cry of the dead. The new encounters of the living happen through the voices of the dead. As Gabriel Sagard, one of the first Western witnesses of the Feast interpreted it in the 1620s: "by means of these ceremonies and gatherings they contract new friendships and unions amongst themselves, saying that, just as the bones of their deceased relatives and friends are gathered together and united in one place, so also they themselves ought during their lives to live all together in the same unity and harmony, like good kinsmen and friends."[12] Thoreau's thought here comes full circle. From his early engagement with Greek lament—with *aei, aei*, a "tireless always"[13] that vitalizes—to his late interest in the Huron *haéé, haé*, that enables the dead to speak through the living, he tells the story of communities of shared losses, of communal ties established and agitated by loss.

Coda: Pythagoras's Birds

A Week's epigraph comes from the first book of Ovid's *Metamorphoses*, where Ovid narrates the birth of the earth out of Chaos and its slow shaping by oceans and rivers, the forces of transformation, dissemination, and mixing. In the last book of the *Metamorphoses* Ovid introduces Pythagoras, charging him with the task of articulating his philosophy of change, the philosophy that had guided Ovid himself throughout the book. In Pythagoras "metamorphosis" is a name given to forces that defy death:

> All is subject to change and nothing to death . . .
> Nothing retains its original form, but Nature, the
> goddess
> Of all renewal, keeps altering one shape into another.
> Nothing at all in the world can perish, you have to
> believe me;
> things merely vary and change their appearance. What
> we call birth
> is merely becoming a different entity; what we call
> death
> is ceasing to be the same. Though the parts may
> possibly shift
> their position from here to there, the wholeness in
> nature is constant.[1]

Much the same as in Thoreau's philosophy, what is called death in Pythagoras never interrupts life but is instead a change, a process of rearticulation of forms whereby parts of a being get differently mingled. For Pythagoras this ongoing change of all beings—what he understands as the generalized "becoming a different entity" of every being—means that each being trespasses into other beings that are, on the surface of it, both formally (in terms of shape and size) and qualitatively (in terms of species or even forms of life) quite unrelated to it. Never a great respecter of taxonomical divides, Pythagoras details how all creatures are en route to forms quite alien to them; thus, "mud has seeds which produce green frogs"; out of a crab, buried in the earth after its "branching claws" are removed, "a scorpion will shortly emerge"; "a war-horse buried in soil will become the source of a hornet";[2] "when a human backbone has decomposed in a well-sealed tomb, / some think that the spinal marrow is then transformed / to a snake." Yet, even though "all these creatures can trace their beginnings to alien forms," it is to birds that Pythagoras affords the privileged status of emblematizing the capacity of all creatures to appear as they are out of something they are not. Thus, just like "the bird of Juno," a peacock that "carries the stars in her tail," and the "eagle" who bears "doves of beautiful Venus," the whole "avian family" proves that life proceeds by metamorphoses, for all birds are born from yolks, that is, from something they obviously are not: "the whole of the avian family—who would suppose they could grow from the yolk of an egg, if one didn't know for a fact that they did?"[3] Birds become exemplars of transformation, because they teach us the secret of metamorphoses, namely, that they are not abrupt and discontinuous, but slow and continuous. Metamorphoses are not about magical and sudden shape-shifting but are the result of long cumulative processes. The alien otherness that creatures emerge from and progress into is hosted by them, just as Juno hosts the stars, eagle doves, or more generally, all eggs host future birds. Each individual is thus always at least divided and even multiple, accommodating incongruous natures that grow in it, taking it slowly to another form and in another direction. This continuous metamorphosis that eventually turns beings into forms alien to them is termed by Pythagoras "winged."[4] "Winged" metamorphosis means precisely that radical otherness is generated slowly and smoothly, as if creatures were flying gently and unassumingly into other forms. A general metamorphic flux of beings ("In the whole of the world there is nothing that stays unchanged. / All is in flux")[5] is thus imagined as a general flight. All creatures are birdlike.

Humans too, for they also have "winged souls" enabling them to fly into other beings: "We too are part of the world . . . we also possess winged souls. We are able to make our / abodes / inside wild beasts and to hide away in the hearts of the cattle."[6] There is no contradiction between Pythagoras's claim that everything is in metamorphosis (that everything changes its form by letting go of its identity), and his claim that everything is in metempsychosis (that all souls fly into other bodies, thus preserving their identity). For as scholars explain and he clearly states in Ovid's rendering, if metempsychosis is not specifically human ("we *also* possess winged souls"), if everything is souled, it is because souls are not imagined here in any post-Platonic, Christian manner, as highly individualized psyches. It is not that a tree, a flower, a dog, and a human have interiority that moves into other beings, persevering in its souled self-sameness while experiencing transformation of the corporeal form. Instead, as Gilbert Simondon puts it, the three souls that Pythagoras differentiates (human, animal, and vegetal)

> are considered to be of the same nature. It is the body and its functions which establish the differences between the various ways of living for a soul incarnated in a human body, the manner of living for a soul incarnated in a vegetal body, or a soul incarnated in an animal body. What emerges out of these first doctrines of the identity of souls and their community in nature is metempsychosis . . . an ancient doctrine that supposes the soul is a living principle not attached to the individuality of one specific existence or another. An animal soul can serve to animate a human body, it can reincarnate itself in a human body, and a soul that has passed through a human body, after a human existence, can perfectly come back into existence in vegetal or animal form.[7]

Souls are thus substantially identical but possess qualitatively different vital principles. What differentiates them as they emerge from the originary community of life into which they always return is a particular incarnation they receive from the body. The body determines the way life will live for a while, before it moves into a different material constellation; that is why an animal soul can "reincarnate itself in a human [or vegetal] body," experiencing different ways of living in a generalized communal transformation of qualities. For that reason, as Simondon explains, the soul in Pythagoras is not "a properly individual reality" but a force that flies and passes through, "individualizing itself for a certain length of time under the guise of a determined existence."[8] Determinate existences are only postures or modes of souls that mask this fundamental fact: that in their communal fusion that is life, souls go on and persevere after all specific determinations have been blurred or cancelled. According to Simondon this is the major heuristic contribution of Pythagoras's doctrine of winged metempsychosis: "one shouldn't neglect the heuristic contribution of such a doctrine . . . because through this belief the possibility of the continuance of life becomes manifest, the reality of the passage of something else, which is more than the individual."[9] Souls—either vegetal, animal, or human in quality—are durable and interchangeable even if individually determined. Individuals come to be out of preindividuated life that is immortal and continues after its individual forms have been shattered or morphologically changed through metamorphosis. It is precisely that idea of preindividual and nonindividuated communal life in Pythagoras's teaching that the West will forget following Christian intervention: "it is this idea of a durability of souls, of the virtual immortality of souls that will be taken back up by the spiritualistic doctrine of Christianity but with an additional innovation that is obviously quite important: the individuality, the personality of the soul."[10]

Pythagoras himself clearly maintains that souls are vital forces rather than personalized minds when he says that in "wander[ing] from place to place," and "crossing over from beast to man" the soul "never perishes wholly."[11] Not to perish "wholly" is precisely to survive otherwise, not as a self-same individual mind but as something else. But by the same token, not to perish "wholly"—and that is the sense of the "continuity" of life that Simondon detects in Pythagoras—means that something of what was remains in what is or in what is going to be. Every single being is disseminated into many others through what is called its "death." It becomes vegetal, animal, and human, and precisely what connects them, the one life that they share, establishing a living community among them. And just as the migration of a single being establishes living communities among many beings, so every single being is a community in itself, for it is made of many ways of life and a variety of former beings. Thus, ways of life in their many incarnations don't obliterate the past but carry it in their future incarnations as remnants, traces of past incarnation of souls, as if life, just as Thoreau said it would, keeps on by commemorating itself. And just as if each living being were a cluster of traces of past beings that have flown into it, each being a little flock of birds and a living memorial.

Appendix I: Freud and Benjamin on Nature in Mourning

THOREAU'S ASTONISHING theory of grief, which brings grief into the vicinity of natural life, fundamentally differs from grieving practices enacted in antebellum America as well as from Freud's analysis of mourning. As Dana Luciano argues, the American nineteenth-century pragmatics of mourning were structured around absorption of affliction into linear time, thus facilitating the "feeling" of the historical. Luciano understands American strategies of mourning to be generated by:

1. An identification of the time of mourning as "afflicted time," by which she means longer, shorter, or lingering sequences of time allotted to mourning, during which the mourner recovers.[1] Afflicted time is thus a period the psyche needs to reinvest its vital energies and to confirm the "essentially narrative organization of the body and its uses."[2]
2. A linking of the "feeling body to history through the incitement of grief," grief being understood as the power to remove "any obstacles to the forward-moving flow of time."[3]
3. A sentimentalization of grieving to freeze it into transcendence. "Sentimentality's inclination toward eternity and transcendence" thus results, in Luciano's understanding, in transforming the time of mourning—the afflicted time—into a "sacred time."[4]

Thoreau's theory of mourning differs from such nineteenth-century practices on all counts. First, in Thoreau only personal grief, which is denigrated and considered almost unethical in its concern for the interest of the survivor, can be identified with afflicted time. In contrast, mourning proper (perpetual grief) is the motion of natural (ahistorical) life, neither a durational segment designated for suffering the loss nor time somehow "sacralized" as eternal. Second, because mourning in Thoreau is an activity performed by nature, grieving is not a phenomenon that is temporarily hosted by a mind capable of self-reflection, which would link it to the body's "internal" history. Instead of being a privileged site for the formation of a personal history, and because it is a function of exteriority, mourning is in Thoreau related to physiology, botany, geology, and geography. Finally, because it is identified with the universal life of all creatures, mourning is not sentimentalized and so cannot be performed by the reification of a finite feeling into a sacred thing. Moreover, if, as Mary Louise Kete suggests, sentimental mourning synecdochically seeks a type of restoration "through the production and circulation of a remembrance or token of the lost person," then Thoreau's understanding of mourning functions in direct contrast, for the agitation of life that grieving represents for him is negated or arrested when it is hosted by a finite object. In other words, in Thoreau, loss cannot be represented, albeit synecdochically, but must instead be lived.[5]

But more important than outlining Thoreau's difference from the antebellum culture of mourning is the task of outlining how his theory of grieving differs from Freud's. Such a task seems to me imperative not just because Freud's understanding of mourning comes to dominate twentieth-century philosophies of grief—rendering us virtually incapable of understanding grief in any other way—but also because, at first sight, Thoreau's theory of grief seems to a large extent to coincide with and so anticipate that of Freud. Freud famously understood mourning as a labor through which the ego tests reality to corroborate its ominous feeling that the loved object has been lost ("reality-testing has revealed that the beloved object no longer exists, and demands that the libido as a whole sever its bonds with that object"). To fulfill this requirement the ego then gives itself time—often an uneconomically long time—to face a new reality structured by loss. That time is identified precisely as the time of mourning: "Normally, respect for reality gains the day. Nevertheless its orders cannot be obeyed at once. They are carried out bit by bit, at the great expense of time and cathectic energy, and in the meantime the existence of the lost object is psychically prolonged. Each single one of the memories and expectations in which the libido is bound to the object is brought up and hypercathected, and detachment of the libido is accomplished in respect of it."[6] Finally, "when this work [of detaching the ego from the loss] has been accomplished" the integrity of the ego is restored, and the boundaries between its psychic space and the lost object are enforced. Thus, it would appear, as in Thoreau, where personal grief consolidates the survivor's self, successful mourning in Freud is about saving the mourner's ego by detaching it from the world of the dead.

Yet despite these similarities, in my view a fundamental difference separates Thoreau's understanding of mourning from Freud's. In Thoreau's understanding, human or personal grief confines a loss to the crypts and pyramids of the psyche, burying it in the psychic cemetery of the mind, where it will continue its spectral life. If this human grief has to part with the loss, in what Thoreau calls its "unfaithfulness," that is not because the lost object is simply let go, but because the psyche can't access it, even if it is centrally present in the psyche and the psyche does not want to let it go. In personal grief a strange relation between the psyche and the lost object is established, in which the psyche is separated from what it contains. According to this model, then, the psyche—far from either denying or dispensing with the loss—knows its loss and desires to bring it closer to itself, but cannot (Just like the mind, in Thoreau's example from *A Week*, sees its own shadow but can't mingle with it).[7] In Freud's mourning, in contrast, the psyche has the access to its loss from which it then desires to distance itself; as Freud puts it, the loss that "persisted in the psyche" has to be disinvested, and the cathectic energy reinvested in another object (Freud identifies this "reinvestment" as the "normal" result of mourning and describes it as the "withdrawal of the libido from this [lost] object and its displacement on to a new one").[8] From that perspective Thoreau's personal grief moves in the opposite direction to Freud's mourning: while for Thoreau the life of the psyche is focused on a loss that it can't reach, and while this focus is what keeps it going, as if revitalizing it, for Freud life (libido) minimizes its focus on the loss by taking its vitality away and giving it to other objects. Thus, whereas for Thoreau the mourner's self is maintained only by manufacturing visible but inaccessible psychic crypts, for Freud it is consolidated only when the shadow of the loss that falls on it is dispelled through the replacement of what was lost by something newly found, a process that Alessia Ricciardi calls Freud's " 'modernist [mourning] machine' perpetually working to replace its objects."[9]

Similarly, a difference in orientation and dynamic separates Thoreau's perpetual grief from Freud's melancholia, even though, at first sight, they also seem to obey a similar logic. Both are stubbornly committed to the loss: melancholia, on Freud's account, testifies to a "rigidity" of the psyche in the process of reality-testing, as if the melancholic were inflexibly refusing to accept the fact of reality altered by the loss. However, this is not to say that the melancholic delusionally refuses to understand that the loved object is lost, rather, that he refuses to leave the loss, as it were, and to shift his life (libidinal energy) to another object in reality. In Freud's account of melancholia, the affects attaching the psyche to the lost object, do not, once that object is lost, get redirected to another object in the world as they do in his account of mourning. They therefore fail to keep the ego invested in the world external to it and instead withdraw from the world into the ego. Having nothing left to bind him to the world, the melancholic thus loses the world. The libidinal energy that used to be attached to the lost object, relocated into the ego and representing for the ego everything that object was, now gets attached inside the psyche to the only object available there, the ego itself. Overwhelmed with and attached to what the lost object represented to it, the melancholic ego identifies with its loss. As Freud understands it, through its identification with the loss, the melancholic ego depersonalizes itself into that loss:

> The object-cathexis proved to have little power of resistance and was brought to an end. But the free libido was not displaced on to another object; it was withdrawn into the ego. There, however, it was not employed in any unspecified way, but served to establish an identification of the ego with the abandoned object. Thus the shadow of the object fell upon the ego, and the latter could henceforth be judged by a special agency, as though it were an object, the forsaken object. In this way an object-loss was transformed into an ego-loss and the conflict between the ego and the loved person into a cleavage between the critical activity of the ego and the ego as altered by identification.[10]

Melancholia, then, is a grief that erodes the grieving ego into what has been lost. Thus, it might be argued, Thoreau's perpetual grief and Freud's melancholia both describe a commitment to loss that leads to a weakening of the mourner's self or, moreover, to a selflessness provoked by the psyche's effort to inhabit the lost object and thus to lose itself in it. As Freud also puts it: "the melancholic displays something else besides which is lacking in mourning . . . an impoverishment of his ego on a grand scale. In mourning it is the world which has become poor and empty, in melancholia it is the ego itself."[11]

But the selflessness of Freud's melancholic, generated by his getting lost in the loss, bears only a superficial resemblance to Thoreau's perpetual grief. The fundamental difference between these two accounts of grief lies in what it is that the mourner loved in the loss, or, what it is that the lost object represents for the mourner. Freud's hypothesis is that, unlike a healthy mourner, the melancholic's choice of loved object is narcissistic: "the disposition to fall ill of melancholia (or some part of that disposition) lies in the predominance of the narcissistic type of object-choice."[12] In other words the ego disposed to melancholia will not invest its libidinal energies in a radically different other, in whom it can't recognize anything that it appreciates about itself; it will instead invest in an other in whom it recognizes itself. What the melancholic loves in the other, then, is precisely himself:

> The narcissistic identification with the object then becomes the substitute for the love-investment. . . . This substitution of

> identification for object-love is a significant mechanism for the narcissistic illnesses. . . . It naturally corresponds to the regression of a type of object-choice to original narcissism. . . . So melancholia derives some of its characteristics from . . . the process of regression from the narcissistic object-choice to narcissism. . . . The libido that has been withdrawn from the external world has been directed to the ego and thus gives rise to an attitude which may be called narcissism.[13]

The melancholic's narcissistic identification with the loved and now lost object; or, the love for himself that he finds in the lost object, elucidates what the melancholic mourns in what he lost. For if what he loved in the lost object was himself, then his melancholic mourning for the loss reveals itself as his mourning for the part of himself he will have lost when he lost the loved object. The melancholic's refusal to let the loss go is thus a refusal to let a part of himself go, his resistance to self-depletion. When he withdraws his "love-investments" from the external world and transfers them into his ego, he is moving what of himself was attached to the other back into himself. To say, as Freud does, that the melancholic then ends up using this "internalized" love-investment to identify with the lost object (the love-investment "was not employed in any unspecified way, but served to establish an identification of the ego with the abandoned object") is to say that he uses his love to identify with himself (what Freud means when he says that melancholia is the "process of regression from the narcissistic object-choice to narcissism"). The "object" that lets its shadow fall on the melancholic ego is, therefore, that ego itself, now regressively united with itself, and the melancholic's self-impoverishment, paradoxically, shows itself to be a rescuing of the self, its falling into one with itself. In Freud's account, then, melancholics are not romantically undone by the love for the lost other but are instead narcissists driven by desire for self-appropriation.

Thoreau navigates perpetual grief in an opposite direction to Freudian melancholia. In contrast to the Freudian melancholic's identification with his own self in what he loves, Thoreau imagines pure grief to resolutely resist any narcissism or self-appropriation and predicates it on the utter insignificance of the survivor's self. Thus, whereas Freud's melancholic is a vampire that sucks life into itself ("the narcissistic identification with the object then becomes the substitute for the love-investment"), Thoreau's "pure griever," if there is such at all, would be a person who abandons itself, giving whatever there is of its own vitality over to the external world. In contrast to Freud's melancholic, who disinvests himself from the loss to revive himself, Thoreau's griever disinvests himself from himself to revive the loss. Differently put, Thoreau's griever is charged with a task of radical self-transformation, which would lead to shrinking the self, as if to make room for the loss to inhabit it. He is charged with enacting a concrete, as if material, metamorphosis through which he would lend his body and a part of his vitality to the loss in order to recall it to life. That is how Thoreau explains it in his *Journal:* "On the death of a friend, we should consider that the fates through confidence have devolved on us the task of a double living-that we have henceforth to fulfill the promise of our friend's life also, in our own, to the world" (J, 1, 114).

The agency of Thoreau's grief by means of material transformations and multiplication of personhoods in a single body suggests that unlike Freud's, his pure or perpetual grief is less a psychological than an ontological phenomenon. As he recognizes in the letter to Emerson, it is natural life's investment of vitality into decayed and dead matter to recover it "without loss," albeit in a different form: "Nature does not recognize it, she finds her own again under new forms without loss." In fact, it is in the identification of pure grief with nature that one finds the major difference between Thoreau's and Freud's theories. For while Freud is interested in what, in the letter to Emerson from March 11, 1842, Thoreau termed the "privacy" of grief, and what should be understood as the psychology of grieving, Thoreau gestures toward its ontology, identifying impersonal life—the natural life that generates individual beings—as mournful and restorative.

Conversely, I suggest, Thoreau's ontology, rather than psychology of mourning, locates the real—indeed, astonishing—similarity between Thoreau and Freud not in Freud's theory of melancholia but in what he identified as the "unconscious affect." This is what

Jean-François Lyotard, to whom we owe perhaps the only systematic analysis of it, explains as a site of the impersonal and memorial life, hence, as a life that comes close to Thoreau's nature in grief. But the term "unconscious" is misleading here (as Lyotard himself insists), for it suggests that the affect in question had once been conscious and could become so again. That is not the sense of "unconscious" here; instead, in question is a strange if not unthinkable meaning for it, referring to memories and "past shocks (recent or long past)"[14] that were never conscious and never repressed. As Lyotard explains this counterintuitive phenomenon, Freud's "unconscious affect" is a formation consisting of a sedimentation of the past that is not lacking in the present, of memories that no psyche revives and no desire wants to return to. Freud's unconscious affect is made of what Lyotard calls a "past that does not haunt the present" and doesn't "signal itself even . . . as a specter, an absence, which does not inhabit [the present] in the name of full reality." Therefore it is a past "which is not an object of memory like something that might have been forgotten and must be remembered."[15] This formation of memories that nobody has forgotten or remembered is so removed from consciousness (which includes its own unconscious) that under no circumstance can it be represented and so appropriated by the subject. This memorial affect exists where "there are no representations, not even disguised, indirect, reworked, reshaped ones like those with which secondary repression endows the forgotten past, the suffering, while the 'psychic apparatus' is in a position to resist them, to adapt to them, and so to accommodate them."[16] The way Freud imagines the "unconscious affect"—as an affective life that is not personal—thus announces, in Lyotard's account, a "complete break from the philosophy of consciousness, even if the term 'unconscious' still refers to it." Freud's metapsychology thus diagnoses the existence of an impersonal affect, which, despite its strangeness, he can't ignore: "Freud was the very first to say to himself: pure nonsense, an affect that does not affect consciousness. How can one say it affects? What is a feeling that is not felt by anyone? What is this 'anyone'? How can I, he asks, even be led on the path of this insane hypothesis if there exists no witness? Is not the affected the only witness to the affect?"[17]

What doesn't belong to any singular psyche, however, cannot strictly speaking be thought of as psychological. That is why Freud calls this affect a metapsychological phenomenon, while Lyotard identifies it as metaphysical: "This is the other metaphysics, the one that does not hinge upon a subject as the focus of all evident vision."[18] For me, it is the ontological status of this affect that counts: it is imagined not as a "sentiment" or "feeling," but closer to a material or natural power, as a series of "forces and conflicts of force (attraction and repulsion), and the results (effects) are assessed quantitatively."[19] In other words, it is imagined as an affective living force, a life that is also a memory or a memorial, similar to Thoreau's understanding of life as self-commemorating ("the living fact commemorates itself" is how he puts it, suggesting that life coincides in every present moment with the memory of itself [W, 154]). If this life affects persons, it does so not as something that is "theirs," coming from the depths of their psyches, but rather as something that comes to them from the exterior, as a life that is everywhere, in excess, an element similar to air or water: "Its 'excess' (or quantity, of intensity), exceeds the excess that gives rise (presence, place, and time) to the unconscious and the preconscious. It is 'in excess' like air and earth are in excess for the life of a fish."[20]

Freud himself wonders why this affective life, which originates and remains in the unconscious without ever reaching the mind, would still be identified as some sort of mental activity, such as the word "unconscious" implies. What rises in the unconscious and stays there might, he realizes, equally logically be called somatic; why then not understand these affects outside of the psychological, as "somatic processes from which something mental can once more proceed?"[21] To that question he offers two answers. The first answer is identified as practical, because the unity of a personhood hinges on it. In other words, understanding the unconscious as mental rather than somatic enables us to maintain the continuity of personal identity, as it allows us to treat processes of which a person is not conscious, from which the individual is as if absent, as bridges connecting conscious moments.[22] The second answer gestures toward a fabulous, if not magical, ontology that brings Freud close to Thoreau in its mixing of mental and affective with the material. To keep the name "unconscious" for what is not gathered into a self, Freud suggests, doesn't necessarily signal its opposition to the somatic, because a sharp distinction between mental and somatic might be merely a philosophical prejudice. Freud reminds his readers of the erstwhile belief in the mental activity of nonhumans and, more precisely, nonselves. The belief that all forms of life hosted mental processes, that material or natural life was also a mental process of which no self was aware, proved to humans the existence of similar processes within them, that similarly escaped their awareness. It suggested to them that they shared with nonhuman living beings not only the same life but also the same "mentality," as it were, so that to apprehend the processes and forces moving other living forms was to apprehend processes that were within them. Humans read the mental processes that are within them but inaccessible to them by presuming plants and animals to be animated by the same unconscious, impersonal mental power. As Freud reconstructs it, such an understanding of the nonhuman as endowed with mental powers not gathered into a self became less trustworthy as the "gap between the ego and the non-ego widened," that is, as the mental and spiritual was increasingly reserved for the self-conscious human ego. Freud, however, adds that at the moment when he writes his essay on the unconscious (1915) the exclusive attachments of consciousness to the human and the mental to a self have weakened; now the possibility of animal consciousness is raised, which suggests to him that, following the same logic, it might not be long before we acknowledge mental, albeit nonself-centered processes, in all life:

> Consciousness makes each of us aware only of his own states of mind; that other people, too, possess a consciousness is an inference which we draw by analogy . . . (It would no doubt be psychologically more correct to put it in this way: that without any special reflection we attribute to everyone else our own constitution and therefore our consciousness as well, and that this identification is a sine qua non of our understanding.) This inference (or this identification) was formerly extended by the ego to other human beings, to animals, plants, inanimate objects and to the world at large, and proved serviceable so long as their

> similarity to the individual ego was overwhelmingly great; but it became more untrustworthy in proportion as the difference between the ego and these 'others' widened. Today, our critical judgment is already in doubt on the question of consciousness in animals; we refuse to admit it in plants and we regard the assumption of its existence in inanimate matter as mysticism. . . . Psychoanalysis demands nothing more than that we should apply this process of inference to ourselves also. . . . If we do this, we must say: all the acts and manifestations which I notice in myself and do not know how to link up with the rest of my mental life must be judged as if they belonged to someone else.[23]

In its metapsychological phase psychoanalysis therefore asks us less to ascribe mental processes to ourselves than to treat them as forms of impersonal life affecting us from the "exterior," to use Lyotard's phrase. Gesturing toward the extraordinary ethical potential of this insight for what it means to care about all living beings, psychoanalysis is in effect asking us nothing less than to acknowledge the existence of life comprising the past—memories, commemorations, and affects—as not exclusively ours or even human, but something with which all humans, in Freud's normative phrasing (reminiscent of Kant's imperatives), are asked to identify. As in Thoreau, this insight suggests that all beings, indeed life as such, has had memorial sufferings that we are called on to take upon ourselves.

Whereas most philosophies of mourning to appear throughout the twentieth century moved nature away from what they were formulating, following Freud, as a highly stylized and personalized mourning, Walter Benjamin followed a path similar to Thoreau's. It is thus possible that in the whole (modern) history of Western thought, only Benjamin, in "On Language as Such and on the Language of Man"—a short, cryptic essay written in 1916 and unpublished in his lifetime—came close to Thoreau's ideas of nature as endless/formless mourning that recreates as it grieves.

Like Thoreau, Benjamin argued that the difference between human and natural life, the realm that included animal, vegetal, and mineral beings as well as things, was determined by humankind's distance from mourning, which was allotted to nature alone. Humankind's inability to abide in (nature's) grief is signaled for Benjamin by the fact that humans are endowed with a language that identifies and names (he calls it the "language of connotations"), which distances humans from both the realm of knowledge (given to God) and the realm of endless metamorphoses and transmutations (which Benjamin understands to be a denotative language only, hence an "undifferentiated" language with no fixed and identifiable meanings, a language that he identified with the grief of nature). In God's language a being or a thing is its name, because this language is the force of creation that names as it forms. Uttering a name is equivalent to generating a being, which makes divine language the language of absolute cognition and truth. That is why Benjamin can say that things have "proper" names only in God, for only in God are their names also their existences. In contrast, human language operates outside the space of Eden (the site of God's creative and cognizant naming). As Benjamin imagines it, language was given to humankind—as specifically human—only after the fall from Eden, hence, when things were already created. It was thus given to humans in the form of words that identify rather than generate existences. Because the identity of things is created by God before humans names them, a name ("human word") never coincides with a thing and so never names accurately (as is the case in God's language), which is also why human language, Benjamin argues, is not cognizant of the world. The most decisive element of what human language will come to be resides in the fact that because things already exist when humankind names them, their naming requires temporality and is performed through an act of preferring (this name rather than that). Human language is thus always a language of choice, hence of judgment, and is accompanied by the guilt generated by the choices it makes. A long passage in Benjamin's essay identifies judgment and guilt not only as immanently human but also as the single most crucial force of personalization:

> The knowledge to which the snake seduces, that of good and evil, is nameless. It is vain in the deepest sense, and this very knowledge is itself the only evil known to the paradisiacal state.

> Knowledge of good and evil abandons name; it is a knowledge from outside, the uncreated imitation of the creative word. Name steps outside itself in this knowledge: the Fall marks the birth of the *human word*, in which name no longer lives intact and which has stepped out of name-language, the language of knowledge. . . . The knowledge of things resides in the name, whereas that of good and evil is, in the profound sense in which Kierkegaard uses the word, "prattle," and knows only one purification and elevation, to which the prattling man, the sinner, was therefore submitted: judgment. Admittedly, the judging word has direct knowledge of good and evil. . . . This judging word expels the first human being from Paradise; they themselves have aroused it in accordance with the immutable law by which this judging word punishes—and expects—its own awakening as the sole and deepest guilt.[24]

In identifying beings they don't generate, human words are marks similar to stains or blemishes, thus something like accusatory yet untrue signs. Because they mark and so affect the identity of what they mark, human words are never arbitrary. But because their essence is not the essence of the marked, they are also never true. Benjamin terms such marks "connotations," because humans make and trade their words as if they were precise meanings, coincident with the factual, yet the human word always marks others, as if accusing them.[25] And because the accusation embedded in all human communication is mistaken for connotation or "truth," the human word or judgment is also always the execution of an unjust punishment (according to the law Benjamin formulates here, punishment is not something that comes once the judgment is pronounced, but is rather a function of the word itself: it is the "judging word [that] punishes"). That unavoidable injustice is what evokes "deepest guilt." Hence, the awful existential double bind of humans: as long as they speak the naming word—as long as they are human—they are guilty of punishing the innocent; but because they all use the judging word, none of them is innocent either.

In opposition to the human word, which names and punishes what it differentiates by identifying it, mourning is the "undifferentiated, impotent expression of language."[26] That is to say, mourning does not use the language of connotation: it does not name and, by naming, identify; hence, it does not judge or punish, and because of its innocence, it is also impotent (incapable of enacting the violence of human names). The language of mourning is thus neither linguistic (it doesn't signify or formalize) nor stabilizing (it doesn't essentialize in the way that God's language does). Benjamin calls it the language of "denotation or translations only"; it operates as pure process, as the constant transition or "continua of transformation"; it is "the communication of matter in magic communion," material reality in the process of its constant reformulation or, as Benjamin also identifies it, life in the process of its concrete translation ("translation passes through continua of transformation, not abstract areas of identity and similarity").[27] Thus, in yet another striking similarity to Thoreau, Benjamin identifies the performance of mourning with the process of the natural, that is, the means through which guiltless, impersonal, and unnamed life restores loss:

> After the Fall, however, when God's word curses the ground, the appearance of nature is deeply changed. Now begins its other muteness, which is what we mean by the "deep sadness of nature." . . . This proposition has a double meaning. It means, first, that she would lament language itself. Speechlessness: that is the great sorrow of nature. . . . This proposition means, second, that she would lament. Lament, however, is the most undifferentiated, impotent expression of language. It contains scarcely more than the sensuous breath; and even where there is only a rustling of plants, there is always a lament. Because she is mute, nature mourns. Yet the inversion of this proposition leads even further into the essence of nature; the sadness of nature makes her mute. In all mourning there is the deepest inclination to speechlessness, which is infinitely more than the inability or disinclination to communicate. That which mourns feels itself thoroughly known by the unknowable. To be named—even when the namer is godlike and blissful—perhaps always remains an intimation of mourning.[28]

The passage establishes an explicit identity between life ("sensuous breath") and mourning. The mourning—Benjamin uses the word synonymously with "sorrow," "lament," and "sadness"—in question thus amounts to a transmutation of matter generated by the proliferation of "undifferentiated" sensations ("continua of transformation"). Far from being psychological, the rhythm of this mourning is thus the pulse of life, beating before or outside words and images. Like Thoreau, Benjamin identifies this pulsation with the rustling of plants: Benjamin: "and even where there is only a rustling of plants, there is always a lament"; Thoreau, describing how "autumn" is grieving: "We heard the sigh of the first autumnal wind. . . . In all woods the leaves were fast ripening for their fall. . . . Our thoughts too began to rustle" (W, 335–336).

Because it is ontological (impersonal, infinite) the mourning of nature has nothing human or personal about it. A Thoreau who claims that nature mourns away, and differently from humans, can be compared with a Benjamin who is explicit in claiming that nature's sadness is not related to humans. For, in Benjamin, it is not that nature mourns because it cannot speak the language of humans, or that trees are sad because they are not persons, or that grass laments because it can't identify, or that animals grieve because they can't name, as some humanist interpretations of Benjamin's text have suggested.[29] Rather, life is mute because it mourns ("the sadness of nature makes her mute"). Benjamin demonstrates that life is identified with mourning on account of the way it is: as pulsation of breath, life continuously interrupts itself—in breathing out and it suffers a break in its continuity—but then recreates itself by breathing itself into itself. Unlike God, who creates what never was, life restores a breath that was lost. In restoring what was lost, life makes itself known to the unknown (to death), and what lives touches what had been discontinued, enacting what Benjamin terms "magic communion." That "magic," according to Benjamin, is the very operation of mourning: "that which mourns feels itself thoroughly known by the unknowable." The suffering of discontinuity, then, requires the complete, or as Benjamin put it, "thorough" surrender of what is to what is formless: no person can visit the lost without coming completely undone, without becoming depersonalized. That is why, to the extent that they remain human, arrested in names and identities, absorbed in the guilt and violence of judgment that forms them, humans are incapable of such undoing and regenerative mourning. As Benjamin suggests, even when naming is gentle, like the touch of an angel, even when the identification is fragile like the bliss of beatitude, humans always experience at best the "intimation of mourning." Furthermore, the human incapacity to grieve with the grief of nature becomes the reason for nature's other, less essential, mourning; its "sadness" he says, means that nature "laments language itself." Nature mourns the language of arrest and judgment, of violence and guilt that blocks mourning. It mourns that humans cannot mourn.

Appendix II: On Thoreau's Grave

On October 10, 1860, Thoreau's *Journal* reported on the success of an experiment he had begun in 1855. It concerned the initiation of life in Sleepy Hollow cemetery, where John had already been buried for thirteen years.

Thoreau leveled the ground there for an artificial pond, hoping, in accordance with Thalesian materialism and his own philosophy of wetlands, that the presence of water would generate the swampy wet-soil, damp-air ecosystem out of which all life originates. The *Journal* entry records how the swamp did in fact form as the presence of water mobilized birds and fish to disseminate seeds, which were then constantly "awake," busy blooming into lilies. By 1860 the pond became a force of life expanding over the entire cemetery, taking it over, appropriating death and turning it into life. This is how Thoreau describes the process:

> In August, '55, I leveled for the artificial pond at Sleepy Hollow. They dug gradually for three or four years and completed the pond last year, '59. It is now about a dozen rods long by five or six wide and two or three deep, and is supplied by copious springs in the meadow. There is a long ditch leading into it, in which no water now flows, nor has since winter at least, and a short ditch leading out of it into the brook. It is about sixty rods from the very source of the brook. Well, in this pond thus dug in the midst of a meadow a year or two ago and supplied by springs in the meadow, I find to-day several small patches of the large

John Thoreau's grave. Courtesy Vesna Kuiken.

yellow and the kalmiana lily already established. Thus in the midst of death we are in life. The water is otherwise apparently clear of weeds. The river, where these abound, is about half a mile distant down the little brook near which this pond lies, though there *may* be a few pads in the ditched part of it at half that distance. How, then, did the seed get here? I learned last winter (*vide* December 23, 1859) that many small pouts and some sizable pickerel had been caught here, though the connection with the brook is a very slight and shallow ditch. I think, therefore, that the lily seeds have been conveyed into this pond from the river immediately, or perchance from the meadow between, either by fishes, reptiles, or birds which fed

Henry David Thoreau's grave. Courtesy Vesna Kuiken.

> on them, and that the seeds were not lying dormant in the mud. (Jn, 14, 109–110)

So it is that, two years after writing this entry, when Thoreau was himself buried at Sleepy Hollow next to his brother John, he was interred in the midst of the life that he helped create.

Notes

Introduction

The book epigraph is from Henry David Thoreau's *Journal* (J, 1, 114).

1. Joan Dayan, *Fables of Mind: An Inquiry into Poe's Fiction* (New York: Oxford University Press, 1987), p. 16.
2. Paul Patton, "Mobile Concepts, Metaphor, and the Problem of Referentiality," in *Deleuzian Concepts: Philosophy, Colonization, Politics* (Stanford, CA: Stanford University Press, 2010), p. 31.
3. Stanley Cavell, *The Senses of Walden* (Chicago: University of Chicago Press, 1972), p. 64.
4. Walter Benjamin, "On Language as Such and on the Language of Man," in *Selected Writings*, vol. 1, *1913–1926*, ed. Marcus Bullock and Michael W. Jennings, trans. Edmund Jephcott (Cambridge, MA: Harvard University Press, 1996), p. 62.
5. Ibid., p. 63.
6. Charles Sanders Peirce, *Collected Papers of Charles Sanders Peirce*, vol. 2: *1931–35* (Cambridge, MA: Harvard University Press, 1932), p. 228, quoted in Eduardo Kohn, "How Dogs Dream: Amazonian Natures and the Politics of Transspecies Engagement," *American Ethnologist* 34, no. 1 (2007): 6.
7. Barbara Johnson, "A Hound, a Bay Horse, and a Turtle Dove: Obscurity in Walden," in *Walden*, Norton Critical Edition, ed. William Rossi (New York: Norton, 1992), p. 450.
8. Cavell, *The Senses of Walden*, p. 64.
9. Donald Worster, *Nature's Economy: A History of Ecological Ideas* (Cambridge: Cambridge University Press, 1998 [first published in 1977]), p. 78.
10. Ibid., p. 77.
11. Stanley Cavell, "Ending the End Game: A Reading of Beckett's *Endgame*," in *Must We Mean What We Say?* (Cambridge, MA: Harvard University Press, 1976 [reprint of the Charles Scriber's Sons 1969 edition]), p. 153.
12. Patton, "Mobile Concepts," p. 25.
13. Ibid.
14. Ibid.

15. Roland Barthes, *The Preparation of the Novel*, trans. Kate Briggs (New York: Columbia University Press, 2011), p. 46.
16. Roland Barthes, *The Neutral*, trans. Rosalind E. Krauss and Denis Hollier (New York: Columbia University Press, 2002), p. 11.
17. Barthes, *The Preparation of the Novel*, p. 51.
18. Ibid., p. 47.
19. Cavell, "Ending the End Game," p. 122.
20. Cavell, *The Senses of Walden*, p. 87.
21. Cavell offers the following examples of Thoreau's "thorough meanings": "Let me recall one or two passages: 'The abdomen under the wings of the butterfly still represents the larva. This is the tidbit which tempts his insectivorous fate. The gross feeder is a man in the larva state.' 'I saw a striped snake run into the water, and he lay on the bottom . . . more than a quarter of an hour perhaps because he had not yet fairly come out of the torpid state. It appeared to me that for a like reason men remain in their present low and primitive condition'" (ibid., pp. 42–43).
22. Cavell's passage on the nonskeptic's maltreatment of the skeptic that serves as a background for my formulations of "affirmative reading" comes from "Knowing and Acknowledging": "The appeal to ordinary language cannot directly repudiate the skeptic . . . by, for example, finding that what he says contradicts what we ordinarily say or by claiming that he cannot mean what he says: the former is no surprise to him and the latter is not obviously more than a piece of abuse" (*Must We Mean What We Say?*, p. 240).
23. I am grateful to Sharon Cameron for helping me refine this argument.
24. Jonathan Lear, *Radical Hope: Ethics in the Face of Cultural Devastation* (Cambridge, MA: Harvard University Press, 2006), p. 90 ("bird-philosopher"); pp. 83–84, 80.
25. Cavell, *Must We Mean What We Say?*, p. 241.
26. Ibid.
27. Gilles Deleuze and Claire Parnet, *Dialogues II*, trans. Hugh Tomlinson and Barbara Habberjam (New York: Columbia University Press, 1987), p. 1: "Questions are invented, like anything else. . . . The art of constructing a problem is very important: you invent a problem, a problem-position, before finding a solution. None of this happens in an interview, a conversation, a discussion. Even reflection, whether it's alone, or between two or more, is not enough. Above all, not reflection. Objections are even worse. Every time someone puts an objection to me, I want to say: 'OK, OK, let's go on to something else.' Objections have never contributed anything. It's the same when I am asked a general question. The aim is not to answer questions, it's to get out, to get out of it. Many people think that it is only by going back over the question that it's possible to get out of it. . . . They won't stop returning to the question in order to get out of it. But getting out never happens like that. Movement always happens behind the thinker's back, or in the moment when he blinks. Getting out is already achieved, or else it never will be."
28. Quoted in Walter Harding, *The Days of Henry Thoreau: A Biography* (Princeton, NJ: Princeton University Press, 1982), p. 193.

29. Henry David Thoreau, *Collected Essays and Poems*, ed. Elizabeth Hall Witherell (New York: Library of America, 2001), p. 595.
30. Sherman Paul, *The Shores of America: Thoreau's Inward Exploration* (Urbana: University of Illinois Press, 1958), p. 104; for a reading that detects crucial differences between Thoreau's poem and "Lycidas," see Linck C. Johnson, *Thoreau's Complex Weave* (Charlottesville: University Press of Virginia, 1986), p. 51.
31. David B. Williams, "Benchmarks: September 30, 1861: Archaeopteryx Is Discovered and Described," *Earth*, http://www.earthmagazine.org/article/benchmarks-september-30–1861-archaeopteryx-discovered-and-described. In some reports the discovery is dated 1860, but I date it 1861, following Richard Owen, who was the first English paleontologist to discuss the fossil at length. See Richard Owen, "On the Archeopteryx of von Meyer, with a Description of the Fossil Remains of a Long-tailed Species, from the Lithographic Stone of Solenhofen," *Philosophical Transactions of the Royal Society of London* 71 (January 1833): 41.
32. Ibid. In some reports discovery is dated 1860, but I date it 1861 following Owen.

I Introduction: Perpetual Grief and the Example of the Fish Hawks

1. Sharon Cameron, *Impersonality: Seven Essays* (Chicago: University of Chicago Press, 2007), p. 66.
2. For *álaston pénthos* in Homeric hymns, see Nicole Loraux, *Mothers in Mourning*, trans. Corinne Pache (Ithaca, NY: Cornell University Press, 1998), p. 100. Thoreau discusses Homeric Hymns in "Greek Classic Poets" (*Early Essays and Miscellanies*, ed. William L. Howarth, The Writings of Henry D. Thoreau [Princeton, NJ: Princeton University Press, 1975], p. 57).
3. *Iliad*, 22, 261. For Achilles's grief, see Laura Slatkin, *The Wrath of Thetis: Allusion and Interpretation in the Iliad* (Berkeley: University of California Press, 1991): "For Achilles in particular, *achos* (grief) is a constant" (p. 86). See also Loraux, *Mothers in Mourning*, p. 101. As Thoreau's essay "Greek Classic Poets," suggests (pp. 50–58), he was familiar with Greek, Latin, German, and French editions of the *Iliad* and was deeply immersed in the secondary literature on Homer.
4. Homer, *The Odyssey* (14.200–201), trans. Robert Fagles, intro. Bernard Knox (New York: Penguin Books, 1996), p. 307; Loraux, *Mothers in Mourning*, p. 100.
5. Loraux, *Mothers in Mourning*, p. 95.
6. Homer, *The Odyssey* (24.468–470), trans. Robert Fagles, p. 481. For systematic references Thoreau makes to the *Odyssey* (books VII, X, XI) as late as 1862, in "Wild Apples," see Lewis Hyde's editorial notes to "Wild Apples," in *The Essays of Henry D. Thoreau*, ed. Lewis Hyde (New York: North Point Press, 2002), p. 365. Loraux also discusses this example in *Mothers in Mourning*, p. 100.
7. Nicole Loraux, *Mothers in Mourning*, p. 101.
8. Ibid., p. 99.

9. Ibid., pp. 100–101.
10. Ibid.
11. Ibid., p. 102.
12. Ibid.
13. Sophocles, *Antigone* (1295–1300), ed. and trans. Andrew Brown (Warminster: Aris & Phillips, 1993), p. 127.
14. Euripides, *The Bacchae* (1218–1227), trans. C. K. Williams (New York: Farrar, Straus and Giroux, 1990), p. 77.
15. Ibid. (1285), p. 80.
16. Loraux, *Mothers in Mourning*, p. 102.
17. Nicole Loraux, *The Mourning Voice: An Essay on Greek Tragedy*, trans. Elizabeth Trapnell Rawlings (Ithaca, NY: Cornell University Press, 2002), p. 27; see also, Loraux, *Mothers in Mourning*, p. 98.
18. Emile Benveniste, "Expression indo-européenne de l'immortalité," *Bulletin de la Société linguistique de Paris* 38 (1937): 109, translation mine.
19. Loreaux, *The Mourning Voice*, p. 27.
20. Emile Benveniste, "Expression indo-européenne de l'immortalité," p. 111, translation mine.
21. Jean-Pierre Vernant, "The Reason of Myth," in *Myth and Society in Ancient Greece*, trans. Janet Lloyd (New York: Zone Books, 1996), p. 225.
22. Ibid., p. 221.
23. Thoreau knew Aeschylus well, being a translator of his *Prometheus Unbound.* But there are many other ways he could have been familiar with the phrase, for classical tradition turned it into an ontological formula that summarized ideas regarding life for many philosophers of vitalistic orientation (Lucretius or Cicero, for instance). (For detailed discussion of Lucretius's and Cicero's appropriation of "anêrithmon gelasma," see: Martin Kellogg, "The Wealth of Words," *Proceedings of the California Teacher's Association*, Session of Year 1897 (San Francisco: Myssell-Rollins, 1898, p. 58.) He could have also read a long discussion in *Blackwood's Edinburgh Magazine* (Volume 61, no. 380, June, 1847, p. 699), always available in Boston, which under the title *Horae Catullianae*, published the "Letter to Eusebius," in which Curate says the following: "CURATE.—I would venture from the last line—'Ridete quidquid est domi cachinnorum—' a remark upon a passage, the celebrated expression in the *Prometheus* of Æschylus, the [Greek: anêrithmon gelasma]. Some call it 'countless dimples.' Now is it not possible Catullus may have thought of this, and as it were translated it by 'quidquid est cachinnorum'? The question then would be, is it meant to speak to the ear or the eye? Is it of sound or vision? I am inclined to think it is the sound, the communicative laughter of the many waves. 'Dimple' is too little for the gigantic conception of Æschylus, but the laughter of the multitudinous ocean-waves is more after his genius. No one could translate 'cachinnus' 'a dimple.' If, therefore, Catullus had in his mind the Greek passage, it shows his idea of the [Greek: anêrithmon gelasma]."
24. The Pythagorean story existed in many versions throughout the classical tradition. Here is Cicero's (M. Tullius Cicero, *De Officiis*, English trans. Walter Miller [Cambridge, MA: Harvard University Press, 1913], sec-

tion 45): "They say that Damon and Pythias, of the Pythagorean school, enjoyed such ideally perfect friendship, that when the tyrant Dionysius had appointed a day for the execution of one of them, and the one who had been condemned to death requested a few days' respite for the purpose of putting his loved ones in the care of friends, the other became surety for his appearance, with the understanding that if his friend did not return, he himself should be put to death. And when the friend returned on the day appointed, the tyrant in admiration for their faithfulness begged that they would enroll him as a third partner in their friendship." In the early first century Valerius Maximus offered a longer version of "The History of Damon and Pythias" in *De Amicitiae Vinculo.*

25. Vernant, "The Reason of Myth," p. 221.

26. Anne Carson, *Grief Lessons* (New York: New York Review Books Classics, 2006), p. 8.

Literal Sounds

1. Jane Bennett, *Thoreau's Nature: Ethics, Politics, and the Wild* (1994; reprint, Boston: Rowman & Littlefield, 2002), p. 26; Lawrence Buell, *The Environmental Imagination: Thoreau, Nature Writing, and the Formation of American Culture* (Cambridge, MA: Harvard University Press, 1995), pp. 144–145.

2. Bennett, *Thoreau's Nature*, p. 27; Buell, *The Environmental Imagination*, p. 12.

3. Ross Posnock, "Fathers and Sons," *Raritan* 331, no. 3 (Winter 2012): 108.

4. Donald Worster, *Nature's Economy: A History of Ecological Ideas* (1977; reprint, Cambridge: Cambridge University Press, 1998), p. 78.

5. Ralph Waldo Emerson, "Nature," in *Essays and Lectures*, ed. Joel Porte (New York: Library of America, 1983, College Edition), p. 25.

6. Henry David Thoreau, "The Service," in Henry David Thoreau, *Collected Essays and Poems*, ed. Elizabeth Hall Witherell (New York: Library of America, 2001), p. 14.

7. Samuel Taylor Coleridge, *Specimens of the Table-Talk of the Late Samuel Taylor Coleridge*, 2 vols. in 1 (New York: Harper, 1835), 2, p. 3. Thoreau's quote is in the *Journal* (J, 1, 97).

8. Thoreau, "The Service," p. 14.

9. The chapter of *Walden* titled "Sounds" lists a whole series of such discordant vibrations as musical, annulling not only the difference between melodious and inharmonious but also between natural and technological. The song of a nightingale is no better than a sound released by a machine, because any sound that is heard "finely" requires a "refreshing" emptying of the ears: "The whistle of the locomotive penetrates my woods summer and winter, sounding like a scream of a hawk" (Wa, 115); "I am refreshed and expanded when the freight train rattles past me" (Wa, 119). Even the sound of starving "rats in the wall" is musical on condition that one empties the ear enough to avoid attaching any preconceived idea to it (ibid., p. 127). Music is thus transformed into an unintended aural objectivity whose literality is not representative of anything.

Homer's Music Box

1. Walter Harding, *The Days of Henry Thoreau: A Biography* (Princeton, NJ: Princeton University Press, 1982), p. 137.

2. Quoted in Jeffrey S. Cramer's editorial comment to *I to Myself: An Annotated Selection from the Journal of Henry D. Thoreau* (New Haven, CT: Yale University Press, 2007), p. 35. The conversation I had with Cramer during my research at The Thoreau Institute at Walden helped me track down and classify the music boxes. It was he who suggested to me that the first music box I've mentioned, which Thoreau listens to in January 1842, was already in the household and therefore, quite possibly belonged to John. I thank Jeffrey Cramer most sincerely for his suggestions.

3. Harding, *The Days of Henry Thoreau*, p. 134. In a letter to her sister, Lucy Jackson Brown, of January 11, 1842, Lydian Jackson reports that on the evening of January 9 [Nathan?] Brooks "came for [Thoreau]," who visited them that day and "said they were alarmed by symptoms of the lockjaw in John." *The Selected Letters of Lidian Jackson Emerson*, ed. Delores Bird Carpenter (Columbia: University of Missouri Press, 1987), p. 99. Robert D. Richardson depicts John's acute pain as unbearably intense: "Emerson wrote later about lockjaw convulsions so violent they 'bent a man backward so that his head almost touched his heels,' probably on the basis of what he heard about John's death." Robert D. Richardson Jr., *Henry Thoreau: A Life of the Mind* (Berkeley: University of California Press, 1986), p. 133.

4. In the *Iliad* Glaucus famously described it as the decaying and budding of leaves ("For, like frail leaves on trees, the state of man / Is ever hanging. / Some Some by driving winds / Are shaken to the ground, and wither there; / Others again the budding boughs produce / In spring's enlivening hour; so of men, / Successive generations rise and fall, / Some flourish, some decay."). I quote here the edition Thoreau used: Homer, *The Iliad of Homer, from the Text of Wolf, with English Notes and Flaxman's Designs*, ed. C. C. Felton (Boston: Hilliard, Gray, and Company, 1833, reprinted 1846), 6.195–200, p. 195. The cyclical time that Thoreau here has in mind is also different from the circle, which in Pythagoras, Empedocles, and Plato regulates the "cycles of metamorphoses" of souls.

5. For the tremendous influence that Orphic theogonies had on Thoreau, see Ethel Seybold, *Thoreau: The Quest and the Classics* (New Haven, CT: Yale University Press, 1951), p. 16. For the concept of Chronos in Orphic theogonies, see Jean-Pierre Vernant, "Mythic Aspect of Memory," in *Myth and Thought among the Greeks*, trans. Janet Lloyd (New York: Zone Books, 2006), pp. 130–132.

6. Vernant, "Mythic Aspect of Memory," p. 130.

7. In mythological terms, if Chronos generates the sky and earth, or, alternatively—as in Homer and Aeschylus—if Chronos is similar to Okeanos, the river that encircles the world, it is because, like water or earth, Chronos embraces what is vanishing while bringing to visibility what is coming to be. Vernant, "Mythic Aspect of Memory," p. 130.

8. Because they represent an effort to spatialize time, calendars are closer to chronic than to linear time. They enable persons to simultaneously see dif-

ferent segments of time, similar to mythology. I address the question of calendars, calendric time, and Thoreau's calendars of transitory phenomena in Part II of this book.

9. For instance, Tiresias in Sophocle's *Antigone* has access to both past and future; see Carol Jacobs, "Dusting Antigone," in *Skirting the Ethical* (Stanford, CA: Stanford University Press, 2008), p. 9. In contrast, Aristotle claims Epimenides's power of divination to be entirely "retrospective" (Aristotle, *Rhetoric*, 3.17.10; quoted in Jean-Pierre Vernant, "Mythic Aspect of Memory," p. 128).

10. Seeing is thus closer to "seeing through" than to envisioning. For instance, Epimenides, who foretells by remembering, sees not an "image" but the totality of *alētheia*. See Marcel Detienne, *The Masters of Truth in Archaic Greece*, trans. Janet Lloyd (New York: Zone Books, 1999), p. 55.

11. Importantly for my discussion of sound, words arriving through such divination are not therefore generated as "measured," or intentionally aestheticized but instead—because divination means fusing the mind into the corporeality of circular time—as "objective" or natural sounds: "The Iliad seems like a natural sound which has reverberated to our days" (J, 1, 31). For a slightly different assessment of Thoreau's understanding of circular time, see H. Daniel Peck, *Thoreau's Morning Work* (New Haven, CT: Yale University Press, 1990), pp. 46–47. Peck understands that memory can be prospective in Thoreau, because it fuses past and future into identity: "In many entries during this period, we find the writer virtually identifying phenomena of past and future, as in phrases such as 'that reminiscence or prophesying of spring' " (p. 47). However, I claim that chronic time doesn't level all instants into sameness but keeps their differences alive, while enabling the seer to encounter them through different temporal orientations (whether by going backward or forward).

12. Harding, *The Days of Henry Thoreau*, pp. 135–136.

13. The editorial comment on Thoreau's manuscript transcribed in 1842 states: "Marbled orange/red/brown boards; brown spine. Thoreau excised the first fifteen leaves (possibly including the free endpaper and flyleaf) on which John Thoreau, Jr., kept records for the brothers' Concord Academy, and then inverted this volume and wrote the Journal text from back to front" (J, 1, 624).

14. Harding, *The Days of Henry Thoreau*, p. 258.

15. Richard Lebeaux, *Thoreau's Seasons* (Amherst: University of Massachusetts Press, 1984), p. 104.

16. Canby convincingly argues that the entire text concerned Lidian Emerson, basing his reasoning on the fact that large portions of it were "parallel" to the love letters Thoreau wrote to her in May and June 1843. But he also notes that the new passages Thoreau added were not quite in tune with the letters sent to Lidian Emerson; the passages in the essay draft are disjointed, as if signaling an experience of disorientation. See Henry Seidel Canby, *Thoreau* (Boston: Houghton Mifflin, 1939), pp. 161–162. And while it is obvious that many of the passages from this draft of an essay are only slightly revised sentences from the letters to Lidian, it is also plausible that some of the disjointedness of Thoreau's

sentences, the befuddlement of his thinking, comes from the fact that, as he writes, feelings for his dying sister are confused with feelings of love that were originally directed toward another woman.

Aeschylus's Paean

1. Nicole Loraux, *The Mourning Voice: An Essay on Greek Tragedy*, trans. Elizabeth Trapnell Rawlings (Ithaca, NY: Cornell University Press, 2002), p. 63. As Loraux additionally notes, the "chorus in the *Agamemnon* remarks that it is not in the nature of this god to embrace the *thrēnos* and indeed that any claim of proximity between the god and the sound of mourning is entirely inappropriate . . . in Euripides' *Suppliant Women* . . . the mothers who form the chorus and grieve endlessly in sobs and mourning *(aei)* for their fallen sons observe that golden-haired Apollo does not accept their songs" (p. 60).

2. Ibid., p. 63. Loraux differentiates various forms of *thrēnos.* (1)*Thrēnos* is the word used for the various forms that grief *(penthos)* assumed at different historical sites and in different literary situations. In epics, it was the "union of grief and the promise of immortality, of penthos and kleos." (2) The initial transformation of *goos* into *thrēnos* eventually turned the *thrēnos* into a more *"composed lament,"* thus engendering the "exclusion of [emotional] lamentation" and privileging the more balanced and measured aspect of the *logos.* This transformation was of central political importance for the construction of a new Athenian public sphere; mourning, as a "shrill cry" was marginalized—if not abandoned—in favor of the political benefits of eulogy: "This equivalence of *thrēnos* and *goos* illuminates in many different ways the funeral oration and its exclusion of lamentation. This exclusion may be seen as a desire to suppress displays of excessive mourning" (Nicole Loraux, *The Invention of Athens: The Funeral Oration in the Classical City*, trans. Alan Sheridan [New York: Zone Books, 2007], p. 79). (3) *Threnos* marks a complex, if not controversial relation between *penthos* and lyric poetry. Initially, Loraux argues, "*thrēnos* referred to the poet's composed lament. Thus, in the Homeric epic, the *thrēnos* sung by the bard over the hero's body is contrasted in a sort of a dialogue with the moans and sobs of the family and crowds. The lyric poets made a gnomic form of it in which the consolation of the living was accompanied by a whole philosophy of life and death" (ibid., pp. 78–79). Loraux argues that what she calls "tragic mourning" refers to yet another experience of the complex *thrēnos,* which resists all the above delineations of it. It is, for instance, less composed and more contradictory than the Homeric "composed lament," hence closer to a cry disorienting the composed presence of the logos.

3. Ibid.

4. Ibid., p. 64.

5. Ibid., p. 65.

6. As K. P. Van Anglen, the editor of Thoreau's Greek translations, reconstructs, "several brief *Journal* quotations suggest that Thoreau was reading the play as early as the end of January 1840, but there is no evidence that he began to translate the entire work until after he had done the *Prometheus Bound* (Kevin P. Van Anglen, "Introduction," in Henry D. Thoreau, *Translations*, ed. Kevin P. Van Anglen, The Writings of Henry D. Thoreau [Princeton, NJ: Princeton

University Press, 1986], p. 168). But whereas the translation was finished only after the work on *Prometheus Bound*, it started, again according to Van Anglen's reconstruction, in 1842: "His [Thoreau's] *The Seven Against Thebes*, which exists in a single and not entirely complete manuscript, also comes from the same general period, having been begun in Concord during 1842 and finished the following year while Thoreau served as tutor in the household of William Emerson on Staten Island" (Kevin P. Van Anglen, "The Sources for Thoreau's Greek Translations," *Studies in the American Renaissance* 27 [1980]: 292). For a detailed discussion of strategies Thoreau adopted in translating Greek texts, see Van Anglen's "Introduction," pp. 201–204. In the discussion of *Seven against Thebes* that follows, I use—unless otherwise specified—Thoreau's translation (ST) to relate his theory of mourning to choices he made in translating Aeschylus. My concern, therefore, is not at all points to evaluate his translations from the point of view of a classicist.

7. Such revision of grieving practices is visible not only in the instance of the paean but in other instances too. First, the Theban maidens who comprise the chorus are so terrified by the noises they hear around the walls announcing the arrival of the enemy army, that, prompted by their ominous fear of death and mourning that will follow, they turn their suppliant prayers into lamenting "sighs" and "shrieks" (ST, 70, 71), so that Eteocles, the Theban king soon to be killed by his brother Polynices (leader of the attacking Argive army, whom he will in turn kill, so that the two brothers die by each other's hand) has to ask for a "better uttered prayer." Second, the chorus's prayer to Apollo and Cypris, the mother of Harmonia, is turned into inharmonious "sharp cries of terror" (*eeee*).

8. Van Anglen determines that Thoreau used *Aeschylus's Tragoedia: Ad exemplar accurate expressae* Editio stereotypa (Leipzig: Tauchnitz, 1817 and/or 1819) (Anglen, "The Sources for Thoreau's Greek Translations," p. 292). Unlike Thoreau's rendering—in which the paean transposes the grief of the living onto the dead and creates incompatible yet fused ontological juxtapositions—other crucial translations register Aeschylus's modification of the paean not by making it simultaneously reside in the two worlds (of living and dead), but instead clearly entrusting it to one world and logically explaining its presence in another. See, for example, Alan H. Sommerstein's translation of *Seven against Thebes* (871–875): "But it is right that we, . . . / should raise the unpleasing sound / of the Fury's hymn, and sing the hateful paean of Hades" (Cambridge, MA: Harvard University Press, 2008, p. 245). Hutchinson's Greek edition of ΕΠΤΑ ΕΠΙ ΘΗΒΑΣ interprets line 870 as insisting on the common "contrast between paeans and laments," later elaborated by Euripidus (Aeschylus, *Septem Contra Thebas*, ed. G. O. Hutchinson [Oxford: Clarendon Press, 1985], p. 192). E.D.A. Morshead's translation fails to register the ambiguity of these lines, and indeed doesn't even translate the paean: "We must awake the rueful strain / To vengeful powers, in realms below." However, Anthony Hecht and Helen H. Bacon's translation (*Seven against Thebes* [New York: Oxford University Press, 1973], p. 59) does register the paean's ambivalence: "But it is right for us / before we hear their voices / to chant the grating hymn, / the harsh hymn of the Fury, / and over the dead to sing / the hate-filled paean of Hades."

9. Sommerstein notes that "it is illogical that this chorus of citizen maidens should speak in the name of the bereaved, mourning palace; but the chorus of Argive elders in Agamemnon do likewise" (Aeschylus, *Agamemnon*, 1482–1483, 1532, 1565–1566) (*Seven against Thebes*, trans. Sommerstein, p. 251). But "speaking for" the dead, which confuses not just two worlds but also the different persons to whom such speaking gestures, is perhaps less illogical if we take into account what other classicists tell us, namely, that to talk about selves in Aeschylus is a very modern and hence inappropriate way of understanding his tragedies. With the possible exception of Eteocles, there is nothing in them that can be identified as an operation of "psychology" in any modern sense, that is, as portraying a private psyche. For detailed discussion of this issue, see Pierre Vidal-Naquet, "The Shields of the Heroes," in *Myth and Tragedy in Ancient Greece*, ed. Jean-Pierre Vernant and Pierre Vidal-Naquet, trans. Janet Lloyd (New York: Zone Books, 1990), p. 273.

10. Tauchnitz doesn't comment on the ὀλολυγμός, but Sommerstein's translation (*Seven against Thebes*, 266–267; p. 181) explains it as the "auspicious ululation of triumph": "*ὀλολυγμός*, a cry of joy uttered . . . at the slaughter of a sacrificial beast." Sommerstein's translation refers to ululation but omits "paean" altogether. He translates: "Listen to my prayer, and then utter the sacred, auspicious ululation of triumph, the customary Helenic cry at sacrifices, to give confidence to our friend" 266–270; p. 181).

11. As Van Anglen explains, the Tauchnitz edition Thoreau used "at a number of points ascribes lines and speeches to different characters than those in the OCT [Oxford Classical Texts]. . . . The same sort of thing may be seen in *The Seven Against Thebes*. Although once again, there are a number of one or two line rearrangements that may be pointed to, the most significant examples come at the end of the play, after the deaths of Eteocles and Polynices, when Antigone and Ismene appear on the stage. The English of the manuscript follows faithfully the ascription of speeches found in the Tauchnitz (an ordering which is radically different from that found in the OCT). An allied kind of textual difference between Tauchnitz and the OCT is in the lineation and division of speeches among parts of the chorus, in which, once more, Thoreau follows his source" (Anglen, "The Sources for Thoreau's Greek Translations," p. 294). The end of the play is disputed in yet another sense. Sommerstein, for instance, argues that the end of the play is apocryphal: "it becomes unlikely that Antigone and her sister Ismene (862) originally figured at all in a play. . . . Clearly the interpolator wanted to have the sisters sing the antiphonal lament 961–1004 (which Aeschylus doubtless wrote for sections, or section-leaders of the chorus) but did not want to bring them on immediately before it and so break up the continuous sequence of lyric lament" (*Seven against Thebes*, trans. Sommerstein, pp. 147–148).

12. Some scholars understand this attribution of the paean-sound to the dead to be metaphorical only. In contrast to Loraux's "literal" reading of the force of the paean, Richard Segal proposes a figural reading: "A later poet like Aeschylus, who vividly recreates the shrill sound of the voice and the thud of breast-beating in ritual lamentation, particularly in the great *kommos* of the *Choephoroe* (306–478), allows the ritual chanting to slip into metaphor. So, for

example, the chorus prays that 'in place of the dirges at the tomb a paean may *bring back* the beloved (Orestes) with new force'" ("Song, Ritual and Commemoration in Early Greek Poetry and Tragedy," *Oral Tradition* 4 [1989]: 342–345).

13. After Antigone and Ismene have entered, the voice of the chorus divides into two semi-choruses, but, as Sommerstein notes, in certain places (877, 895, 933) "the text makes it clear that this lament is sung by two (groups of) voices in alternation, and many mss. mark the change-points fairly regularly (though none marks them all)" (*Seven against Thebes*, trans. Sommerstein, p. 245). Thoreau's translation registers the alternation according to Tauchnitz's edition.

14. Loraux, *Mothers in Mourning*, p. 103. On many occasions in the play the grief transcends persons, becoming an acoustic phenomenon traversing the city: "Surely an echo of the houses against / Them / Sends forth piercing lamentation, / Groaning for itself, suffering-for itself" (ST, 103), or: "And A groan has gone through / The city, the towers groan, / Groans the man-loving plain" (ST, 103).

15. "Head-beating, a common gesture of mourning" (*Seven against Thebes*, trans. Sommerstein, p. 242). Charles Segal's reading suggests that this could be a tragic figure he identifies as a paradoxical "motif of 'negated song' or 'unchorused dance'" (Segal, *Song, Ritual and Comemoration in Early Greek Poetry and Tragedy*, p. 343).

16. For an excellent discussion of the various ship figures in the play, see H. D. Cameron, *Studies on the Seven against Thebes of Aeschylus* [The Hague: Mouton, 1971], pp. 58–73; for discussion of "black sails," see p. 69.

17. Sommerstein translates: "Grieving has spread right through the city: / the walls groan, and so / does the soil that loved these men; their property / awaits [new owners], / that property over which a dreadful fate came to them, /over which came strife and death as its end" (*Seven against Thebes*, trans. Sommerstein, 900–905; p. 249). Tauchnitz's edition renders: δι ών νεικος έβα / καί θανάτου τέλος (905–906; p. 118).

Cemeteries of the Brain

1. In a remarkably rigorous way Thoreau's trope here references both ancient Greek and contemporary astronomy. From Anaximander to contemporary science, researchers have analyzed how the bronze color of the moon results from its eclipse. For instance, in John Farrar, *An Elementary Treatise on Astronomy*, 2nd ed. (Boston: Hilliard, Gray & Co., 1834, p. 235)—a work that Thoreau studied in Joseph Lovering's course on natural philosophy at Harvard—a lunar eclipse is said to result from the moon's entering the space between the "opaque body [of the earth]" and its shadow, while Asa Smith's *Illustrated Astronomy* (Boston: Chase & Nichols, 1848, p. 31, arguably the most popular American astronomy handbook of the nineteenth century, reiterated Farrar's theories of opaque shadows almost verbatim, offering a series of illustrations to show the surface of the moon during the eclipse as precisely bronze in color. Using Farrar's terminology in a passage on which Thoreau draws, Smith defines a lunar eclipse as the moment when the moon is caught

in the shadow cast by the earth. And just as in Thoreau, that shadow is represented as an enclosed pyramid constructed by the casting of shadows: "All opake bodies cast a shadow when the rays from any luminous body fall upon them. Every primary and secondary planet in the solar system casts a shadow towards that point of the heavens which is opposite to the sun. . . . A total eclipse of the moon occurs when the whole of the moon is immersed in the earth's shadow" (ibid., p. 31).

2. It might appear that Sharon Cameron's "Representing Grief" detects in Emerson's "Experience" modes of mourning similar to what I am describing as Thoreau's "discovery." Her central claim is that the "Waldo" talked about explicitly at the beginning of the essay ostensibly disappears from it only to organize and dominate it by being introjected—encrypted or enshrined—into it (Sharon Cameron, "Representing Grief," in *Impersonality: Seven Essays* [Chicago: University of Chicago Press, 2007], p. 75). But her reading never claims that encryption of loss within the mind was Emerson's explicit theory. Rather, she detects crypts in Emerson's essays by following the psychoanalysis of Nicolas Abraham and Maria Torok, as well as Jacques Derrida's account of it.

3. John Farrar, *An Experimental Treatise on Optics, Comprehending the Leading Principles of the Science (The Third Part of A Course of Natural Philosophy)* (Cambridge, MA: Hilliard and Metcalf, 1826), pp. 221–222. For the classes Thoreau took at Harvard, see Kenneth Walter Cameron, *Thoreau's Harvard Years* (Hartford, CT: Transcendental Books, 1966), p. 17.

Pindar's Doubles

1. Wai Chee Dimock, *Through Other Continents: American Literature across Deep Time* (Princeton, NJ: Princeton University Press, 2006), p. 13.

2. Henry David Thoreau, "Fragments of Pindar," in *Translations*, ed. Kevin P. van Anglen, The Writings of Henry D. Thoreau (Princeton, NJ: Princeton University Press, 1987), p. 124.

3. As classicists explain, the mythology of Pindar's Odes in many respects reclaimed Homer while simultaneously innovating by integrating Pythagorean and Orphic doctrines. Thus, as Erwin Rohde remarked, Pindar's Odes not only proposed the heroic existence to be an "exalted state of being, after [hero's] departure from the earth," but more radically suggested that even "when death has occurred . . . elevation to a higher life remains possible." Because humans are in that way able to live eternally, in Pindar the barrier between finite and infinite or "between men and gods is not insuperable" (*Psyche: The Cult of Souls and the Belief in Immortality among the Greeks*, trans. W. B. Hillis [New York: Routledge, 1925; reprint 2001], p. 414).

4. Martin Heidegger, *Parmenides*, trans. André Schuwer and Richard Rojcewicz (Bloomington: Indiana University Press, 1992), p. 199.

5. Ibid., p. 76.

6. Martin Heidegger, Parmenides, trans. Andre Schuwer and Richard Rojcewicz (Bloomington: Indiana University Press, 1998), p. 74.

7. Steven Fink, "Variations on the Self: Thoreau's Personae in *A Week on the Concord and Merrimack Rivers*," *ESQ* 28, no. 1 (1982): 24–35, at 29.

Odysseus, the Lyric, and the Call of the Dead

1. Henry David Thoreau, "The Funeral Bell," in *Collected Essays and Poems*, ed. Elizabeth Hall Witherell (New York: Library of America, 2001), pp. 578–579.

2. Thus, in 1938, in what is still perhaps the most elaborate discussion of Thoreau's philosophy of poetry, Fred W. Lorch proposes that Thoreau conceived of poetry as an "organic outgrowth of the poet's character," so that the "outer form," the form of a poem, can only "approximate the inner form" of the self, which is always more perfect and stable than art ("Thoreau and the Organic Principle in Poetry," *PMLA* 53, no. 1 [March 1938]: 286–302; see pp. 287, 290). Similarly, while recognizing "Thoreau's prickly reaction against anything proposed by Emerson," F. O. Matthiessen argued in 1941 that Emerson and Thoreau shared an "organic" or "expressive" philosophy of art—organicism being understood as the "slow growth" of the artistic form "unfolding under the care of the poet's patient hands," through a craftsmanship that required a constant presence of the self, and "expressivity" as the externalization of a stable interiority (*American Renaissance, Art and Expression in the Age of Emerson and Whitman* [New York: Oxford University Press, 1941], pp. 153, 156). More recently, Meredith McGill has renewed the idea that we should understand Thoreau's poetry through Emerson's poetics of expression, though at the same time she importantly expands the meaning of expression to include more receptive experiences, such as hearing and reading ("Common Places: Poetry, Illocality, and Temporal Dislocation in Thoreau's *A Week on the Concord and Merrimack Rivers*," *American Literary History* 19, no. 2 [Summer 2007]: 357–374; see pp. 363–365).

3. Frank B. Sanborn, *The Life of Henry David Thoreau* (Boston: Houghton Mifflin, 1917), p. 63.

4. Lawrence Buell, *The Environmental Imagination: Thoreau, Nature Writing, and the Formation of American Culture* (Cambridge, MA: Harvard University Press, 1995), p. 145. It is, of course, unlikely that Thoreau ever read Keats's letter to Richard Woodhouse, of October 27, 1818, where Keats comes closest to Thoreau's idea of the self's nothingness as conditioning poetic utterance, describing the poet as unpoetical, untalented, without substance (with attributes only), and thus selfless: "A Poet is the most unpoetical of any thing in existence; because he has no identity—he is continually in for—and filling some other Body—The Sun, the Moon, the Sea and Men and Women who are creatures of impulse are poetical and have about them an unchangeable attribute—the poet has none; no identity" (*The Letters of John Keats, 1814–1821*, vol. 1, ed. Hyder Edward Rollins [Cambridge, MA: Harvard University Press, 1972], p. 387). Yet, it is not impossible that in his poetics Thoreau was indeed influenced by Keats. For the fourth issue of *The Dial* (July 1842–April 1843)—which published Thoreau's translation of Anacreon—also brought a letter by James Freeman Clarke on Keats's "American brother and remarks on Milton by Keats" (*The Dial* 3, no. 4 [April 1843: 495–505] [Boston: E. P. Peabody, 1843]). In a letter to the editor Clarke explained that George Keats "had preserved and highly praised John's letters and unpublished verses, the

copy of Spenser filled with this work . . . and a Milton in which were preserved in a like manner John's marks and comments. From a fly-leaf of this book, I was permitted to copy the passages I now send you," Clarke writes to the editor. Thus, even though *The Dial* didn't publish all of John Keats's comments on Milton, some of those that could have influenced Thoreau were found there. For instance, there was Keats's observation on Milton's genius as not being genius at all. Keats notes that while a genius would be satisfied with the luxury of his mind allowing him to write pleasurable poems, Milton—or, more accurately, some "sort of thing" operating in him, escaping his control—reached out into "otherness" and "breaking through the clouds" accessed the exteriority that Satan occupies ("he had an exquisite passion for what is properly, in the sense of ease and pleasure, poetical Luxury; and with that it appears to me he would fain have been content, if he could, so doing, have preserved his self-respect and feel of duty performed; but there was working in him as it were that same sort of thing as operates in the great world to the end of a Prophecy's being accomplish'd: therefore he devoted himself rather to the ardours than the pleasures of Song" [p. 497]). Keats also comments on a stanza from *Paradise Lost* in which Milton describes thoughts as powers capable of migrating from mortal to immortal mind, and from beings to things ("With dread of death to flight or foul retreat; Nor wanting power to mitigate and swage / With solemn touches troubled thoughts, and chase / Anguish and doubt and fear and sorrow and pain / From mortal or immortal minds" [p. 503]). His attitude at that point is close to Thoreau's, and identifies the thingness of thoughts, their impersonality, as something that simply "is" and about which scarcely anything can be said: "The light and shade—the sort of black brightness—the ebon diamonding—the Ethiop Immortality—the sorrow, the pain . . . —the thousand Melancholies and Magnificences of this Page—leaves no room for anything to be said thereon but 'so it is' " (p. 502).

5. See Vladimir Jankélévitch, *Music and the Ineffable*, trans. Carolyn Abbate (Princeton, NJ: Princeton University Press, 2003), p. 20.

6. Walter Harding, *The Days of Henry Thoreau: A Biography* (Princeton, NJ: Princeton University Press, 1982), p. 136.

7. "We sighing said, 'Our Pan is dead; / His pipe hangs mute beside the river; / Around it wistful sunbeams quiver, / But Music's airy voice is fled. / Spring came to us in guise forlorn; / The bluebird chants a requiem; / The willow-blossom / waits for him;— / The Genius of the wood is gone. / Then from the flute, untouched by hands, / There came a low, harmonious breath: / For such as he there is no death;— / His life the eternal life commands; / Above man's aims his nature rose. / The wisdom of a just content / Made one small spot a continent, / And turned to poetry life's prose." (Louisa May Alcott, originally appeared in *Atlantic*, 12 [Summer 1863]: 280.)

8. Jankélévitch, *Music and the Ineffable*, p. 5.

9. Plato, "Republic," in *The Collected Dialogues*, ed. Edith Hamilton and Huntington Cairns, trans. Paul Shorey (Princeton, NJ: Princeton University Press, 1961), 399d, e.

10. Nicole Loraux, *The Mourning Voice: An Essay on Greek Tragedy*, trans. Elizabeth Trapnell Rawlings (Ithaca, NY: Cornell University Press, 2002), p. 61.

11. Jankélévitch, *Music and the Ineffable*, p. 5.
12. Plato, "Republic," 399d.
13. Loraux, *The Mourning Voice*, p. 61.
14. Ibid., p. 67.
15. Robert Graves, *The Greek Myths*, vol. 1 (New York: Penguin Books, 1960), p. 112.
16. Ibid.
17. Ibid.
18. Ibid.
19. Ibid.
20. Jane Bennett, *Thoreau's Nature: Ethics, Politics, and the Wild* (Lanham, MD: Rowman & Littlefield, 2002), pp. 19, 97.
21. Clive Scott, *The Poetics of French Verse: Studies in Reading* (Oxford: Clarendon Press, 1998), pp. 102–103.
22. Ibid. Similarly, in proposing a phenomenology of prosody, Simon Jarvis suggests that "rhyme is part of what allows a subjectivity thus conceived to sustain itself. . . . Rhyme marks off the time of innerness against the world's time. Rhyme through the return of similar sounds, does not merely lead us back to those sounds. It leads us back to ourselves. Rhyme is part of what makes Christian and ultimately modern subjects possible" ("Musical Thinking: Hegel and the Phenomenology of Prosody," *Paragraph* 28, no. 2 [2005]: 57–71, p. 64).
23. Review of *Walden*, *New York Times*, September 22, 1854, p. 3 (reprinted in *Emerson and Thoreau: The Contemporary Reviews*, ed. Joel Myerson [Cambridge: Cambridge University Press, 1992], p. 389); Elizabeth Barstow Stoddard, "Letter from a Lady Correspondent," *Daily Alta California*, October 8, 1854, p. 5 (reprinted in Myerson, *Emerson and Thoreau*, p. 394); James Russell Lowell, "Thoreau," in *Literary Criticism of James Russell Lowell*, ed. Herbert F. Smith (Lincoln: University of Nebraska Press 1969), pp. 225–226.
24. Joseph Wood Krutch, *Henry David Thoreau* (New York: William Morrow & Company, 1974; originally published in 1948), pp. 95, 86, 97. Krutch related these deficiencies to the "absence of balance and wholeness of [Thoreau's] personality," and considered that Thoreau's powerlessness to maintain the basic demands of prosody was somehow innate to his thinking (p. 101).
25. J. Lyndon Shanley, *The Making of "Walden"* (Chicago: University of Chicago Press, 1957), pp. 84, 85, 87.
26. Ibid., pp. 75–76.
27. Walter Benn Michaels, "Walden's *False Bottoms*," in *Walden and Resistance to Civil Government*, 2nd ed., ed. William Rossi (New York: Norton, 1992), pp. 420, 411. (Originally published in *Glyph* 1 [1977]: 132–149.)
28. Perry Miller, *Consciousness in Concord* (Boston: Houghton Mifflin, 1958), p. 35.
29. Deborah Elise White, *Romantic Returns: Superstition, Imagination, History* (Stanford, CA: Stanford University Press, 2000), p. 66.
30. Sharon Cameron, *Choosing Not Choosing: Dickinson's Fascicles* (Chicago: University of Chicago Press, 1992), p. 15.

31. That is why everything in Thoreau's *Journal* hinges on quantity rather than quality, or why it can't be summarized or abbreviated, and thus ideologized. If no edition satisfies—except of course the Princeton Edition that reproduces it in its entirety—that is not the failure of its editors. For each abbreviated edition subjectivizes the *Journal* according to the editor's preferences, precisely repressing what the *Journal* is about, namely, an the impersonal objectiveness that resists figuration or summation, something that Thoreau tried to engender from the time of his early experimentation in emptying the mind.

Coda: Melville's Seafowls

1. Herman Melville, "The Encantadas, or Enchanted Isles," in *The Piazza Tales*, ed. Harrison Hayford, Alma A. MacDougall, and G. Thomas Tanselle (Evanston, IL: Northwestern University Press, 1996), pp. 135, 127.
2. Herman Melville, *"Mardi" and "A Voyage Thither,"* ed. Harrison Hayford, Hershel Parker, and G. Thomas Tanselle (Evanston, IL: Northwestern University Press, 1998), p. 157.
3. Herman Melville, "Benito Cereno," in *The Piazza Tales*, pp. 48, 46.
4. Herman Melville, *Moby-Dick*, ed. Hershel Parker and Harrison Hayford, Norton Critical Edition (New York: W. W. Norton & Company, 2002), p. 427.
5. Herman Melville, *Billy Budd, Sailor*, ed. Harrison Hayford and Merton M. Sealts Jr. (Chicago: University of Chicago Press, 1962), p. 126.
6. Ibid., p. 127.
7. Ibid.

II Introduction: Harvard Vitalism and the Way of the Loon

1. Robert D. Richardson Jr., *Henry Thoreau* (Oakland: University of California Press, 1988), p. 71.
2. *The Dial* 3, no. 3 (January 1843): 363.
3. Jean-Pierre Vernant, *Myth and Society in Ancient Greece*, trans. Janet Lloyd (New York: Zone Books, 1990), p. 188.
4. Ibid., p. 191.
5. Robert Hunt, *The Poetry of Science or Studies of the Physical Phenomena of Nature* (1848; reprint, London: Henry G. Bohn, 1854), p. 35.
6. Ibid., pp. 1, 7, 6.
7. Ibid., p. 3.
8. Ibid., p. 38.
9. Jean-Baptiste Lamarck, *La Flore française, ou la description succinte de toutes les plantes qui croissent naturellement en France*, vol. 1 (Paris: De L'Imprimerie Royale, 1778), p. 1. The translation is mine.
10. Lamarck, *La Flore française*, vol. 1, rev. ed. (Paris: H. Agasse, 1805), p. 61. I am grateful to David Wills for drawing my attention to this difference in the editions and for assisting me with the translation.

11. William Ellery Channing, *Thoreau, the Poet-Naturalist: With Memorial Verses* (Boston: Charles E. Goodspeed, 1902), p. 75.
12. Maurice Merleau-Ponty, *Nature: Course Notes from the Collège de France*, ed. Dominique Séglard, trans. Robert Vallier (Evanston, IL: Northwestern University Press, 2003), p. 3.
13. Noam Chomsky and Michel Foucault, "Human Nature: Justice versus Power," in *Foucault and His Interlocutors*, ed. Arnold I. Davidson (Chicago: University of Chicago Press, 1997), p. 110.
14. Ralph Waldo Emerson, "Nature," in *Essays and Poems*, ed. Joel Porte (New York: Library of America, 1983), p. 7.
15. Martin Heidegger, *Aristotle's Metaphysics Θ 1–3, On the Essence and Actuality of Force*, trans. Walter Brogan and Peter Warnek (Bloomington: Indiana University Press, 1995), p. 57. Heidegger's discussion is arguably the most sustained philosophical analysis of this tradition of vitalism, from Aristotle via Leibniz to Kant.
16. Ibid., p. 81.
17. Ibid., p. 86.
18. Ibid.
19. Muriel Combes, *Gilbert Simondon and the Philosophy of the Transindividual*, trans. Thomas LaMarre (Cambridge, MA: MIT Press, 2012), p. 1.
20. Ibid.
21. Henry David Thoreau, "Natural History of Massachusetts," in *Collected Essays and Poems*, ed. Elizabeth Hall Witherell (New York: Library of America, 2001), pp. 38–39.
22. Anne Sauvagnargues, "Crystals and Membranes: Individuation and Temporality," in *Gilbert Simondon: Being and Technology*, ed. Arne De Boever, Alex Murray, Jon Roffe, and Ashley Woodward, trans. Jon Roffe (Edinburgh: Edinburgh University Press, 2012), pp. 56, 69.
23. Jane Bennett, "A Vitalist Stopover on the Way to New Materialism," in *New Materialisms: Ontology, Agency, and Politics*, ed. Diana Coole and Samantha Frost (Durham, NC: Duke University Press, 2010), p. 63.
24. Gilles Deleuze, quoted in Eugene Thacker, *After Life* (Chicago: University of Chicago Press, 2010), p. 187.
25. Thacker, *After Life*, p. 135.
26. I have here in mind Esposito's understanding of the Dionysian in Nietzsche: "The Dionysian is life itself in absolute (or dissolute) form . . . abandoned to its original flow . . . [it] not only escapes determinacy, but is also the greatest power of inderminacy." Roberto Esposito, *Bios, Biopolitics and Philosophy*, trans. Timothy Campbell (Minneapolis: University of Minnesota Press, 2008), p. 89.
27. Emerson, "Nature," in "Essays, Second Series," in *Essays and Poems*, ed. Joel Porte (New York: Library of America, 1983), p. 547.
28. Ibid.
29. William James, *Essays in Radical Empiricism* (Lincoln: University of Nebraska Press, 1997), p. 27.
30. Ibid., p. 41.
31. Ibid., p. 37.

32. Ibid., p. 23.
33. Gilles Deleuze, *Difference and Repetition*, trans. Paul Patton (New York: Columbia University Press, 1994), p. 41.
34. Ibid., p. 45.
35. Gillian Beer, *Darwin's Plots: Evolutionary Narrative in Darwin, George Eliot and Nineteenth-Century Fiction* (Cambridge: Cambridge University Press, 2009), p. 104.
36. For Lyell's understanding of the return of the species, see Stephen Jay Gould, *Time's Arrow, Time's Cycle: Myth and Metaphor in the Discovery of Geological Time* (Cambridge. MA: Harvard University Press, 1987), pp. 101–102.
37. Roland Barthes, *Michelet*, trans. Richard Howard (New York: Hill & Wang, 1987), p. 32.
38. Emerson, "The Method Nature," in Porte, *Essays and Poems*, p. 120.
39. Samuel Taylor Coleridge, *Hints towards the Formation of a More Comprehensive Theory of Life*, ed. Seth B. Watson (London: John Churchill, 1848), p. 47.
40. Ibid. p. 48.
41. Ibid., p. 22.
42. Ibid., p. 40.
43. Ibid., p. 49.
44. Ibid.
45. David Amigoni, *Colonies, Cults and Evolution: Literature, Science and Culture in Nineteenth-Century Writing* (Cambridge: Cambridge University Press, 2007), p. 41.
46. Coleridge, *Hints*, p. 49.
47. Amigoni, *Colonies*, p. 41.
48. Robert Sattelmeyer and Richard A. Hocks, "Thoreau and Coleridge's 'Theory of Life,'" *Studies in the American Renaissance* (1985): 269–284, at p. 280.
49. Coleridge, *Hints*, p. 33.
50. Thoreau, "Natural History of Massachusetts," p. 39.
51. Gilles Deleuze, *Francis Bacon: The Logic of Sensation*, trans. Daniel W. Smith (Minneapolis: University of Minnesota Press, 2003), pp. 52–53.
52. Sattelmeyer and Hocks, "Thoreau and Coleridge's 'Theory of Life,'" p. 270.
53. William Smellie, *The Philosophy of Natural History: Introduction and Various Additions and Alterations by John Ware* (Boston: William J. Reynolds and Company, 1851), p. v.
54. Kenneth Walter Cameron, *Thoreau's Harvard Years* (Hartford, CT: Transcendental Books, 1966), p. 18.
55. Smellie, *The Philosophy of Natural History*, p. 2.
56. Ibid., p. 14.
57. Quoted in Howard A. Kelly and Walter L. Burrage, *American Medical Biographies*, vol. 1 (Baltimore: Norman, Remington, 1920), p. 1191.
58. Giorgo Agamben understands Bichat's distinction between vegetal and animal life as the difference between bare life, which is without qualities, and qualitative, conscious life, arguing that "still today, in discussion about the definition *ex lege* of the criteria for clinical death, it is a further identification of this bare life—detached from any brain activity and, so to speak, from any

subject—which decides whether a certain body can be considered alive or must be abandoned to the extreme vicissitude of transplantation." Giorgo Agamben, *The Open: Man and Animal*, trans. Kevin Attell (Stanford, CA: Stanford University Press, 2002), pp. 14–15. Although this distinction is certainly the premise of many contemporary legal decisions regarding clinical death, Bichat was far from understanding vegetal life as "bare" or without qualities just because it was unconscious. To the contrary, for Bichat unconscious vegetal life is qualified. It is so many legal fictions, as well as Agamben's thesis itself, that have abandoned it to what is without qualities.

59. Xavier Bichat, *Physiological Researches upon Life and Death*, trans. Tobias Watkins (Philadelphia: Smith & Maxwell, 1809), p. 5.

60. Ibid.

61. Arnold Guyot, *The Earth and Man: Lectures on Comparative Physical Geography, in Its Relation to the History of Mankind*, trans. C. C. Felton (Boston: Gould and Lincoln, 1849), p. 93.

62. Ibid., pp. 94, 95.

63. Among other observations, Thoreau copies this from Guyot: "compared with the Old World, the New World is the humid side of our planet, the *oceanic, Vegetative* world, the passive element awaiting the excitement of a livelier impulse from without" (J, 3, 183).

64. Robert M. Thorson, *Walden's Shore: Henry David Thoreau and Nineteenth-Century Science* (Cambridge, MA: Harvard University Press, 2014).

65. Esposito, *Bios*, pp. 63–65.

66. Ibid., p. 27.

67. Ibid., p. 110.

68. Ibid., p. 157.

69. Ibid., p. 158.

70. Ibid., p. 160.

71. Ibid., p. 180.

72. Ibid., p. 192.

73. Ibid., p. 194.

74. Ibid.

Birds

1. "Barzillai Frost's Funeral Sermon on the Death of John Thoreau Jr.," ed. Joel Myerson, *Huntington Library Quarterly*, 57, no. 4 (Autumn 1994): 367–376, at p. 369.

2. Ibid., pp. 370, 372.

3. Ibid., p. 373.

4. Ibid., p. 371.

5. Ibid.

6. Ibid., p. 370.

7. Henry D. Thoreau, "Nature and Bird Notes: Journal and List of Birds, kept by Sophia, John and H. D. Thoreau," manuscript, Henry W. and Albert A. Berg Collection of English and American Literature, The New York Public Library, Astor, Lenox and Tilden Foundations.

8. Ibid.

9. Christoph Irmscher, *The Poetics of Natural History: From John Bartram to William James* (New Brunswick, NJ: Rutgers University Press, 1999), p. 218.

10. John James Audubon, *Audubon and His Journals*, ed. Maria R. Audubon, 2 vols. (New York: Dover, 1986), vol. 2, pp. 524–525 (quoted in Irmscher, *The Poetics of Natural History*, p. 218).

11. Audubon's mortification of nature makes his famous bird drawings appear as faces of death, with thick, continuous lines forever imprisoning the birds in their forms, as if those lines were thus themselves an immutable category enabling the setting of taxonomies.

12. Michel Foucault, *The Order of Things: An Archaeology of the Human Sciences* (New York: Vintage, 1994), p. 130.

13. The move from experience to description is also present in Audubon's description of a Snowy Owl that he had observed in Cambridge in 1833, a commentary that John references in the notebook. But the example is emblematic of Audubon's procedures more generally—moving from the observation of the bird's habits, to discussion of the most efficient ways to kill it, and listing detailed information obtained from a Maine hunter as to what kind of traps work best. In Audubon's ornithological accounts the life of a creature appears through a network of naturalistic and cultural descriptions—through histories, precisely—hence as what is merely a representation of what is alive. John James Audubon, "Ornithological Biography," in *Writings and Drawings*, ed. Christoph Irmscher (New York: Library of America, 1999), pp. 333–336.

14. "Audubon and Nuttall would jointly publish a description of 12 new species under Townsend's name." Jeff Holt, "The Discovery of the Common," *Cassinia, A Journal of Ornithology of Pennsylvania, New Jersey and Delaware*, (Philadelphia), no. 70 (2002–2003): 31, www.dvoc.org/CassiniaOnLine/Cassinia70/C70_30_32.pdf.

15. Thomas Nuttall, *The Genera of North American Plants*, 2 vols. (Philadelphia: D. Heartt, 1818), vol. 1, p. VI.

16. Monique Allewaert, *Ariel's Ecology: Plantations, Personhood, and Colonialism in the American Tropics* (Minneapolis: University of Minnesota Press, 2013), p. 63.

17. Nuttal, *The Genera of North American Plants*, p. 3.

18. Ibid., p. V.

19. Thomas Nuttall, *A Manual of the Ornithology of the United States and of Canada* (Cambridge, MA: Hillard and Brown, 1832), p. 3.

20. Ibid.

21. Ibid., p. 13.

22. Ibid., pp. 19–20. The thrush also developed a system of gestures to express pain, anger, or disagreement. Nuttall's description of the bird is so detailed that, in his own words, it verges on "the actual biography of the [bird]," suggesting that the only "true" classification is the history of a singularity.

23. Ibid., p. 20: "A second balanced itself on the head, with its claws in the air. A third imitated a milkmaid going to market. . . . The sixth was a cannonier, with a cap on its head and acted as wounded when needed."

24. Ibid.

25. Ibid.

26. Foucault, *The Order of Things*, p. xv.
27. For instance, in the order of "rapacious" birds—which comes first in both John's and Nuttall's accounts—John will list only the Falcon and "Nocturnal Birds of Prey or Owls," reserving a page for each, but the pages are left blank, and no description is provided.
28. Thoreau, "Nature and Bird Notes."
29. Second entry: "May 1842 A female Oriole has her nest near the house, and comes constantly to the fir-trees by the window for the sweet drops ejected by the aphides (as do the Blackpoll warblers—the Chipping sparrows and another bird whose name I can not learn) with a head nearly yellow—and only streaked with dark." See photograph: Thoreau, Baltimore Oriole, in "Nature and Bird Notes."
30. "Great Horned or Cat Owl—apparently a young female—22 inches by 4 feet—shot in Lincoln woods by Mr. Geo. Farrar—sitting upon a tree at 3 o'clock P.M. May 31st 1842. The rest of the family seen near the spot. The half of a hare left by them. June 3d 1842. Visited the nest of the owl—or his hunting lodge—near the top of a large white pine tree. Formerly the haunt of a male raccoon—in a thick swamp between two hills." See photograph: Thoreau, Nocturnal Birds, in "Nature and Bird Notes."
31. Thoreau, "Nature and Bird Notes."
32. I quote Aristotle from "Metaphysica," in *The Basic Works of Aristotle*, ed. Richard McKeon, trans. W. D. Ross (New York: Random House, 1941), p. 693 (983b). Thoreau had already studied Aristotle while at Harvard; see Kenneth Walter Cameron, *Thoreau's Harvard Years* (Hartford, CT: Transcendental Books, 1966), p. 95. The editors of Thoreau's *Journal* specify that Thoreau's discussion of Thales is based on his readings of Ralph Cudworth, *The True Intellectual System of the Universe*, 4 vols. (London: J. F. Dove, for Richard Pristley, 1820). The first part of that book is dedicated to first philosophers, which Cudworth discusses via Aristotle's first book of Metaphysics, from which I quoted the section on Thales. Cudworth treats the first philosophers—Thales, Anaximander, and Anaxagora—as a group, without distinguishing their respective theories. Because all of Cudworth's discussions are guided by one goal only—to give infinite privilege to the Christian-Platonic principle of "spiritual" creation—throughout the book he identifies the first philosophers as "atheists," disqualifying them as holders of the "vulgar opinion" "that matter is the only substance," "that there is no other substance besides body and matter" (p. 77); this materialistic orientation culminated with Anaxogora, who, "looking at the maxims of Italic philosophers that nothing could be physically made out of nothing, or no real entity generated or corrupted . . . concluded that therefore in generations and alterations, these were not made out of nothing, and animated into nothing, but that every thing was naturally made" (p. 85). On the basis of his reading of Cudworth, then, Thoreau wouldn't have mistaken the materialistic orientation of the first philosophers. It is therefore important to register that unlike Cudworth, who dismisses such an orientation as "vulgar," Thoreau praises its materialism and vitalism—the idea that nothing "animated vanishes into nothing"—as a philosophy that the last philosopher has yet to reach.

33. Martin Heidegger, *Basic Concepts of Ancient Philosophy*, trans. Richard Rojcewicz (Bloomington: Indiana University Press, 2008), p. 29.
34. Ibid., p. 28.
35. Aristotle, "De Anima," in *The Basic Works of Aristotle*, ed. Richard McKeon, trans. J. A. Smith (New York: Random House, 1941), p. 541 (405a20).
36. Ibid, p. 553 (411a9).
37. Heidegger, *Basic Concepts*, p. 181.
38. In Thoreau's strange taxonomy a cat—a specimen—assumes the role of a whole species by acting as a missing link between birds and quadrupeds, precisely because it is a quadruped whose run resembles flight and draws a continuous waving line; all cats are therefore beings of passage, quadruped-birds.

Fossils

1. Most readers of Thoreau's discourse on fish read it anthropocentrically, as suggestive of human ethics. Sherman Paul (*The Shores of America: Thoreau's Inward Exploration* [Urbana: University of Illinois Press, 1958]) was among first to offer a sustained analysis of Thoreau's "fish principle." In his idealizing reading, the "fish principle" is understood not to concern fish at all but rather practices of thinking, whereas nature is dematerialized into a symbol of "spirit": "Speaking of fishes, Thoreau was in effect speaking of thoughts, of a mode of the spirit. . . . The fishes, indeed, were a catalog of thoughts" (p. 213). Lawrence Buell's influential *Literary Transcendentalism* (Ithaca, NY: Cornell University Press, 1973) revised Paul's claim; while not denying the symbolic aspect of the discourse, it signaled nevertheless its scientific importance: "Here Thoreau's 'scientific' side comes to the fore more than anywhere else in the book, but in a casual way he also uses ichthyology for metaphorical purposes" (p. 211). More recently Philip Cafaro ("Thoreau's Environmental Ethics in *Walden*," *Concord Saunterer, Journal of the Thoreau Society* 10 [2002]: 17–63) has offered a fascinating reading of Thoreau's fish discourse, understanding it, through Locke's epistemology, not as a series of propositions about ichthyology or life more generally, but as a "challenge to conventional ethics" whose importance derives from its moving away "from facts to values and from description to prescription" to end with "demands of interspecific justice" (pp. 19, 20). Interestingly, readers invested in Thoreau's scientific discourse tend to disregard his discourse on fish. Nina Baym's "Thoreau's View of Science" (*Journal of the History of Ideas* 26, no. 2 [1963]: 221–234) doesn't mention it at all, whereas Laura Dassow Walls (*Seeing New Worlds: Henry David Thoreau and Nineteenth-Century Natural Science* [Madison: University of Wisconsin Press, 1995]) treats it as epistemological in its orientation; it proposed "knowing as an activity of sympathy," and related how Thoreau "had the capacity to be 'touched' by fish" (p. 46).
2. I here differ from Richardson, who believes Thoreau was exposed to Cuvier only in 1851 and after reading Darwin; see Robert D. Richardson Jr., *Henry Thoreau: A Life of the Mind* (Berkeley: University of California Press, 1989), p. 254. It is difficult to know exactly when Thoreau was first introduced

to Cuvier. But taking into account how influential the book became among Harvard professors and how invested in questions of natural history John Thoreau was, it is probable that Thoreau knew some of Cuvier's arguments even before attending Harvard. The English translation of Cuvier's book was commissioned by Robert Jameson, professor of natural history at Edinburgh and tutor of Charles Darwin. For the history of English translations and publications of the "Preliminary Discourse," see Martin J. S. Rudwick, "Preface," in *George Cuvier: Fossil Bones, and Geological Catastrophes* (Chicago: University of Chicago Press, 1997), p. xi.

3. Kenneth Walter Cameron, "Tabular View of Thoreau's Courses," in *Thoreau's Harvard Years* (Hartford, CT: Transcendental Books, 1966), p. 20.

4. William Smellie, *The Philosophy of Natural History, Revised by John Ware* (Boston: William J. Reynolds, several editions, 1824–1851). John Ware, *The Philosophy of Natural History* (Boston: Tragard and Thompson, 1866); the quotation is from p. 16.

5. Georges Cuvier, "Preliminary Discourse," in Rudwick, *George Cuvier*, p. 183.

6. Ibid., p. 186.

7. Ibid., p. 205.

8. Ibid., pp. 186, 190.

9. François Jacob, *The Logic of Life: A History of Heredity*, trans. Betty E. Spillmann (Princeton, NJ: Princeton University Press, 1973), p. 110; Edward Lurie, *Louis Agassiz: A Life in Science* (Baltimore: Johns Hopkins University Press, 1988), p. 61.

10. As Martin J. S. Rudwick remarks for Cuvier "there was no evidence to support a transformist view of life: human and animal mummies showed there had been no organic change since the time of the ancient Egyptians. There was also no evidence of gradual change between fossil and living animal species; on the contrary, all the evidence pointed to the reality of extinction. In particular, no human fossils were known." (Rudwick, *George Cuvier*, p. 76.)

11. Moreover, as François Jacob argues, if there are similarities among branches, species, or genera, they are only superficial: "It is on the surface that small continuous series of numerous variations can be found. In depth there can be only radical changes, only leaps from one plane to another." (Jacob, *The Logic of Life*, p. 111.)

12. Louis Agassiz, *Recherches sur les poissons fossiles*, vol. 1 (Neuchâtel, Switzerland: Petitpierre, 1833). In 1833 the *American Journal of Science*—available to Thoreau through the Boston Society of Natural History—reprinted from the *Edinburgh Journal of Science* excerpts from Agassiz's first volume: Robert Jameson, ed., "Poissons Fossiles par L'Agassiz—Great Work of Prof. Agassiz on Fossil Fishes," *American Journal of Science* 28 (1833): 193–236.

13. Agassiz, *Recherches sur les poissons fossiles*, p. 19.

14. See Martin J. S. Rudwick's discussion of how attitudes toward marine fossils changed during the eighteenth century (*The Meaning of Fossils: Episodes in the History of Paleontology* [Chicago: University of Chicago Press, 1976], pp. 76–90). Rudwick locates the change in Woodward and Scheuchzer: "In his

Complaints and Claims of the Fishes (1708), written probably to counter Karl Lang's use of Lhwyd's theory for the interpretation of Swiss fossils, Scheuchzer made the fossil fish of Oeningen speak up to defend their own organic origin against those who denied they had ever been alive, and to emphasize their witness to reality of the Deluge" (p. 87).

15. Georges Cuvier, "Special Importance of the Fossil Bones of Quadrupeds," in Rudwick, *George Cuvier*, p. 207.

16. Agassiz, *Recherches sur les poissons fossiles*, pp. 20–21. For a history of discussion regarding the mobility of strata before Agassiz, especially in Hutton, see Stephen Jay Gould, *Time's Arrow, Time's Cycle: Myth and Metaphor in the Discovery of Geological Time* (Cambridge, MA: Harvard University Press, 1987), pp. 83–91.

17. "Visit of Prof. Agassiz to Mr. Mantell's Museum at Brighton," Editorial, quoting Agassiz, *American Journal of Science* 28 (1935): 194.

18. As Johnson reconstructs, the first draft of *A Week* was written in 1845, the second in 1846–1847. It was "revised and expanded at intervals during the following two years," finally being published in 1849 (Linck C. Johnson, *Thoreau's Complex Weave: The Writing of A Week on the Concord and Merrimack Rivers* [Charlottesville: University Press of Virginia, 1986], p. xi). As Harding reconstructs, Thoreau worked particularly hard on the *A Week* manuscript in Spring 1847. On March 12, 1847, he thought it was nearing the completion and "had Emerson write Duyckinck," to recommend the book. "Duyckinck replied immediately that he would be willing to consider the manuscript, but it was May 28 before Thoreau got around to mailing it. There were corrections and revisions he found that he wanted to make and they took time. And Duyckinck had the manuscript only two weeks when Thoreau asked for it back again to make still further corrections. He returned it to them on July 3." (Walter Harding, *The Days of Henry Thoreau: A Biography* [Princeton, NJ: Princeton University Press, 1982], p. 244.) But plans for publication fell through, so Thoreau kept working on the second draft, completing it in 1847: "Unable to find a publisher, he continued to work on the manuscript until 1849," when he published it at his own expense (Johnson, *Thoreau's Complex Weave*, p. xi).

19. "During the winter of 1846–47 thousands of Bostonians crowded into the Tremont Temple, on some evenings as many as five thousand of them, to hear the foreign professor. So great was public interest that Agassiz had to repeat his lectures each day to a second audience" (Lurie, *Louis Agassiz*, p. 126). The talks were published as: Louis Agassiz and A. A. Gould, *Principles of Zoology: Touching the Structure, Development, Distribution, and Natural Arrangement of the Races of Animals, Living and Extinct*, Part 1, "Comparative Physiology" (Boston: Gould, Kendall and Lincoln, 1848). I quote from this edition.

20. Ibid., p. xiv.

21. Ibid., pp. xiv–xv.

22. Thoreau's "Long-Book," containing notes for *A Week*, suggests that he only gradually overcame his opposition to Agassiz's political ichthyology. Until August 1844 he claimed there were "steady revolutions of nature," but

by the fall of 1844 he abandoned that idea to accept fish and human life as contemporaneous (J, 2, 89, 102, 108). For the Agassiz-Darwin controversy, see Harding, *The Days of Henry Thoreau*, p. 383.

23. Agassiz, *Recherches sur les poissons fossiles*, p. 20.

24. Thoreau read the following editions. Charles Darwin, *Journal of Researches into the Natural History and Geology of the Countries Visited during the Voyage of H. M. S. Beagle round the World, under the Command of Capt. Fitz Roy, R.N.*, 2 vols. (New York: Harper & Brothers, 1846). Immediately after reading Darwin, Thoreau read Cuvier for the first time: Baron Georges Léopold Chrétien Frédéric Dagobert Cuvier, *The Animal Kingdom Arranged in Conformity with Its Organization . . . with Additional Descriptions of All the Species Hitherto Named, and of Many Not before Noticed, by Edward Griffith . . . and Others*, 16 vols. (London: Printed for Geo. B. Whittaker, 1827–1832).

25. Edward Tuckerman, *An Enumeration of North American Lichenes* (Cambridge, MA: John Owen, 1845), p. 1. The term "affinity" belongs to the arsenal of classical chemistry and botany (that of Goethe and Linnaeus) and signifies attractions enacted not only by agreement of functions or similarity of forms but also by oppositional differences.

26. Ibid., p. 4.

27. For "affinities" in Darwin, see Beer: "For Darwin, the old conviction that nature does not make leaps opened into some of his most radical insights, leading him away from the idea of the chain of being or the ladder, with its hierarchical ordering of rungs, towards the ecological image of the 'inextricable web of affinities'. These affinities he perceives sometimes as kinship networks, sometimes as tree, sometimes as coral, but never as a single ascent with man making his way upward" (Gillian Beer, *Darwin's Plots: Evolutionary Narrative in Darwin, George Eliot and Nineteenth-Century Fiction* [Cambridge: Cambridge University Press, 2009], pp. 18–19). The radicalism of the Tuckerman-Thoreau position consists in its devaluing the boundaries of "kinship." They understand affinities as a force transcending kinships toward singular branches and then even further toward disconnected particulars.

28. Tuckerman, *North American Lichenes*, p. 4. Botanists and zoologists of Tuckerman's time recognized that Linnaeus's classifications were too artificial, and that vegetal life was more mobile and relational than Linnaeus ever believed. Thus more and more naturalists were leaning toward a democratization of natural history, engaging in debate about methods of classifying. From that point of view, Tuckerman was participating in, rather than initiating, a discussion. However, it was less Tuckerman's democratization of taxonomic practices than his suggestion that, philosophically speaking, life might be composed of "promiscuous" connections that struck Thoreau as offering radical possibilities for rethinking the living. For his own part, realizing that what he was saying about the nature of life would subvert even minimal scientific efforts to classify botanic families, Tuckerman decided not to follow his own insight. For an analysis of the state of botanical and zoological classification between 1825 and 1840 in Europe, and the problems that unstable nomenclature caused in classifying genera and species, see M. Eulàlia

Gassó Miracle, "On Whose Authority? Temminck's Debates on Zoological Classification and Nomenclature: 1820–1850," *Journal of The History of Biology* 44, no. 3 (Spring 2011): 445–481.

29. As Georges Canguilhem explains, nineteenth-century sciences of life collapsed the distinction between abnormal and anomalous: "Lalande's *Vocabulaire* [*philosophique*] shows that confusion of an etymological nature has helped draw anomaly and abnormal closer together. . . . In a strictly semantic sense 'anomaly' points to a fact, and is a descriptive term, while 'abnormal' implies reference to a value and is an evaluative, normative term; but the switching of good grammatical methods has meant a confusion of the respective meanings of anomaly and abnormal. 'Abnormal' has become a descriptive concept and 'anomaly,' a normative one." (Georges Canguilhem, *The Normal and the Pathological*, trans. Carolyn R. Fawcett [New York: Zone Books, 1991], pp. 131–132.)

30. Geoffroy Saint-Hilaire divided anomalies into "varieties, structural defects, heterotaxy and monstrosities," all of which represented more or less complex deviations of an individual within a relatively stable "specific type"; see Canguilhem, *The Normal and the Pathological*, pp. 133–138; for Darwin, see Beer, *Darwin's Plots*, pp. 59, 73. Darwin's far less intense insistence on the rights of the anomalous was received by many Victorian readers as bordering on the fictional. As Jonathan Smith reconstructs, Ruskin for instance, voiced the deep disturbance Darwin produced in the mainstream Victorian mind when Ruskin reproached him for supporting the possibility that plants could become animals: "Speaking of an unpublished pamphlet sent by an unnamed author, on 'every sort of plant that looked or behaved like an animal, and every sort of animal that looked or behaved like a plant,' Ruskin makes clear his sense of the pointlessness of such an exercise: 'He gave descriptions of walking trees, and rooted beast; of flesh-eating flowers, and mud-eating worms; of sensitive leaves, and insensitive persons'" (Jonathan Smith, *Charles Darwin and Victorian Visual Culture* [Cambridge: Cambridge University Press, 2006], p. 167). But as Gillian Beer points out, although in Darwin "deviation, not truth to type, is the creative principle," his method nevertheless "placed great emphasis upon congruities" (Beer, *Darwin's Plots*, pp. 59, 73).

31. Harding, *The Days of Henry Thoreau*, p. 288. Thoreau's insistence on anomalous singularities will only intensify. In 1860 he reads Darwin's *On the Origin of Species* but remains underwhelmed. Of 280 pages of his common place book on "Natural History," composed in late 1858 to 1860, only six contain notes from Darwin's book, inserted between extensive notes from Aristotle's *History of Animals* and Pliny's *Natural History* (comprising more than half of the book), as if to reassert that Pliny's anomalous and fantastic taxonomies were still more appealing than Darwin's branching, with its attendant subordination. Moreover, most of the quotes Thoreau recorded from Darwin are not related to the essence of Darwin's theory (inheritance, struggle for existence, and natural selection). Instead, he notes only what Darwin himself registers as accidental, anomalous, and even incredible: the existence of mammals on oceanic islands, the anomalous presence of frogs in the mountains of New Zealand (for "frogs, toads, newts have never been found on any of the many islands with which the

great oceans are studded"); the even greater anomaly of New Zealand's diversity of flowering plants (which Darwin finds inexplicable: "we must, I think, admit that something quite independently of any difference in physical conditions has caused so great a difference in number"); or the anomalous case of Northern America, where nonindigenous, naturalized plants "are of a highly diversified nature." Reporting Asa Gray's findings, Darwin states: "in the last edition of Dr. Asa Gray's Manual of the Flora of the Northern United States 260 naturalised plants are enumerated, and these belong to 162 genera. We thus see that these naturalized plants are of a highly diversified nature. They differ, moreover, to a large extent from the indigenes, for out of the 162 genera, no less than 100 genera are not there indigenous, and thus a large proportional addition is made to the genera of these States" (Darwin, *The origin of species by means of natural selection, or the preservation of favoured races in the struggle for life*, new ed. [New York: D. Appleton, 1860], 343, 340, 108).

32. Michel Foucault, *The Order of Things: An Archeology of the Human Sciences*, trans. unacknowledged (New York: Vintage Books, 1994), p. 387.

33. In that respect Darwin appears to reconcile Thoreau's and Agassiz's opposing positions. For although he doesn't predicate his theories on categorical stability, neither does he abandon it. As Gillian Beer argues, in Darwin "categorization, classification, description, must all be understood to be implicated in movement, process, and time" (Beer, *Darwin's Plots*, p. 59).

34. Walls, *Seeing New Worlds*, p, 33; see also p. 39, where Walls suggests Thoreau's investment to be in the "specificity of actual experience."

35. Henry David Thoreau, "Natural History of Massachusetts," in *Collected Essays and Poems*, ed. Elizabeth Hall Witherell (New York: Library of America, 2001), p. 41.

36. Believing that Agassiz should, like him, be invested in knowing by endlessly particularizing—and judging by Agassiz's surprise at the diversity Thoreau presented to him, as Cabot testifies, Agassiz didn't think the amassing of "particular instances" was necessary—Thoreau inquires what it is that Agassiz specifically wants to see ("What of the above does M. Agassiz particularly wish to see?" [C, 294]). He then sends large numbers of specimens, instructing Cabot/Agassiz to let his perception follow the nuances: "I sent you 15 pouts, 17 perch, 13 shiners, 1 larger land tortoise and 5 muddy tortoises. . . . Also 7 perch, 5 shiners, 8 breams, 4 dace. . . . Observe the difference between those from the pond, which is pure water, and those from the river" (C, 302). Furthermore, because differentiation is the business of life, Thoreau's specimens are living: "Red-finned Minnows, of which I sent you a dozen alive" (C, 294); "I send you . . . one black snake, alive, and one dormouse? . . . The tortoises were all put in alive, the fishes were alive yesterday" (C, 302). In emphasizing the mobility of "particular instances" and the evasion of naming as crucial to Thoreau's understanding of life, I differ radically from John Hildebidle's argument that Thoreau's thinking on nature both starts from and is grounded in the classifications of natural sciences: "Most centrally, perhaps, he relies on that great work of naming which had been the fundamental contribution of the eighteenth century to the sciences. . . . Naming is to Thoreau always a complex and transcendent task; and these names are at

least a place to start, a common if not always steady ground" (John Hildebidle, *Thoreau, A Naturalist's Liberty* [Cambridge, MA: Harvard University Press, 1983], pp. 86–87).

37. The impossible name Thoreau desires would not only apply to a single being but would have to be contrived at the moment that the being was observed. Moreover, to remain faithful to the mobility of the being it names, the name itself would have to be in process, so that names would be transformed into nuanced sensorial phenomena themselves, manufactured by a refinement of attention, with no noun being applicable to either a continuity of one being or a generality of multiple "instances." Thoreau's desire to invent a scientific nomenclature that would be a flow of sensorial materiality only intensified, signaling that the empiricism of the fish discourse was not a whimsical digression. As late as October 27, 1858, trying to describe leaves, he works on a language in which the difference between the materiality of what is named and the ideality of naming would collapse into the animated flow of what is perceived in the instant: "It is impossible to describe the infinite variety of hues, tints, and shades, for the language affords no names for them, and we must apply the same term monotonously to twenty different things. If I could exhibit so many different trees, or only leaves, the effect would be different. . . . And it is these subtle differences which especially attract and charm our eyes. . . . To describe these colored leaves you must use colored words. How tame and ineffectual must be the words with which we attempt to describe that subtle difference of tint, which so charms the eye? Who will undertake to describe in words the difference in tint between two neighboring leaves on the same tree? or of two thousand?—for by so many the eye is addressed in a glance. In describing the richly spotted leaves, for instance, how often we find ourselves using ineffectually words which merely indelicate faintly our good intentions, giving them in our despair a terminal twist toward our mark,—such as redd*ish*, yellow*ish*, purpl*ish*, etc. We cannot make a hue of words, for they are not to be compounded like colors, and hence we are obliged to use such ineffectual expressions as reddish, brown, etc. They need to be ground together" (Jn, 11, 255–256).

38. Four days later Thoreau generalizes this insight to "all science," claiming that because they are severed from life, scientific "descriptions" don't have access to truth: "The scientific differs from the poetic or lively description somewhat as the photographs which we so weary of viewing from paintings and sketches, though this comparison is too favorable to science. All science is only a makeshift, a means to an end which is never attained. After all, the truest description, and that by which another living man can most readily recognize a flower, is the unmeasured and eloquent one which the sight of it inspires. No scientific description will supply the want of this, though you should count and measure and analyze every atom that seems to compose it" (Jn, 14, 117).

39. Philip Cafaro, "Thoreau on Science and System," *Paideia* Archive, p. 5, http://www.bu.edu/wcp/Papers/Envi/EnviCafa.htm.

40. The phrase is from Walls, *Seeing New Worlds*, p. 123.

41. Ibid., p. 153.

42. Gillian Beer, *Open Fields: Science in Cultural Encounter* (Oxford: Clarendon Press, 1996), p. 14.

43. As Cafaro puts it: "Against philosophy's tendency to equate real knowledge with general scientific knowledge, Thoreau asserts: (1) the existence and value of particular knowledge; and (2) the existence and value of acquaintance, irrespective of knowledge" (Cafaro, "Thoreau on Science and System," p. 3).
44. Beer, *Darwin's Plots*, p. 74.
45. Mary B. Hesse, *Models and Analogies in Science* (Notre Dame, IN: University of Notre Dame Press, 1966), pp. 82–83. As Hesse explains, Cuvier's "structural homologies" tried to isolate what is "common" and then idealize it into an essence shared by what appears to be irreducibly different: "The logical argument rests on the presumption that if AB is connected with D in the model, then there is some possibility that B is connected with D, and that this connection will tend to make D occur with BC in the explicandum" (p. 85).
46. Beer, *Darwin's Plots*, p. 74.
47. Mary B. Hesse, *Models and Analogies in Science*, p. 78.
48. Beer, *Darwin's Plots*, p. 78.
49. Thoreau, "Natural History of Massachusetts," p. 210.
50. Walls, *Seeing New Worlds*, p. 40.

Stones

1. Charles Lyell, *Principles of Geology*, 5th ed., 4 vols. (London: John Murray, 1837), vol. 3, p. 309: "It has been stated that the . . . primary series sometimes passes by intermediate gradations into strata of mechanical origin. The formation of intermediate character [of rocks] was "termed by Werner 'transition rocks.' . . . Some of the shales, for example, associated with these [primary] strata, often passed insensibly into clay slates, undistinguishable from those of the granitic series."
2. Ibid., vol. 1, pp. 106–107.
3. Arnold Guyot, *The Earth and Man: Lectures on Comparative Physical Geography in Its Relation to the History of Mankind*, trans. C. C. Felton (Boston: Gould and Lincoln, 1849), pp. 22, 23, 24. In his third lecture, Guyot discusses the "law of life" and applies it to metals: "I say, further, place side by side two plates of the same metal, but unequally heated, and there is established between them an interchange of temperature. . . . Thus everywhere a simple difference, be it of matter, be it of condition, be it of position, excites a manifestation of vital forces, a mutual exchange between the bodies, each giving to the other what the other does not possess" (ibid., p. 95).
4. "Nonprogressionism," another aspect of Lyell's geology, might have contributed to this claim. As Stephen Jay Gould explains, "nonprogressionism" meant that "change is continuous, but leads nowhere" (Gould, *Time's Arrow, Time's Cycle: Myth and Metaphor in the Discovery of Geological Time* [Cambridge, MA: Harvard University Press, 1987], p. 123). Change is neither theological nor teleological; life is neither guided by some "spiritual" powers nor progresses toward a higher goal but on the contrary (Thoreau must have particularly appreciated this aspect of Lyell's theory because it was in accord with the Greek philosophy of cyclical time) can sometimes "return" to its previous states by reversing its processes. Because life is continuous—because nothing

had vanished definitively as the result of a catastrophe—what has been is never absolutely extinct but can be recovered for "the renovating as well as the destroying causes are unceasingly at work" (Lyell, *Principles of Geology*, vol. 2, p. 358). Importantly, for Lyell, life renovates itself not only geologically ("the repair of the land being as constant as its decay" [ibid.]) but also biologically. Thus, when the earth, which is now in geological winter, enters summer again, extinct species will return: "We might expect, therefore, in the summer of the 'great year.' . . . That there would be a great predominance of tree-ferns and plants allied to palms and arborescent grasses. . . . Then might those genera of animals return, of which the memorials are preserved in the ancient rocks of our continents" (ibid., vol. 1, p. 123). The idea that memorials will come alive and the dead will return—a much commented on and ridiculed thesis, which, on Gould's account, Lyell nevertheless considered as the "incarnation of rationality" (Gould, *Time's Arrow*, p. 143)—must have interested Thoreau not just for its geological consequences but also because of what it suggested concerning relics and monuments. In reading Lyell as "nonprogressionist," I follow Gould and differ from Rossi, who sees Lyell's geology as predicated on the idea of progress and "natural theology" (William Rossi, "Poetry and Progress: Thoreau, Lyell, and the Geological Principles of *A Week*," *American Literature* 66, no. 2 [June 1994]: 275–299, p. 277).

5. Lyell, *Principles of Geology*, vol. 3, p. 306.

6. Jane Bennett, "Systems and Things," *New Literary History* 43, no. 2 (2012): 225–233, at p. 227.

7. Henry David Thoreau, Houghton Library, Harvard University, Cambridge, Massachusetts, MH 15, J. A version of this passage appears in *A Week* (W, 169).

8. Jacob Bigelow, "On the Burial of the Dead," in *Nature in Disease and Miscellaneous Writings* (Boston: Ticknor and Fields, 1854), pp. 171–172.

9. Stanley French, "The Cemetery as Cultural Institution: The Establishment of Mount Auburn and the 'Rural Cemetery' Movement," in *Death in America*, ed. David E. Stannard (Philadelphia: University of Pennsylvania Press, 1975), p. 79.

10. Bigelow, "On the Burial of the Dead," p. 173.

11. Ibid., p. 174.

12. Ibid., p. 189.

13. Ibid., p. 193.

14. French, "The Cemetery as Cultural Institution," p. 83.

15. Bigelow, "On the Burial of the Dead," p. 175.

16. French, "The Cemetery as Cultural Institution," p. 83.

17. Cited in Eduardo Cadava, *Words of Light: Theses on the Photography of History* (Princeton, NJ: Princeton University Press, 1997), p. 8.

18. Ibid.

19. Linck C. Johnson, *Thoreau's Complex Weave: The Writing of A Week on the Concord and Merrimack Rivers* (Charlottesville: University Press of Virginia, 1986), p. 459.

20. Ibid., p. 458.

21. Jacques Derrida, *Glas*, trans. John P. Leavey Jr. and Richard Rand (Lincoln: University of Nebraska Press, 1990), p. 144.
22. Ibid.
23. Johnson, *Thoreau's Complex Weave*, p. 474. A version of that passage also appears in the *Journal* (J, 2, 70), and the first sentence in my quote is from that version.
24. Andrew L. Christenson, "The Identification and Study of Indian Shell Middens in Eastern North America: 1643–1861," *North American Archaeologist* 6, no. 3, (1985): 227–243. My analysis here closely follows Christenson.
25. Ibid, p. 229.
26. Ibid, p. 233: "Geologist C. T. Jackson first described these massive deposits, which were 6m high and 50m long, and had well-preserved shell stratified into regular layers."
27. Ibid., pp. 233–234.
28. Jeffries Wyman, "An Account of Some Kjoekkenmoeddings, or Shell-Heaps, in Maine and Massachusetts," *American Naturalist* 1, no. 11 (January 1868): 71.
29. Christenson, "The Identification and Study of Indian Shell Middens," p. 235. Christenson specifies that "Clamshell Hill," around Concord was understood to be "a freshwater mussel midden" because " 'clam' was commonly used for freshwater mussel at the time" (ibid.).
30. Henry David Thoreau, "The Natural History of Massachusetts," in *The Major Essays of Henry David Thoreau*, ed. Richard Dillman (Albany, NY: Whitston Publishing, 2001), p. 210. For an excellent discussion of Thoreau's research into other shell middens, specifically those located on Cape Cod, see Christenson, "The Identification and Study of Indian Shell Middens," pp. 235–236.
31. Thoreau wasn't right just about the historical value of middens and mounds. He was also right when he contradicted the scientific estimate that "potholes" at the Amoskeag were artificial and constructed by Native Americans, by claiming that they were in fact natural. Relying on Cotton Mather's opinion, the Royal Society of London estimated in its *Transactions* that the "pot holes' " were artificial. In *A Week*, that estimate became the object of Thoreau's explicit critique: "The Indians who hid their provisions in these holes, and affirmed 'that God had cut them out for that purpose," understood their origin and use better than the Royal Society, who in their Transactions, in the last century, speaking of these very holes, declare that 'they seem plainly to be artificial.' Similar "potholes," may be seen at the Stone Flume on this river, on the Ottaway, at Bellows' Falls on the Connecticut, and in the limestone rock at Shelburne Falls on Deerfield river in Massachusetts, and more or less generally about all falls. Perhaps the most remarkable curiosity of this kind in New England is the well known Basin on the Pamigewesset" (W, 247). The quote demonstrates more than Thoreau's extraordinary knowledge of New England geology; it also testifies to his extraordinary understanding of Native American culture. For in understanding potholes as geological formations, Thoreau also presumes that they were appropriately used by Native Americans to preserve food, which he again interprets in a vitalistic manner. He describes the "current"

washing down a stone, which then "revolves as on a pivot where it lies, gradually sinking in the course of centuries deeper and deeper into the rock, and in new freshets receiving the aid of fresh stones" (W, 247). Fresh stones then mix with "rocky shells of whirlpools," "as if, by force of example and sympathy after so many lessons, the rocks, the hardest material, had been endeavoring to whirl or flow into the forms of the most fluid" (W, 247–248). In that whirlpool of living stones and vibrant shells Native Americans would then deposit "their provisions," understanding that the "real purpose" of the earth is to mix with the humans, to "conceal" and preserve (W, 246), that is, to archive. For the sources of Thoreau's references to Mather and the Royal Society, see William Brennan, "An Index to Quotations in Thoreau's 'A Week on the Concord and Merrimack Rivers,'" *Studies in the American Renaissance* (1980): 259–290. As Brennan notes, Thoreau found his reference in Jeremy Belknap, *The History of New-Hampshire* (Dover, NH: Printed for O. Crosby and J. Varney, by J. Mann and J. K. Remick, 1812). It was Belknap who remarked that "the account was 'formerly sent to the Royal Society' (actually it was derived from letters that Cotton Mather wrote to John Woodward and Richard Waller), which published it. Thoreau errs in attributing the opinion quoted here to the Royal Society itself; most likely it was Mather's" (Brennan, "An Index to Quotations," p. 275).

32. Peter Nabokov, *A Forest of Time: American Indian Ways of History* (New York: Cambridge University Press, 2002), p. 42.

33. Ibid., p. 131.

34. Alfonso Ortiz, "Some Concerns Central to the Writing of 'Indian' History," *Indian Historian* 10 (Winter 1977): 17–22, at 20 (quoted in Nabokov, *A Forest of Time*, p. 132).

35. Ibid.

36. Sam Bingham and Janet Bingham, *Between Sacred Mountains: Navajo Stories and Lessons from the Land* (Tucson: University of Arizona Press, 1984), p. 1 (quoted in Nabokov, *A Forest of Time*, p. 133). Nabokov disqualifies as an "eco-pious cliché," the idea that Native American cultures knew how to live in "harmony" with nature, for it reinforces the very divide between nature and culture, place and history, that Native American cultures put in question: "Sweeping claims about the hallowed nature of Indian environments demand localized refinement lest defining distinctions between tribal traditions get submerged beneath romantic or eco-pious clichés about Indians and nature" (ibid., p. 133).

37. "Sermons in stones . . . streams" is a misquotation from Shakespeare, *As You Like It*, II.i.15–16: "books in the running brooks, / Sermons in stones" (Brennan, "An Index to Quotations," p. 275). For an excellent discussion of the ways in which stones testify in Thoreau, see Shari Goldberg, *Quiet Testimony: A Theory of Witnessing from Ninetieth Century American Literature* (New York: Fordham University Press, 2013). Goldberg precisely notices that "Thoreau reads the stones as containing voice and annals of the past" (ibid., p. 83).

38. The argument that I've been advancing throughout this chapter views Thoreau's readings of Native American ways of archivization and the complex relations they established between the natural and historical as culturally sensitive and archaeologically important, for it was he who made certain Native American practices legible to nineteenth-century Americans. It runs counter

to that proposed by a series of critiques. Most notably and explicitly (and despite the estimate of contemporary archeologists, who recognize the extraordinary accuracy of Thoreau's archaeology), Eric Sundquist—to whose otherwise brilliant reading my own Thoreau owes much—claims it to be an "abstrusely theoretical kind of archaeology" (Eric Sundquist, *Home as Found: Authority and Genealogy in Nineteenth-Century American Literature* [Baltimore: Johns Hopkins University Press, 1979], p. 49). While correctly observing that Thoreau's Native American archaeology integrates "cultural artifact with nature," Sundquist estimates such integration to represent an "imperialist archaeology" that aims to determine that "arrowheads" as always "natural," such that they will "never be properly measured or understood" by the cultural (ibid., p. 48). For Sundquist, Thoreau's archaeology therefore renders Native American relics "illegible." He finds ultimate support for his position in the following *Journal* entry: "[arrow-head] is a stone fruit. Each one yields me a thought. I come nearer to the maker of it than if I found his bones. His bones would not prove any wit that wielded them, such as this work of his bones does. It is humanity inscribed on the face of the earth, patent to my eyes as soon as the snow goes off, not hidden away in some crypt or grave or under a pyramid. Not disgusting mummy, but a clean stone, the best symbol or letter that could have been transmitted to me. The Red Man, his mark. At every step I see it. . . . It is no single inscription on a particular rock, but a footprint—rather a mind-print—left everywhere, and altogether illegible. No vandals, however vandalic in their disposition, can be so industrious as to destroy them. . . . Indifferent they to British Museums, and, no doubt, Nineveh bulls are old acquaintances of theirs, for they have camped on the plains of Mesopotamia, too, and were buried with the winged bulls. They cannot be said to be lost nor found. Surely their use was not so much to bear its fate to some bird or quadruped, or man, as it was to lie here near the surface of the earth for a perpetual reminder to the generations that come after. As for museums, I think it is better to let Nature take care of our antiquities" (Jn, 12, 91–92). Sundquist reads as a claim that the "arrowhead is most at home when lost" (Sundquist, *Home as Found*, p. 48), and he understands Thoreau's insistence on "losing" arrowheads as an attempt to turn them into "ghosts," dematerialized illegible specters. In interpreting the same *Journal* entry, Joan Burbick's otherwise excellent reading, similarly suggests that "by making the arrowhead illegible, Thoreau silences the documents of the past" (Joan Burbick, *Thoreau's Alternative History* [Philadelphia: University of Pennsylvania Press, 1987], p. 149). To my mind the arrowheads can't be lost or found because nothing is ever lost in the earth's archives, for the earth itself remembers and recovers it; and similarly, nothing is ever really found for nothing is ever completely new, nothing is ever generated in and of itself without being made of what has been. The most nuanced reading of the entry is arguably that of Robert E. Abrams, for whom Thoreau's use of the word "illegible" in the entry in question refers to the same phenomenon as "letter," and "meaning," which means that the "illegibility" of the stone is the entrance to its meaning, his way of recovering the world it signifies, rather than silencing it (Robert E. Abrams, *Landscape and Ideology in American Renaissance Literature: Topographies of Skepticism* [New

York: Cambridge University Press, 2002], p. 49). None of these critics, however, seems to be aware of the possibility that Thoreau's sentence about the arrowheads's illegibility is a tacit reference to Thackeray: "What matter for the arrow-head, illegible stuff? Give us the placid grinning kings, twanging their jolly bows over their rident horses, wounding those good-humored enemies, who tumble gayly off the towers, or drown, smiling, in the dimpling waters, amidst the anerithmon gelasma of the fish" (William Makepeace Thackeray, "John Leech's Pictures of Life and Character," *Quarterly Review* 191 [December 1854]: 3). Thoreau obliquely referenced the same line in the context of analyses of "anerithmon gelasma" of ocean waves and fish-hawk flight, which I discussed in the first section of this part of the book.

39. For an extended discussion of the calendars, see Kristen Case, "Thoreau's Radical Empiricism: The Calendar, Pragmatism, and Science," in *Thoreauvian Modernities: Transatlantic Conversations on an American Icon*, ed. François Specq, Laura Dassow Walls, and Michel Granger (Athens: University of Georgia Press, 2013), pp. 189–192. See also H. Daniel Peck, *Thoreau's Morning Work* (New Haven, CT: Yale University Press, 1990), p. 96. Peck rightly suggests that "all observations about the status of these charts, including their relative completeness, must remain speculative until further evidence about them surfaces" (ibid.). My own interpretation is similarly speculative.

40. Ibid., p. 163.

41. Peck understands "calendars" as Thoreau's effort to insert and impose categories on the changeable. On his reading, the "calendars" thus register his "struggle to determine which natural phenomena are central and discrete enough to serve as viable categories" (ibid., p. 166). My argument throughout this chapter is that Thoreau works toward destabilizing separation of phenomena into discrete units.

Photographs

1. Robert Hunt, *The Poetry of Science or Studies of the Physical Phenomena of Nature* (London: Henry G. Bohn, 1854), p. 168.

2. Ibid.

3. Ibid., p. 174.

4. Louis Jacques Mandé Daguerre, "Daguerreotype," in *Classic Essays on Photography*, ed. Alan Trachtenberg (New Haven, CT: Leete's Island Books, 1980), p. 12.

5. Oliver Wendell Holmes, "The Stereoscope and the Stereograph," in Trachtenberg, *Classic Essays on Photography*, p. 81.

6. Hunt, *The Poetry of Science*, p. 169.

7. Ibid.

8. Ibid., p. 174.

9. Hunt judges as both erroneous and unfortunate the idea that heliography—whether Niépce's or Daguerre's—became known as "photography": "Photography is the name by which the art of sun-painting will be for ever known. We regard this as unfortunate" (ibid., p. 170).

10. Ibid., p. 175.
11. Robert Hunt, *Researches on Light* (London: Longman, Brown, Green & Longmans, 1844, 2nd ed. 1854), p. 32. References are to the second edition.
12. Ibid., p. 30.
13. Ibid., p. 27.
14. Graham Clarke, for instance, suggests that the "crude format and poor quality" of Niépce's image "declare it not so much an image as an archaeological fragment" (Graham Clarke, *The Photograph* [Oxford: Oxford University Press, 1997], p. 12).
15. Hunt, *The Poetry of Science*, p. 354. "Alternatively, beings believed to be extinct are fossilized into a stone, as in the case of ammonites or snake-stones, which Christians believed to be snakes transformed by a saint's prayers into a coil, but which current science regards as a regular geological process through which life archives forms to 'reuse' them at a later date" (ibid.).
16. Amanda Jo Goldstein, "Obsolescent Life: Goethe's Journals on Morphology," *European Romantic Review* 22, no. 3 (June 2011): 405–414. Goldstein offers a beautiful reading of Goethe's late biology that is useful for estimating the distance separating post-Lyell vitalists from early German romantics. According to her reconstruction, Goethe's late biology is materialistic but not vitalistic. In it, "in *dis*organizing acts of going to dust, vapor, or droplets, things leave their shapes on neighboring surfaces and scatter their influence into the air" (ibid., p. 411), but the traces so left are not living repositories themselves. Rather, they are commemorative correspondences, traces that the living has left "in dust . . . rather than in flesh" (ibid., p. 409).
17. Hunt, *The Poetry of Science*, p. 354.
18. Ibid., p. 356.
19. For instance, Thoreau went for a night walk on September 1, 1850, but that walk had nothing to do with light, or with the idea of recovery. Instead, it was to hear "the sound of the whistle & the rattle of the cars" (J, 3, 110). In fact, until November 25, 1850, Thoreau believed that sunlight was restorative, a belief he would abandon after reading Hunt: "Just as the sun shines in to us warmly & serenely—our creator breathes on us & re-creates us" (J, 3, 125).As a result he believes the creator capable of "recreating" us to be moonlight.
20. Walter Harding, *The Days of Henry Thoreau: A Biography* (Princeton, NJ: Princeton University Press, 1982), p. 303. Of all readers of Thoreau, Gordon V. Boudreau has perhaps paid most attention to what he calls Thoreau's "Deep Cut" *Journal* passages and his observations of sand and vegetal foliage made during daytime and nighttime walks; see Gordon V. Boudreau, *The Roots of Walden and the Tree of Life* (Nashville: Vanderbilt University Press, 1990). But Boudreau reads those passages through "Freudian and Eriksonian critics . . . regarding them as evidence of an obsessive and rejective personality" (ibid., p. 2) or, alternatively, metaphorically, as a "means of drawing aside the curtain from the mystery of artistic creation" (ibid., p. 3). By focusing on science, my reading follows a different path.
21. Henry David Thoreau, *Faith in a Seed*, ed. Bradley P. Dean (Washington, DC: Island Press, 1993), p. 101.

22. Henry David Thoreau, *Night and Moonlight* (New York: Hubert Rutherford Brown, 1921), p. 6 (first published in *Atlantic Monthly Magazine*, November 1863, pp. 579–583). My claim that Thoreau is invested in photography understood in Hunt's or Niépce's sense—as a laboratory of nocturnal light erasing the inscriptions of sunlight—runs counter to Sean Ross Meehan's interesting reading, which is predicated on Talbot. Meehan posits that Thoreau's understanding of photography as "writing" or "inscribing," is motivated by the latter's wish to reach some precise representation of nature, and then to preserve it and fix it: "Thoreau's language approaches the remediating terms of photographic portraiture; like Talbot before him, Thoreau seems to engage here the representational desire to capture the 'pictures of nature's painting' with something more permanent and unmediated than mere painting or sketching" (Sean Ross Meehan, "Pencil of Nature: Thoreau's Photographic Register," *Criticism* 48, no. 1 [2007]: 7–38, at p. 27).

Swamps, Leaves, Galls

1. William Bartram, "Travels through North and South Carolina, Georgia, East and West Florida, the Cherokee Country, the Extensive Territories of the Muscogulges or Creek Confederacy, and the Country of the Chactaws," in *Travels and Other Writings*, ed. Thomas P. Slaughter (New York: Library of America, 1996), p. 106.

2. "In the swamp, clay has its colloidal properties as well, but they are there molecule by molecule, or grain by grain; this is prior to the form, and it is what later will maintain the homogeneous and well molded brick. The quality of the matter is the form's origin" (Gilbert Simondon, "Simondon and the Physico-Biological Genesis of the Individual," trans. Taylor Adkins, October 2007, Fractal Onology site, https://fractalontology.wordpress.com/2007/10/03/translation-simondon-and-the-physico-biological-genesis-of-the-individual/.

3. Regarding swamps as processual sites of generation, Thoreau thus eventually affords them a role in earthly life similar to that played by blood in the human organism: swamp "is the most living part of nature. This is the blood of the earth" (Jn, 13, 163). Thus, as Rodney James Giblett puts it, for Thoreau the "wetland feeds and holds together the skeleton of the body of nature. Without the wetland there would be nothing to replenish the skeletal system of the dry land" (Rodney James Giblett, "Henry Thoreau: A 'Patron Saint' of Swamps," in *Postmodern Wetlands: Culture, History, Ecology* [Edinburgh: Edinburg University Press, 1996], p. 234).

4. Laura Dassow Walls, "Walking West, Gazing East, Planetarity on the Shores of Cape Cod," in *Thoreauvian Modernities: Transatlantic Conversations on an American Icon*, ed. François Specq, Laura Dassow Walls, and Michel Granger (Athens: University of Georgia Press, 2013), p. 37.

5. Giblett, "Henry Thoreau," p. 229.

6. I owe my general information on cryptogams to the Edinburgh Royal Botanical Garden's instructive website, http://web.archive.org/web/20071111-8072522/http://www.rbge.org.uk/rbge/web/hort/crypt02.jsp, as well as to Wikipedia, http://en.wikipedia.org/wiki/Cryptogam.

7. Ben Woodard, *Slime Dynamics: Generation, Mutation, and the Creep of Life* (Winchester, VA: Zero Books, 2012), p. 28.
8. "The mollusk is a being - almost a—quality. It has no need of any frame - / . . . It isn't then a simple bit of spit, but most precious reality. / The mollusk is endowed with a mighty energy for keeping itself shut / up." Francis Ponge, *The Nature of Things*, trans. Lee Fahnestock (New York: Red Dust, 2011), p. 26. My discussion of Thoreau's fungi is influenced by Elissa Marder's excellent discussion of Ponge's "Mollusk" in her "Human, None Too Human" (doctoral dissertation, Yale University, 1989), p. 111.
9. As the editors of Thoreau's *Journal* note, Thoreau read Augustin Candolle and Kurt Sprengel's *Elements of the Philosophy of Plants* (trans. from the German) (Edinburgh: W. Blackwood, 1821); see the *Journal* (J, 1, 534).
10. Thoreau, "Autumnal Tints," in *Collected Essays and Poems*, ed. Elizabeth Hall Witherell (New York: Library of America, 2001), p. 381.
11. Aristotle, "De Anima," in *The Basic Works of Aristotle*, ed. Richard McKeon, trans. J. A. Smith (New York: Random House, 1941), p. 553.
12. Ibid., p. 557.
13. Ibid., p. 555.
14. Ibid., p. 562.
15. Ibid., p. 555.
16. Ibid., p. 554.
17. John Turberville Needham, *Observation upon the Generation: Composition and Decomposition of Animal and Vegetable Substances* (London, 1749), pp. 31, 30.
18. Ibid., p. 35.
19. Candolle and Sprengel, *Elements of the Philosophy of Plants*, p. 199.
20. Ibid., p. 202.
21. "So some human beings in the November of their days exhibit some fresh radical greenness-which . . . confirms their essential vitality" (J, 7, 166).
22. William Rossi, "Thoreau's Multiple Modernities," in *Thoreauvian Modernities*, p. 63.
23. Gordon V. Boudreau (*The Roots of Walden and the Tree of Life* [Nashville: Vanderbilt University Press, 1990], pp. 125–130) offers a detailed summary of critical readings of Thoreau's preoccupation with vegetal foliage up to 1990. While the summary shows that critics were mostly preoccupied with Thoreau's observations of the earth's self-organization into vegetation and organic life in "Spring" and were less concerned with what he wrote about decay and generation throughout the *Journal* and in "Autumnal Tints," for instance, the commentary is nevertheless instructive as it documents critics' almost unanimous insistence on either psychologizing or metaphorizing his thinking on leaves. More recent criticism—especially work by Lawrence Buell, Laura Dassow Walls, and William Rossi—significantly departs from that previous orientation. But while these readers put emphasis on ecological aspects of Thoreau's botany, or look in it either for the signs of a confluence with pre-nineteenth-century scientific developments influenced by Humbolt or for its anticipation of Darwin, I am interested in tracing a path that connects Thoreau with contemporary scientific developments in the Paris of the 1830s, via Harvard scientific circles.

24. Rossi, "Thoreau's Multiple Modernities," p. 62.
25. Reginald Heber Howe Jr., "Thoreau, the Lichenist," quoted in Ray Angelo, "Thoreau as Botanist: An Appreciation and a Critique," *Arnoldia* 45 (Summer 1985): 13–23, at p. 17.
26. Angelo, "Thoreau as Botanist," p. 17. On Angelo's understanding, Thoreau's difficulties in gaining expertise concerning certain plant groups might also be explained by a lack both of good manuals and of competent botanists, because "aside from Asa Gray, virtually all other botanists in New England at this time were amateurs" (ibid.).
27. Ibid.
28. Georges Canguilhem, *A Vital Rationalist*, ed. François Delaporte, trans. Arthur Goldhammer (New York: Zone Books, 1994), p. 68.
29. Ibid.
30. Georges Canguilhem, *The Normal and the Pathological*, trans. Carolyn R. Fawcett (New York: Zone Books, 1991), pp. 55, 56.
31. Ibid., p. 56.
32. Ibid.
33. Canguilhem argues that in the nineteenth century it was Auguste Comte who intensified the normative aspect of the "normal": "In the case of Comte, the vagueness of the notions of excess and deficiency and their implicit qualitative and normative character is even more noticeable" (ibid., p. 56).
34. Ibid., p. 61.
35. Michel Foucault, *The Birth of the Clinic: An Archaeology of Medical Perception*, trans. A. M. Sheridan Smith (New York: Vintage Books, 1994), p. 153.
36. Canguilhem, *The Normal and the Pathological*, p. 196.
37. As Esposito persuasively argues, the crucial category of Nazi biopolitics, its "most salient feature," was "the enfolding of pathology into abnormality. What characterizes the degenerate above all is his distance from the norm. . . . The biological abnormality is nothing but the sign of a more general abnormality that links the degenerate subject to a condition that is steadily differentiated with regard to other individuals of the same species" (Roberto Esposito, *Bios, Biopolitics and Philosophy*, trans. Timothy Campbell [Minneapolis: Minnesota University Press, 2008], p. 119). Insofar as it renders the pathological normal, Thoreau's understanding of life also functions as a radical critique of all thanato-political drifts.
38. On the "French phase" in antebellum medicine, see Russell M. Jones, "American Doctors and the Parisian Medical World, 1830–1840," *Bulletin of History of Medicine* 47 (1973): 40–65.
39. John Duffy, *From Humors to Medical Science: A History of American Medicine* (Urbana: University of Illinois Press, 1993), p. 73.
40. Ibid.
41. Jacob Bigelow, "Treatment of Disease," in *Nature in Disease* (Boston: Ticknor and Fields, 1854), p. 75.
42. Joan Burbick, *Healing the Republic: The Language of Health and the Culture of Nationalism in Nineteenth-Century America* (Cambridge: Cambridge University Press, 1994), pp. 47–48.
43. Jacob Bigelow, "On Self-Limited Diseases," in *Nature in Disease*, p. 37.

44. Ibid.
45. Ibid.
46. On Holmes's claim that Bigelow's treatise "had more influence on medical practice in America than any other similar brief treatise," see Duffy, *From Humors to Medical Science*, p. 73.
47. Elisha Bartlett, *An Essay on the Philosophy of Medical Science* (Philadelphia: Lea & Blanchard, 1844), pp. 85–86.
48. Herman Melville, *The Confidence-Man: His Masquerade*, ed. Hershel Parker (New York: Norton, 1971), p. 69.
49. Margaret Redfern, *Plant Galls* (London: HarperCollins, 2011).
50. H. Peter Loewer, *Thoreau's Garden: Native Plants for the American Landscape* (Mechanicsburg, PA: Stackpole Books, 1996), p. 199.
51. Foucault, *The Birth of the Clinic*, pp. 152–153.
52. Here I echo contemporary discussions of life proposed by Jacques Derrida and Roberto Esposito. Although the nature and scope of my arguments doesn't allow me to engage detailed discussion of their philosophies of life, I nevertheless want to signal, for those invested in contemporary debates, how Thoreau's understanding of life merits consideration by virtue of its originality.
53. Esposito, *Bios*, p. 105.

Coda: Antigone's Birds

1. G. W. F. Hegel, *Phenomenology of Spirit*, trans. A. V. Miller (Oxford: Oxford University Press, 1977), p. 271 (par. 453).
2. Ibid., p. 269 (par. 451).
3. Ibid., p. 271 (par. 452).
4. Ibid., p. 270 (par. 452).
5. Ibid., p. 271 (par. 452).
6. Sophocles, *Antigone*, ed. and trans. Andrew Brown (Warminster, UK: Aris & Phillips, 1993), pp. 41, 43 (250ff.).
7. Carol Jacobs, "Dusting Antigone," in *Skirting the Ethical* (Stanford, CA: Stanford University Press, 2008), pp. 13–14.
8. Sophocles, *Antigone*, p. 55 (410ff.).
9. Jacobs, "Dusting Antigone," p. 15.
10. Ibid.
11. Sophocles, *Antigone*, pp. 55, 57 (421ff.).
12. Jacobs, "Dusting Antigone," p. 23.
13. Ibid., p. 21.
14. Ibid., p. 22.

III Introduction: On Embodied Knowledge and the Deliberation of the Crow

1. Perry Miller, *Consciousness in Concord* (Boston: Houghton Mifflin, 1958), p. 33.
2. Sherman Paul, *The Shores of America: Thoreau's Inward Exploration* (Urbana: University of Illinois Press, 1958), p. 301.

3. Walter Benn Michaels, "Walden's False Bottoms," in *Walden*, ed. William Rossi, Norton Critical Edition (New York: W. W. Norton & Company, 1992), pp. 407, 410, 411.

4. Alfred I. Tauber, *Henry David Thoreau and the Moral Agency of Knowing* (Berkeley: University of California Press, 2001), p. 209; James D. Reid, Rick Anthony Furtak, and Jonathan Ellsworth, "Locating Thoreau, Reorienting Philosophy," in *Thoreau's Importance for Philosophy*, ed. Rick Anthony Furtak, Jonathan Ellsworth, and James D. Reid (New York: Fordham University Press, 2012): "[In *Walden*] the metaphysician will find suggestive remarks on the nature of reality and our fascination with appearances; Kantians will discover an idealist who praises the life of the knowing subject and her conceptual contribution to the perceived world" (ibid., p. 4).

5. Stanley Cavell, *The Senses of Walden* (Chicago: University of Chicago Press, 1992), p. 102; Lawrence Buell, *The Environmental Imagination: Thoreau, Nature Writing, and the Formation of American Culture* (Cambridge, MA: Harvard University Press, 1995), p. 146; Sharon Cameron, *Writing Nature: Henry Thoreau's Journal* (New York: Oxford University Press, 1985), p. 40. H. Daniel Peck, *Thoreau's Morning Work* (New Haven, CT: Yale University Press, 1990), p. 72. Russell Goodman, "Thoreau and the Body," in *Thoreau's Importance for Philosophy*, ed. Reid, Furtak, and Ellsworth, p. 33.

6. Cora Diamond, "The Difficulty of Reality and the Difficulty of Philosophy," in *Philosophy & Animal Life* (New York: Columbia University Press, 2008), p. 46.

7. Ibid., p. 73.

8. Ibid., p. 44.

9. Ibid., p. 57.

Toward Things as They Are

1. Stanley Cavell, *The Senses of Walden* (Chicago: University of Chicago Press, 1992), p. 73.

2. Ibid., p. 72.

3. Ibid., p. 98.

4. Ibid., p. 94.

5. Ibid.

6. Ibid., p. 95.

7. Ibid., p. 106.

8. Ibid. (emphasis mine).

9. Ibid., p. 95. Thoreau's commitment to the recovery of materiality is also what distinguishes him in a fundamental way from Emerson. Whereas Emerson suggested that "an Idea of Reason is an Idea to be aspired to," Thoreau suggests on the contrary that disobedience to such an idea, the suspension of the mind, and nearing objects is to be aspired to, which in fact leads him to formulate a materialist project that radically contradicts Emerson's idealistic preoccupations.

10. Ibid., p. 106.

11. Ibid., pp. 107–108.

12. Ibid., p. 99.

Thinking with Geological Velocity

1. I use the text of the *Gita* in the version Thoreau read: *The Bhagvat-Geeta: Or Dialogues of Kreeshna and Arjoon, in Eighteen Lectures; With Notes*, trans. Charles Wilkins (London: C. Nourse, 1785, reprint 1867). I reference the reprint from 1867.

2. "How often, asleep at night, am I convinced of just such familiar events—that I am here in my dressing-gown, sitting by the fire—when in fact I am lying undressed in bed! Yet at the moment my eyes are certainly wide awake when I look at this piece of paper; I shake my head and it is not asleep; as I stretch out and feel my hand I do so deliberately, and I know what I am doing. All this would not happen with such distinctness to someone asleep." René Descartes, "Meditations on First Philosophy," in *The Philosophical Writings of Descartes*, vol. 2, trans. John Cottingham, Robert Stoothoff, and Dugald Murdoch (Cambridge: Cambridge University Press, 1984), p. 13.

3. *The Bhagvat-Geeta*, p. 34.

How to Greet a Tree?

1. *The Bhagvat-Geeta: Or Dialogues of Kreeshna and Arjoon, in Eighteen Lectures; With Notes*, trans. Charles Wilkins (London: C. Nourse, 1785, reprint 1867), p. 46. I reference the reprint from 1867.

2. As Sharon Cameron puts it: "[contemplated] feelings, mind states, and objects of mind . . . are empty phenomena, whose insubstantiality could not constitute individuality. What body, feelings, mind states, and mind objects are empty of is personal identity" (Sharon Cameron, *Impersonality: Seven Essays* [Chicago: University of Chicago Press, 2007], p. 5).

3. Robert Kuhn McGregor, "Henry David Thoreau: The Asian Thread," in *Thoreau's Importance for Philosophy*, ed. Anthony Furtak, Jonathan Ellsworth, and James D. Reid (New York: Fordham University Press, 2012), p. 210; Paul Friedrich (*The Gita within Walden* [Albany: State University of New York Press, 2009) convincingly reconstructs the ways Thoreau camouflaged Krishna's concepts in *Walden*, changing for instance "the field" *(kshetra)* into the "bean-field" (ibid., pp. 47–48) or Krishna's "tree Ashwatta, whose root is above and whose branches are below, and whose leaves are the Vedas" into the "upside-down yellow pine" (ibid., pp. 45, 43).

4. Cavell, *The Senses of Walden* (Chicago: University of Chicago Press, 1992), pp. 118–119.

5. Ibid., p. 116.

6. Ibid., pp. 116, 119.

7. Bill Brown, *A Sense of Things: The Object Matter of American Literature* (Chicago: University of Chicago Press, 2003), pp. 24, 14.

8. Martin Heidegger, *What Is Called Thinking?*, trans. J. Glenn Gray (New York: Harper & Row, 1968), pp. 43–44.

9. Ibid., p. 45.

Contemplating Matter

1. Wilkins's translation renders as *"Brahm"* what Thoreau sometimes renders as *"Brahma"* and what more modern translations render as *"Brahman."* However, Thoreau's own renderings vary; most often he transliterates "Brahm/Brahman" as "Brahma," explicitly stating that by this transliteration he understands the "impersonal god" of Hinduism (Nirguna Brahman), rather than its personification (Īshvara). For instance, in an 1846 *Journal* entry he establishes a difference between an impersonal Brahma and its personifications: "No race of thinkers occupies so lofty a platform-such a table land- They do not quite fail when they personify Brahma and speak as from the mouth of Divinity" (J, 2, 257). More explicitly, in that entry, Thoreau comments on Krishna's philosophy of absorption of the self in the "divine" being, which Wilkins translates in the following way: "Such a one, whose soul is thus fixed upon the study of *Brahm*, enjoyeth pleasure without decline. . . . The man who is happy in his heart, . . . one devoted to God . . . obtaineth the immaterial nature of Brahm" (*The Bhagvat-Geeta: Or Dialogues of Kreeshna and Arjoon, in Eighteen Lectures; With Notes*, trans. Charles Wilkins [London: C. Nourse, 1785, reprint 1867], p. 46). Thoreau's commentary suggests that he doesn't distinguish between Brahm and Brahma: "The end and object is an immense consolation-to be absorbed in Brahma" (J, 2, 259). But in the passages from the *Gita* that Thoreau quotes in his *Journal*, Wilkinson sometimes renders *"Brahm"* as "almighty," as if to Christianize the Hindu divinity. Hindu philosophy's technical vocabulary was not yet standardized, and various references to the same concept create interpretative difficulties.

2. Ellen M. Raghavan and Barry Wood, "Thoreau's Hindu Quotations in *A Week*," *American Literature* 51 (March 1997): 94–98. I quote the following editions: Īsvara Krishna, *The Sānkhya Kárika, Or, Memorial Verses on the Sānkhya Philosophy, Translated from the Sanscrit by Henry Thomas Colebrooke, Esq.; Also The Bhashya or Commentary of Guarapáda; Translated, and Illustrated by an Original Comment by Horace Hayman Wilson* (Oxford: Oriental Translation Fund of Great Britain and Ireland, 1837; reprint, London: A. J. Valpy, 1887) (I reference the 1887 edition). As Hegel explicates in his famous lectures on Indian philosophy, "it was these texts that in fact introduced Hinduism to western philosophers." Hegel is referring to Colebrooke's commentaries of the Sānkhya system as the only systematic philosophical interpretation of any Indian philosophy that the West had at its disposal around 1800: "It is quite recently that we first obtained a definite knowledge of Indian Philosophy; in the main we understand by it religious ideas, but in modern times men have learned to recognize real philosophic writings. Colebrooke, in particular, communicated abstracts to us from two Indian philosophic works, and this forms the first contribution we have had in reference to Indian Philosophy. What Frederick von Schlegel says about the wisdom of the Indians is taken from their religious ideas only. He is one of the first Germans who took up his attention with Indian philosophy, yet his work bore little fruit because he himself read no more than the index to the Ramayana." (G. W. F. Hegel, "Oriental Philosophy," in *Lectures on the History of Philosophy*, vol. 1, trans. E. S. Haldane [Lincoln: University of Nebraska Press, 1995], pp. 127–128.)

3. Īsvara Krishna, *The Sānkhya Kárika*, p. 109.
4. Ibid., p. 111.
5. Ibid., p. 109.
6. Ibid., p. 112.
7. Ibid., p. 110.
8. Ibid., p. 111.
9. In the history of Western philosophy, Wilson argues, this infinite and materialized vitality can be found where Thoreau found it also, in the Pythagoreans and in Aristotle's *Physics*, where the existence of an infinite, formless, and yet sensitive matter is suggested. Wilson specifically mentions Aristotle's *Physics*, III, 6, 200b–208a, as a text that can corroborate his claim that Hindu *Prakriti* is a version of the Western materialist understanding of matter. Yet, he also attaches such uncertainty to his claim not to be quite certain, for although *Prakriti* is distinctly material and identical to "nature," Aristotle "'nowhere declares of this nature of his, whether it be corporeal or incorporeal, substantial or accidental" (Īsvara Krishna, *The Sānkhya Kárika*, p. 112).
10. Ibid., pp. 104–105.
11. Ibid., p. 105.
12. Ibid.
13. More accurately than Colebrooke, Swami Bhaskarananda specifies that in the Sānkhya system "*Purusha* is pure consciousness or pure sentience," which is in fact opposed to *Prakriti*, and can't dissolve into it, as Thoreau suggests it can (Swami Bhaskarananda, *The Essentials of Hinduism* [Seattle: Viveka Press, The Vedanta Society of Western Washington, 2002], pp. 167–168).
14. Thus, as McGregor puts it, it is precisely Hindu philosophy that helps Thoreau "develop a view of nature unique in the Western society of his time. The prevailing American and European view of nature was to hold that only God and humanity possessed a spirit, that nature was spiritually dead. This view, promoted by much of the Judeo-Christian tradition and reinforced by the science of Descartes, Newton, and others, promoted the objectification of the natural world. Emerson's heresy was to suggest that the world of God and the spirit did extend to plants and animals, but he had done so in a framework that denied the value of the objective, sensate world—the bodies of animals and people alike. Thoreau rejected Emerson's heresy—only to embrace a far more heretical view of his own: The spiritual and the natural world were inseparable and present all around him. Spirit not only existed in nature; its existence was there to be understood, if one was mentally prepared to do the work." (Robert Kuhn McGregor, "Henry David Thoreau: The Asian Thread," in *Thoreau's Importance for Philosophy*, ed. Rick Anthony Furtak, Jonathan Ellsworth, and James D. Reid [New York: Fordham University Press, 2012], p. 214.)

The Noontime of the Mind

1. There are numerous other instances in the *Journal* where Thoreau reasserts that environmental conditions generate mental conditions. For instance, on August 31, 1851, he records the influence of the air on thought: "The greater stillness-the serenity of the air-its coolness & transparency the mistiness being condensed-are favorable to thought. (The pensive eve.)" (J, 4,

22–23). Even more explicitly, on August 7, 1853: "The stillness & the shade enable you to collect & concentrate your thoughts" (J, 6, 289). Additionally, not only environmental conditions but nutrition also generate or alter thoughts: "It is abstinence from loading the belly anew until the brain and divine faculties have felt their vigor-not till some hours does the my food invigorate my brain-ascendeth into the brain" (J, 4, 23).

2. I cite from F. A. Flückiger and J. E. Gerock, "Contributions to the Knowledge of Catha Leaves," *Pharmaceutical Journal and Transaction, Pharmaceutical Society of Great Britain*, third series, 18 (1887–1888): 221–224, at p. 222, which quotes in its entirety Botta's testimony from 1837.

3. As Mark Van Doren suggests, Thoreau kept intensifying his effort to prolong this interval during which, in Van Doren's remarkable phrase, "consciousness is caught in the trap of the specific," becoming the impression the senses receive. In Van Doren's account, the desirability of prolonging the captivity of consciousness in this selfless interval was effected by Thoreau's understanding of impression not as a subjective but rather an objective phenomenon: "For [Thoreau] thought meant 'impression' and 'impression' meant 'reality' " (Mark Van Doren, *Henry David Thoreau: A Critical Study* [Boston: Houghton Mifflin, 1916], p. 82). Thoreau thus understood suspending thought in an impression as equivalent to transforming it into something real, like an object or a thing.

Thinking with the Body 1

1. Similarly, in *A Week*, the flux of animated matter somehow takes thought into itself—the "rivers of sap and woody fibre, . . . rivers of stars, and milky-ways . . . rivers of rock on the surface of earth" are said to "flow and circulate [our] thoughts" into the same animated current (W, 331).

2. William Hazlitt, "On Going a Journey," in *Selected Writings*, ed. Ronald Blythe (New York: Penguin Books, 1982), p. 139.

3. Ibid., p. 137.

4. Immanuel Kant, "What Is Orientation in Thinking?," in *Political Writings*, trans. H. B. Nisbet, ed. H. S. Reiss (New York: Cambridge University Press, 1985), pp. 238–239.

5. Jacques Derrida, *The Beast & the Sovereign*, vol. 2, trans. Geoffrey Bennington, ed. Geoffrey Bennington and Peggy Kamuf (Chicago: University of Chicago Press, 2011), p. 60.

6. Susan Scott Parrish, *American Curiosity* (Chapel Hill: University of North Carolina Press, 2006), p. 17. See also Rebecca Solnit, *Wanderlust: A History of Walking* (New York: Penguin, 2000), p. 49.

7. Michel de Certeau, *The Practice of Everyday Life*, trans. Steven Rendall (Berkeley: University of California Press, 1984), pp. 93, 97.

8. Ibid.

9. Michel Henry, *I Am the Truth: Toward a Philosophy of Christianity*, trans. Susan Emanuel (Stanford, CA: Stanford University Press, 2003), p. 55.

10. The idea of walking as a passage over Lethe is signaled by Thoreau's comparison (in "Walking") between individual and "national" or collective walks.

Thoreau suggests that the pilgrims' crossing of the Atlantic could have been one such collective walk, requiring of them a complete forgetting of their previous selves, as well as of the worlds from which they started out, all they knew about Europe, its values, and institutions: "The Atlantic is a Lethean stream, in our passage over which we have had an opportunity to forget the Old World and its institutions. If we do not succeed this time, there is perhaps one more chance for the race left before it arrives on the banks of the Styx; and that is in the Lethe of the Pacific, which is three times as wide" (Wk, 170). The success of the walk, then, is in the completeness of the oblivion of the previous self/life that it generates.

11. As Catherine Malabou reminds us "In tales, metamorphosis is always at once an effect of the other . . . and something addressed to it. . . . What sense would there be to undergoing metamorphosis all alone?" (Catherine Malabou, *The Heidegger Change: On the Fantastic in Philosophy*, trans. Peter Skafish [Albany: State University of New York Press, 2011], p. 152.)

Thinking with the Body 2

1. David Scott explains that this reference to a northern Indian poet is in fact to a Persian rather than a Hindu tradition: "Such esoteric materials from northern India point in turn toward Persia, the hearth of Mughal court language, its poetry, and its varied Sufi orders. . . . The Persian dimension in Thoreau is often rather overlooked and yet one that is quite substantive" (David Scott, "Re-Walking Thoreau and Asia: 'Light From the East' for 'A Very Yankee Sort of Oriental,'" *Philosophy East & West* 57, no. 1 [January 2007]: 14–39, at p. 21). Scott offers a fine account of Persian influences on Thoreau's thought.

2. Roland Barthes, *The Neutral*, trans. Rosalind E. Krauss and Denis Hollier (New York: Columbia University Press, 2005), p. 185.

3. Charles F. Duffy, "Honest, Hard-Working, but Shiftless: Henry David Thoreau on the Irish Immigrants," *The Kouroo Artist Project*, http://www.kouroo.info/Thoreau/ThoreauOnTheIrish.pdf, p. 6. Duffy also provides an extensive discussion of scholarship on Thoreau's relations with Irish immigrants.

4. St. Augustine, *Confessions*, trans. Henry Chadwick (Oxford: Oxford University Press, 1991), p. 171.

5. René Descartes, "Meditations on First Philosophy," in *The Philosophical Writings of Descartes*, vol. 2, trans. John Cottingham, Robert Stoothoff, and Dugald Murdoch (Cambridge: Cambridge University Press, 1984), p. 15.

6. Robert Pogue Harrison, *The Dominion of the Dead* (Chicago: University of Chicago Press, 2003), p. 44.

7. Walter Harding, *The Days of Henry Thoreau: A Biography* (Princeton, NJ: Princeton University Press, 1982), p. 183.

8. On cold days, when the windows had to remain closed, Thoreau worked hard to keep the house transparent. For instance, he never put curtains in his house ("it costs me nothing for curtains, for I have no gazers to shut out but the sun and the moon, and I am willing that they should look in" [Wa, 67]), and because the windows were low and large, anybody could indeed "look in."

Everything that was "in" was exposed to the gaze of passersby and outsiders. Unlike Collins's shanty, whose interiority wasn't visible from the outside, the interiority of Thoreau's house is transparent. There are no secrets in it, nothing personal or private; one might in fact say that it has no interior.

9. Emerson, "Thoreau," in *Essays and Poems*, ed. Joel Porte (New York: Library of America, 1983), p. 1018.

Personhood

1. Plotinus, *Enneads*, III, 8, "On Nature and Contemplation and the One," trans. A. H. Armstrong (Cambridge, MA: Harvard University Press, 1967), p. 361.
2. Ibid., pp. 363, 369.
3. Pierre Hadot, *Plotinus or The Simplicity of Vision*, trans. Michael Chase (Chicago: University of Chicago Press, 1993), p. 44.
4. Gilles Deleuze and Félix Guattari, *What Is Philosophy?*, trans. Hugh Tomlinson and Graham Burchell (New York: Columbia University Press, 1994), p. 212.
5. Plotinus, *Enneads*, IV, 4, 2, 3–8. Quoted in Pierre Hadot, *Plotinus or the Simplictiy of Vision* (Chicago: University of Chicago Press, 1998), p. 32.
6. Deleuze and Guattari, *What Is Philosophy?*, p. 212.
7. Ibid., pp. 363, 381.
8. Ibid., p. 27.
9. Ibid., p. 24.
10. Cavell here argues that the change of the object of contemplation—switching God for a natural creature in order to know that creature such as it is in itself—completely changes the position of the contemplator and his world: "The idea of God is that of a relation in which the world as a whole stands; call it a relation of dependency, or of having something 'beyond' it. The idea of the thing-in-itself is the idea of a relation in which we stand to the world as a whole; call it a relation of the world's externality" (Stanley Cavell, *The Senses of Walden* [Chicago: University of Chicago Press, 1992], p. 107).
11. Thoreau's understanding of the nature and generation of the world differs not only from that of the mystics but also from Kant. Kant argues that "reason needs to assume reality as given before it can conceive the possibility of anything," because only the idea of such a reality, which he identifies with an "unlimited being," "archetypal being," or simply God, enables us to offer an intelligible explanation of the universal purposiveness of all beings. Without presuming the existence of an infinite and intelligent creator preceding all creation, we can explain only by "complete absurdities" the purpose of all life on earth (Immanuel Kant, "What Is Orientation in Thinking?," in *Political Writings*, ed. H. S. Reiss, trans. H. B. Nisbet [Cambridge: Cambridge University Press, 1991], pp. 241–242). But it doesn't occur to Kant that perhaps life and beings exist without purpose and that life is not teleological; that trees and grass are in fact not directed toward an invisible goal when they grow, a goal known only to the being that created them for that purpose; and therefore, that the concept of absolute being preceding the world is not necessary. The

beauty and radicalism of Thoreau's position lies in his removing the purposiveness of life. There is life, there is nature, and its "there is" is all there is to it: grass doesn't grow so that animals can eat it, animals are not created so that humans can eat them. No form of life is purposively made to serve higher lives.

12. René Descartes, "Meditations on First Philosophy," in *The Philosophical Writings of Descartes*, vol. 2, trans. John Cottingham, Robert Stoothoff, and Dugald Murdoch (Cambridge: Cambridge University Press, 1984), p. 10. Kant had only to potentialize and interiorize the Cartesian premise. Deleuze's economic formulation is: "[In Kant] the manifold would never be referred to an object if we did not have at our disposal objectivity as a form in general ('object in general', 'object=x'). Where does this form come from? The object in general is the correlate of the 'I think' or of the unity of consciousness; it is the expression of the cogito, its formal objectivation. Therefore the real (synthetic) formula of the cogito is: I think myself and in thinking myself, I think the object in general to which I relate a represented diversity. *The form of the object does not derive from the imagination but from the understanding*" (Gilles Deleuze, *Kant's Critical Philosophy*, trans. Hugh Tomlinson and Barbara Habberjam [Minneapolis: University of Minnesota Press, 1999], pp. 15–16). For Descartes, the world was out there, outside the mind, but this outside-ness was inconsequential for the mind. For Kant, the outside-ness is transported into the mind so long as it fits the form of objecthood that the mind generates by taking itself as the model of all objects.

13. Here I am relying on Deleuze's formulation: contemplative contraction "is by no means a memory, nor indeed an operation of the understanding: contraction is not a matter of reflection. Properly speaking, it forms a synthesis of time. . . . This synthesis contracts the successive independent instants into one another, thereby constituting the lived, or living, present" (Gilles Deleuze, *Difference and Repetition*, trans. P. Patton [New York: Columbia University Press, 1994], pp. 70–71). If starting a new life at Walden Pond marked Thoreau's distancing from his previous ways, then as William Rossi argues in a remarkable essay on mourning in *Walden*, it also signals his departure from previous practices of mourning. To start a new life at the pond meant, Rossi suggests, to start grieving anew, as if living some different and better grief: "Dated 5 July 1845, the day after he moved to the pond, the [Journal] entry begins, famously, 'Yesterday I came here to live.' . . . Considering that John Thoreau would have been thirty years old on 5 July, as Henry . . . would not have failed to recall, this famous entry also resonates through the grief it contains. . . . This gesture strives to incorporate John's life . . . the life he would have had and perhaps has still. . . . The inscription on the shore of Walden Pond draws John into 'here,' the present time and place" (William Rossi, "Performing Loss, Elegy, and Transcendental Friendship," *The New England Quarterly* 71, no. 2 [June 2008]: 252–277, at pp. 267–268). What Rossi describes looks much like the grieving that preceded Thoreau's life at Walden. For it appears that Thoreau's grief is less about helping him to separate from John than about his desire to "draw" John's past life into the present by fusing it with his own, as if it were through such a "gesture of incorporation," as Rossi

calls it, that the personhoods of the two brothers could be made indistinguishable, thereby producing John's survival. Yet Rossi is right in suggesting that *Walden* testifies to a different grief from the one recorded in *A Week*. The difference is not, however, about *what* grief should achieve; it is about *how* it will achieve. As I argued in Part I of this book, Thoreau's early response to the possibility of annulling the "encroachment of time" to infuse the past into what is, was mythological. It led him to explore ancient cosmogonies, tales, and tragedies, all of which suggested a porous self; and all of which afforded the self a possibility of radical transformations, thanks to its inhabiting a circular temporality in which it gets forwarded into the future, eventually to arrive in the past. *Walden* offers a different answer to the same question of how to organize time other than by the linear succession of past, present, and future. Thoreau's astonishing revision of both Eastern and Western practices of contemplation—a revision that is nothing short of a reinvention—was such an answer. It suggests that the power of grief to reorganize time—to generate a synthesis of time and so enable the survival of those who were in those who are—could be enacted only at a distance from mythology, by means of a rigorous epistemology requiring disciplined mental and corporeal practices. For it is through contemplation, as Thoreau discovers at Walden, that Sadi survives not just in him but as him.

Coda: Benjamin's Seagulls

1. Walter Benjamin, "On Some Motifs in Baudelaire," in *Selected Writings*, vol. 4, *1938–1940*, ed. Howard Eiland and Michael W. Jennings, trans. Harry Zohn (Cambridge, MA: Harvard University Press, 2003), p. 338.
2. Ibid., p. 339.
3. Walter Benjamin, *The Arcades Project*, trans. Howard Eiland and Kevin McLaughlin (Cambridge, MA: Harvard University Press, 1999), p. 416.
4. Ibid., p. 418.
5. Ibid., p. 416.
6. Ibid., p. 447.
7. Ibid., p. 417.
8. Ibid.
9. Sylviane Agacinski, *Time Passing: Modernity and Nostalgia*, trans. Jody Gladding (New York: Columbia University Press, 2003), p. 55.
10. Benjamin, *The Arcades Project*, p. 448.
11. Ibid., p. 449.
12. Walter Benjamin, "Seagulls," in Samuel Weber, *Benjamin's—abilities*, trans. Samuel Weber (Cambridge, MA: Harvard University Press, 2008), pp. 325–326. For an excellent discussion of Benjamin's article, see Weber's commentary "Seagulls," in *Benjamin's—abilities*, especially pp. 316–320.
13. Agacinski, *Time Passing*, p. 56.
14. Benjamin, *The Arcades Project*, p. 416.

IV Introduction: The Ethics of Communal Mourning and the Flight of the Turtle Dove

1. Roberto Esposito, *Communitas: The Origin and Destiny of Community*, trans. Timothy Campbell (Stanford, CA: Stanford University Press, 2010), p. 11.

2. Ibid.

3. Mitchell Breitwieser, "Henry David Thoreau and the Wrecks on Cape Cod," in *National Melancholy: Mourning and Opportunity in Classic American Literature* (Stanford, CA: Stanford University Press, 2007), p. 152.

4. Stanley Cavell, *The Senses of Walden* (Chicago: University of Chicago Press, 1992), p. 51.

5. Barbara Johnson reconstructs the history of readings of the passage in the following way: "It should come as no surprise that the hound, the bay horse, and the turtle dove are almost universally seen as symbols by Thoreau's readers. The questions asked of this passage are generally, What do the three animals symbolize? And Where did the symbols come from? The answers to these questions are many and varied: for T. M. Raysor, the animals represent the 'gentle boy' Edmund Sewall, Thoreau's dead brother John, and the woman to whom he unsuccessfully proposed marriage, Ellen Sewall; for Francis H. Allen, the symbols represent 'the vague desires and aspirations of man's spiritual nature'; for John Burroughs, they stand for the 'fine effluence' that for Thoreau constitutes 'the ultimate expression of fruit of any created thing.' Others have seen in the symbols 'a mythical record of [Thoreau's] disappointments' (Emerson), a 'quest . . . for an absolutely satisfactory condition of friendship' (Mark Van Doren), the 'wildness that keeps man in touch with nature, intellectual stimulus, and purification of spirit" (Frank Davidson), and a 'lost Eden' (Alfred Kazin)." (Barbara Johnson, "A Hound, a Bay Horse, and a Turtle Dove: Obscurity in Walden," in *A World of Difference* [Baltimore: Johns Hopkins University Press, 1987], p. 53.)

6. Ibid.

7. Ibid.

8. Cavell, *The Senses of Walden*, p. 51.

9. Ibid.

10. Ibid., p. 53

11. Ibid.

Deathways of Things

1. Martin Heidegger, "The Thing," in *Poetry, Language, Thought*, trans. Albert Hofstadter (New York: HarperCollins, 2001), p. 170.

2. I quote from the revised version. The first version of this meditation is recorded on Sunday, September 24, 1843, in the *Journal* (J, 1, 465).

3. Carolyn Steedman, *Dust: The Archive and Cultural History* (New Brunswick, NJ: Rutgers University Press, 2002), p. 68.

4. Miguel Tamen, *Friends of Interpretable Objects* (Cambridge, MA: Harvard University Press, 2001), p. 59.

Ragged Fragments and Vital Relics

1. Walter Harding, *The Days of Henry Thoreau: A Biography* (Princeton, NJ: Princeton University Press, 1982), p. 70.

2. [Henry David Thoreau], obituary for Anna Jones, *Yeoman's Gazette* (Concord, MA), November 25, 1837.

3. The obituary influenced Hawthorne so much that he drew on it "in describing the search for the body of Zenobia in *The Blithedale Romance*." The *Concord Freeman* in fact republished the obituary from the *Christian Register*, June 26, 1845, p. 119. For the obituary, as well as for a historical contextualization of it, see Leslie Perrin Wilson, "Martha Hunt in Her Own Words," *Thoreau Society Bulletin* 255 (Summer 2006): 10–11. I am grateful to Leslie Perrin Wilson, curator of the Concord Free Public Library's special collection, for drawing my attention to this.

4. For the surviving drafts, their condition, and differences among them, see the editorial note to Henry David Thoreau, "Early Essays and Miscellanies," in *The Writings of Henry D. Thoreau*, ed. Joseph J. Moldenhauer and Edwin Moser, with Alexander C. Kern (Princeton, NJ: Princeton University Press, 1975), p. 378.

5. Ibid.

6. Ibid.

7. Henry David Thoreau, "Against Burial of the Dead," unpublished manuscript, Houghton Library, Harvard University, MH 15J. Newton is mention in another context in *A Week:* "Let not us sailors of late centuries take upon ourselves any airs on account of our Newtons and our Cuviers" (W, 366). The sentence suggests that Thoreau feared that to claim how great events are effects of all lives, as he did in the omitted fragment I quoted, could be mistakenly understood as his advocating that there is no need for individual reform and moral perfecting.

8. Harding, *The Days of Henry Thoreau*, p. 277.

9. Ralph Waldo Emerson, "Emerson, to Marcus Spring, July 23, 1850," in *The Selected Letters of Ralph Waldo Emerson*, ed. Joel Myerson (New York: Columbia University Press, 1997), p. 358. On many other occasions Thoreau performed similar practices of mourning and continued to take care of the dead. For instance, in 1853, when Emerson's mother died, "it was Thoreau who journeyed over to Littleton to bring home Emerson's mentally retarded brother Bulkeley and to watch him through the funeral" (Harding, *The Days of Henry Thoreau*, p. 302). When Bulkeley himself died on May 27, 1859, "in Littleton, Massachusetts, where he had for years been in the care of a local farmer, it was Thoreau who took over handling all the arrangements for both the funeral and the burial" (ibid., p. 410). Similarly, on March 22, 1859, Thoreau "had been one of the few outside the family asked to attend the private funeral service for the seventeen-year-old Theodore Parker Pratt, son of Minot Pratt, who had been buried under an oak near the family homestead" (ibid.).

10. As Thoreau specifies in a letter to Emerson from July 25, 1850, he had taken Channing with him, and Arthur Fuller arrived on the Island on that day (C, 262–263).

11. As the *Journal* editors note, from the first manuscript (May 12–September 19) "thirty-two leaves were removed, none of which has been located," whereas from the fourth manuscript (September 25–December 2) "forty-six leaves were removed, none of which has been located" (J, 3, 502, 503).

12. Steve Grice, "A Leaf from Thoreau's Fire Island Manuscript," *Thoreau Society Bulletin* 258 (Spring 2007): 1. Grice combined the fragment he recovered, and which is in his possession, with another one from the Pierpont Morgan Library: "This fragment [the one in his possession] is the top half of another two-sided, half page manuscript fragment which in 1938 was removed from Manuscript Edition set 518 to become page 18 of MA 920 in the Pierpont Morgan Library collection. Both fragments are on very thin, lined paper with a greenish hue, and the widths match as does the color of the ink. Brought together they become a leaf from an obviously longer, unpublished description by Thoreau of his July 1850 trip to Fire Island, New York."

13. Thoreau in fact had a similar experience of observing the dead off the coast of Cape Cod the previous year. In 1849, he walked with Channing the Cape Cod beach after the wreck of the St. John. As in the Fire Island experience, he saw clothes and bones mixed—a mixture made, as he put it, of "shattered fragments" (CC, 7)—and described the new mixtures with a disturbing accuracy, wondering how it is—through which force—that water shatters forms: "A little further along the shore we saw a man's clothes on a rock; further, a woman's scarf, a gown, a straw bonnet, the brig's caboose . . . I was even more surprised at the power of the waves, exhibited on this shattered fragment. . . . I saw many marble feet and matted heads as the cloths were raised, and one livid, swollen and mangled body of a drowned girl . . . to which some rags still adhered, with a string, held concealed by the flesh, about its swollen neck; the coiled up wreck of a human hulk, gashed by the rocks or fishes, so that the bone and muscle were exposed, but quite bloodless- merely red and white-with wide-open staring eyes" (CC, 5–6, 14).

14. Harding, *The Days of Henry Thoreau*, p. 70.

15. Mitchell Breitwieser, "Henry David Thoreau and the Wrecks on Cape Cod," in *National Melancholy: Mourning and Opportunity in Classic American Literature* (Stanford, CA: Stanford University Press, 2007), p. 155.

The Huron-Wendat Feast of the Dead

1. Quoted in Robert F. Sayre, *Thoreau and the American Indians* (Princeton, NJ: Princeton University Press, 1997), p. 105. Sayre recognizes that there is dispute over whether 1850 was the year when Thoreau started the notebooks. Already in 1847 Thoreau had extensive notes on Native American cultures that he used in *A Week*. His dating and enumeration start from the 102-page notebook on Canada, which Thoreau began in November 1850.

2. For the list of topics, see Richard F. Fleck, ed., *The Indians of Thoreau: Selections from the Indian Notebooks of Henry David Thoreau* (Albuquerque, NM: Hummingbird Press, 1974), as well as Sayre, *Thoreau and the American Indians*, p. 113.

3. Sayre, *Thoreau and the American Indians*, p. 113.

4. Ibid., p. 112.

5. Ibid., p. 113.
6. All quotes that follow from Thoreau's Indian Notebooks are from "Indian Notebooks," Pierpont Morgan Library, Literary and Historical Manuscripts; the notebooks are catalogued in twelve individual records, MA 595–606. This grouping includes the first "Canadian Notebook" (MA 595), which, however, is not always grouped with the other eleven. I quote from MA 602, Notebook 6. The manuscript was written from November 1852 to January 1855. Sayre (*Thoreau and the American Indians*, p. 110) reconstructs how Thoreau consulted Jean de Brébeuf's *Jesuit Relations* for 1636 at the Harvard library on October 5, 1852.
7. Jean de Brébeuf, "The Huron Feast of the Dead," in *Jesuit Relations*, ed. Allan Greer (New York: Bedford/St. Martin's, 2000), p. 63.
8. On this distinction see the excellent discussion in Erik R. Seeman, *The Huron-Wendat Feast of the Dead: Indian-European Encounters in Early North America* (Baltimore: Johns Hopkins University Press, 2011), pp. 8–9. My understanding of the ritual throughout is indebted to Seeman's book.
9. I here quote from de Brébeuf himself ("The Huron Feast of the Dead," pp. 64–65). Thoreau had shortened de Brébeuf's description and substituted "something that 2 souls remain with the body" for de Brébeuf's "I gathered from his conversation" ("Indian Notebooks," Notebook 6, MA 602).
10. Seeman, *The Huron-Wendat Feast of the Dead*, p. 62.
11. "Indian Notebooks," Notebook 6, MA 602, with interpolations from de Brébeuf, "The Huron Feast of the Dead," p. 66.
12. Quoted in Seeman, *The Huron-Wendat Feast of the Dead*, p. 61.
13. Nicole Loraux, *Mothers in Mourning*, trans. Corinne Pache (Ithaca, NY: Cornell University Press, 1998), p. 98.

Coda: Pythagoras's Birds

1. Ovid, *Metamorphoses*, trans. David Raeburn (London: Penguin Classics, 2004), pp. 602, 606.
2. Ibid., p. 611.
3. Ibid., p. 612.
4. Ibid., p. 615.
5. Ibid., p. 602.
6. Ibid., p. 615.
7. Gilbert Simondon, *Two Lessons on Animal and Man*, trans. Drew S. Burk (Minneapolis, MN: Univocal Publishing, 2011), p. 33.
8. Ibid., p. 34.
9. Ibid.
10. Ibid.
11. Ovid, *Metamorphoses*, p. 602.

Appendix I

1. Dana Luciano, *Arranging Grief: Sacred Time and the Body in Nineteenth-Century America* (New York: New York University Press, 2007), p. 16.
2. Ibid., p. 27.

3. Ibid. pp. 11, 14–15.
4. Ibid., p. 19.
5. Mary Louise Kete, *Sentimental Collaborations: Mourning and Middle-Class Identity in Nineteenth-Century America* (Durham, NC: Duke University Press, 2000), p. 63.
6. Sigmund Freud, "Mourning and Melancholia," in *On Metapsychology*, trans. James Strachey, ed. Angela Richards (New York: Penguin, 1991), p. 253.
7. Additionally, the fact that for Thoreau loss is encrypted without being denied by the psyche seems to me to distinguish his account of grief from that described in Nicolas Abraham and Maria Torok, *The Shell and the Kernel*, ed. and trans. Nicholas T. Rand (Chicago: University of Chicago Press, 1994). Abraham and Torok differentiated among two types of mourning: (1) Introjection, similarly to Freud's mourning, is a "gradual process," through which the psyche recognizes the fact of loss and eventually comes to terms with it by reinvesting its libidinal energy in other objects. (2) Incorporation "results from those losses that for some reason *cannot be acknowledged* as such" (ibid., p. 130). Denying the loss, the psyche buries it in what Abraham and Torok call an "intrapsychic tomb." In this unhealthy mourning, loss becomes the "carrier of a crypt" (ibid., p. 131). In Thoreau's account of mourning, in contrast, loss necessarily gets encrypted, but the crypt isn't denied. In other words, for Thoreau, it is not that the loss is hidden from the sight of the psyche, merely, as if in a bad dream, that the psyche can't reach it.
8. Freud, "Mourning and Melancholia," p. 209.
9. Alessia Ricciardi, *The Ends of Mourning: Psychoanalysis, Literature, Film* (Stanford, CA: Stanford University Press, 2003), p. 18. This insight is complicated in Freud's "Analysis Terminable and Interminable" (trans. Joan Riviere), *International Journal of Psycho-Analysis* 18, no. 4 (1937): 373–405, where the very labor of analysis seems to be unstoppable, because the past resists its passing. By generalizing the work of analysis into work as such, Freud suggested, as Derrida interpreted, any work can be the work of mourning: "I have tried to show elsewhere that the work of mourning is not one kind of work among others. It is work itself, work in general, the trait by means of which one ought perhaps to reconsider the very concept of production—in what links it to trauma, to mourning, to idealizing iterability of exappropriation, thus to the spectral spiritualization that is at work in any *technē*." (Jacques Derrida, *Specters of Marx: The State of the Debt, the Work of Mourning, & the New International*, trans. Peggy Kamuf [New York: Routledge, 1994], p. 97). Such an interpretation—that mourning is endless—less confuses the distinction between mourning and melancholia (for melancholia is the absence of work), than opens mourning to the potential of the infinite, thus bringing Freud closer to Thoreau's understanding of pure grief.
10. Freud, "Mourning and Melancholia," pp. 257–258.
11. Ibid., p. 254.
12. Ibid., p. 259.
13. Here I rely on Shaun Whiteside, whose translation of this passage seems to me more accurate than Strachey's: Sigmund Freud, "Mourning and Melancholia," in *On Murder, Mourning and Melancholia*, trans. Shaun Whiteside (London: Penguin, 2005), pp. 209, 210.

14. Jean-François Lyotard, *Heidegger and "the Jews,"* trans. Andreas Michel and Mark Roberts (Minneapolis: University of Minnesota Press, 1990), p. 11.
15. Ibid.
16. Ibid.
17. Ibid., p. 12.
18. Ibid.
19. Ibid., p. 11.
20. Ibid., p. 12.
21. Sigmund Freud, "The Unconscious," in *On Metapsychology: The Theory of Psychoanalysis*, trans. James Strachey, ed. Angela Richards (New York: Penguin, 1991), p. 169.
22. "To this we may reply that the conventional equation of the psychical with the conscious is totally inexpedient. It disrupts psychical continuities, plunges us into the insoluble difficulties of psycho-physical parallelism" (ibid., p. 169).
23. Ibid., pp. 170–171.
24. Walter Benjamin, "On Language as Such and on the Language of Man," in *Walter Benjamin: Selected Writings*, vol. 1: *1913–1926*, ed. Marcus Bullock and Michael W. Jennings (Cambridge, MA: Belknap Press of Harvard University Press, 2004), p. 71.
25. "Anyone who believes that man communicates his mental being by names cannot also assume that it is his mental being that he communicates, for this does not happen through the names of things—that is, through the words by which he denotes a thing. And, equally, the advocate of such a view can assume only that man is communicating factual subject matter to other men, for that does happen through the word by which he denotes a thing. This view is the bourgeois conception of language, the invalidity and emptiness of which will become increasingly clear in what follows. It holds that the means of communication is the word, its object factual, and its addressee a human being" (ibid., pp. 64–65).
26. Ibid., p. 73.
27. Ibid., p. 70.
28. Ibid., pp. 72–73.
29. The most complex, nuanced, and rigorous interpretation of the Benjamin essay is Irving Wohlfarth, "On Some Jewish Motifs in Benjamin," in *The Frankfurt School: Critical Assessments*, vol. 2, ed. J. M. Bernstein (New York: Routledge, 1994), pp. 38–83 (reprinted from A. Benjamin, *The Problems of Modernity: Adorno and Benjamin* [New York: Routledge, 1989], pp. 157–215). For the humanist interpretation of nature's mourning in Benjamin, see Graeme Gilloch, *Walter Benjamin: Critical Constellations* (Cambridge: Polity Press, 2002), p. 62. Gilloch sees human language itself as the source of nature's mourning: "The source of this sadness is fallen human language" (ibid.). For an interpretation that oscillates between humanistic and posthumanistic, see Beatrice Hanseen, *Walter Benjamin's Other History: Of Stones, Animals, Human Beings, and Angels* (Berkeley: University of California Press, 1998), pp. 160–161. Despite her vacillation, Hanseen concludes by arguing that "Benjamin implied that humanity's divinely ordained task did not amount to mere anthropomorphism" (ibid., p. 161).

Acknowledgments

Thoreau's world slows down almost to the point of stillness. In such a zone, where everything is gradual, he enlarges what is minute and sees movement in what the impatient view as immobile. To enter his world requires deceleration, which coerces its visitor into a fabulous patience. For preventing this slowness from becoming an isolating drift I have to thank the colleagues and friends who helped me in a variety of ways to finish the book, especially by constantly reminding me that sometimes closure is not a bad idea.

Much of the argument of the book developed as I worked my way through the Thoreau archive. I am grateful to the curators and staff of the Concord Free Public Library, Concord, Massachusetts, and the Huntington Library, San Marino, California, who were a great resource. I am also grateful to Linda Briscoe Myers of the Harry Ransom Center at the University of Texas at Austin; Jeffrey S. Cramer of the Thoreau Institute, the Walden Woods Project, Lincoln, Massachusetts; Emily Walhout of the Houghton Library at Harvard University, Cambridge, Massachusetts; and Maria Isabel Molestina of the Pierpont Morgan Library, New York, for their knowledgeable and generous assistance with manuscripts, microfilms, and photo duplication. Above all I thank Anne Garner of the New York Public Library. It was there, in the old-fashioned reading room of the Berg Collection that she first presented me with the Nature and Bird Notebook, written both by John and Henry Thoreau and filled with Sophia Thoreau's pressed and by now almost pulverized flowers; and it was there, spending months, with Anne's encouragements, desperately trying to decipher Henry's handwriting—which in that notebook intersperses with John's and their sister's botanical collection—that I first caught a glimpse of how my argument might be organized and positioned. I also thank Ivan Lupić for assisting me with transcriptions of some particularly difficult passages from the Thoreau manuscripts.

I am very grateful to the great Mexican artist Graciela Iturbide for giving me permission to reprint her *Birds of Mourning* (1978) on the cover.

Lindsay Waters is the best editor one can hope for. I also thank Amanda Peery, editorial assistant at Harvard University Press for always staying on top of everything and making the production process so smooth.

Over the years that this book was in the making, portions of it were presented to scholars at Cornell University, Dartmouth College, Johns Hopkins University, Le Moyne College, London Graduate School, Pennsylvania State University, Princeton University, State University of New York Binghamton, Stanford University, Sussex University, Syracuse University, the University of Florida at Gainesville, the University of Pennsylvania, the University of Western Ontario, and Williams College. My understanding of what I wanted to say regarding Thoreau gained in clarity thanks to the pressure applied to my argument on those occasions by the following people: Peter Balaam, Joshua Bartlett, Chris Castiglia, Theo Davis, Elizabeth Maddock Dillon, Paul Grimstad, Jennifer Gurley, Lucas Hardy, Michael Jonik, Gregg Lambert, Samuel Otter, Donald Pease, Tilottama Rajan (special thanks for the squirrels), Nancy Ruttenburg, Ivy Schweitzer, William Spanos, Kate Stanley, Eric Sundquist, Elisa Tamarkin, Antoine Traisnel, Johannes Voelz, Brigitte Weltman-Aron, and Barbara Will.

I also thank the graduate students in the American Transcendentalism and Nature and Law classes that I taught at Columbia, especially Mary Grace Albanese, Benjamin Barasch, Paula Hopkins, Nicholas Mayer, and Valeria Tsygankova.

Andrew Delbanco has been a generous and supportive colleague, and for that I am very grateful. Edward Mooney offered much-appreciated input on several occasions and also shared his own excellent work on Thoreau and mourning. Lloyd Pratt gave a sustained and precise reading of the manuscript that helped encourage it toward its final version. Cary Wolfe helped considerably by discussing with me on many occasions the questions related to animal life that I raise in the book. Colin Dayan's crucial critical intervention made me see how to reorganize the book and what to leave out. Stuart Burrows was an attentive interlocutor, and his support was a constant encouragement. Elissa Marder was and remains a great, generous friend. I benefited enormously from discussing with her mollusk life in Ponge, womanly life in the Pandora box myth, and many other things about which she knows so much. I thank Geoff Bennington, Timothy Bewes, and Nicholas Royle not just for the attention they paid my work but above all for having such intense minds capable of turning the most trivial things into a philosophical adventure filled with good humor. Ross Posnock is not only a remarkable colleague who has sustained me often, on and off the Columbia campus. He also remains an example of unparalleled dedication to thinking. While cultivating in me appreciation of the negative, his acute remarks about so many things American, but especially pragmatist, were transformative, sending me to the right books and into new insights.

I owe profound gratitude to Sharon Cameron for her attention to this book. The manuscript was revised several more times than I hoped, because she rightly insisted on it. Her criticism was always accurate and constructive, even if she often led me to believe that the right formulation would never come my way. If in the end it has, it is thanks to her.

Thoreau slows one down in other ways, not just by disciplining one's perception into nuance. He is so crazy well educated that any effort to go through his reading list is likely to take years. And in struggling to figure out what he is saying about life and mourning, you find yourself, early in your research, moving through his early and late notebooks and commonplace books filled with commentaries in Greek, ad hoc translations of Aristotle, gnomic remarks on Aeschylus and Sophocles, and obscure references to birds in ancient mythology. All of that sends you back to school, makes you take classes in Greek, and leaves you despairing over its grammar. For seeing my research on Thoreau and the Greeks to fruition, I have Liana Theodoratou to thank. Not only is her extraordinary knowledge of things Greek so contagious that it made me start plotting mad plans to become a classicist, but on top of that she transliterated many of the words I was struggling with and explained the meaning of innumerable references. I also thank her for being such a remarkable friend; it is wonderful that we have been traveling together for so many years through life and Greece. Eduardo Cadava, too, is a kindred soul and extraordinary friend. His unfailing support of my work and attention to my thinking has taught me so many things for which I am grateful to him, about trees and deserts, poetry and photography, listening and sharing.

Vesna and Kir Kuiken are great people and exemplary friends. I thank them for always being there for me, for all the love and support, as well as tolerance and patience, that they decided to dispense in such generous quantities. Lucky as I am to have many people to thank, my biggest debt is to my husband David Wills. He read every single word of this book and always found time to discuss my ideas, even when preoccupied with his own writing. Without his love and support I am certain that this book would have never seen completion.

Index

Abraham, Nicolas, 441n7
Abrams, Robert E., 420n38
Aeschylus, 38, 81, 394n7, 398n9; and Thoreau's translation of, 35, 60–70, 117, 131, 392n23, 397n8
Affirmative reading, 15–19, 390n22. *See also* Cavell, Stanley
Agacinski, Sylviane, 320–322
Agamben, Giorgio, 406n58
Agassiz, Louis, 24, 132, 134, 138–139, 163–172, 178, 180, 183, 189, 412n19, 412n22, 415n36. *See also* Harvard University
Alcott, Louisa May, 91, 299, 402n7
Allen, Francis H., 437n5
Allewaert, Monique, 147–148
Anaxagora, 409n32
Anaximander, 399n1
Angelo, Ray, 234, 426n26
Animation, 7, 60, 67, 122–124, 142, 161–162, 171, 178, 288–289
Arago, Dominique François Jean, 219
Archives, 200–201, 206, 333, 337, 339, 352; body as, 343; bones as, 164; photography as, 215; things as, 334–336. *See also* Museums
Aristotle, 5, 20, 60, 121–122, 157–158, 230–232, 244, 282, 409n32
Audubon, John James, 144–146, 408n11, 408n13
Augustine, Aurelius, 296

Barthes, Roland, 12–13, 128, 293
Bartlett, Elisha, 239
Bartram, John, 203
Bartram, William, 203, 223, 332–333
Baudelaire, Charles, 321–322
Baym, Nina, 410n1
Beckett, Samuel, 11, 13
Becoming, 9–10, 11–12, 29–30, 34–35, 36–40, 78, 82–84, 101, 123–129, 136, 157–158, 176, 198, 274, 284, 286, 307–309, 312, 314, 316–317, 321–332, 335, 364, 399n14; and birds, 14, 17–18, 23, 112, 138, 144, 151, 156, 159, 247–248
Beer, Gillian, 128, 183, 185–186, 413n27, 414n30, 415n33
Benjamin, Walter, 215–216; and birds, 321–322, 436n12; and cemeteries, 197–199; and the flâneur, 319–322; and the language of things, 6–7, 334; and modernity, 252, 318–322; and nature in mourning, 379–384
Bennett, Jane, 19, 41, 95, 124, 192
Benveniste, Émile, 33–35
Bergson, Henry, 140
Bhaskarananda, Swami, 431n13
Bichat, Xavier, 135, 235–237, 241, 406n58
Bigelow, Jacob, 21, 135–138, 194–197, 236–240. *See also* Harvard University
Bingham, Sam and Janet, 206
Biology, 118, 121, 131, 134, 141–142, 162, 172, 188–189, 314, 417n4, 423n16
Biopolitics, 140–142, 168, 426n37
Birds, 1, 15, 17, 22–26, 36–38, 123–124, 138–139, 143–162, 167–168, 178, 186, 247–248, 257, 289, 303–307, 364–367, 408n11, 410n38; and Baltimore Oriole, 150–152, 409n29; and becoming, 14, 17–18, 23, 112, 138, 144, 151, 156, 159, 247–248; and bittern, 15, 22, 156, 159–161; and Book of Isaiah, 160–161; and chickadee, 17–18; and crows, 23,

Birds *(continued)*
145, 257; and fish-hawks, 37–38, 151, 152, 153, 393n9, 420n38; and Herman Melville, 110–113; and John Thoreau, 143–156, 409n27; and loon, 13–14, 23, 128; and memory, 23, 26, 160–161; and metamorphosis, 14, 23, 25, 36–37, 160, 364, 367; and owls, 144, 151, 153, 308, 354, 408n13, 409n27, 409n30; and partridges, 152, 154, 156; and sparrows, 178–179, 409n29; and turtle doves, 23, 326–329, 437n5; and Walter Benjamin, 321–322, 436n12; and whippoorwills, 307. *See also* Fossils; Ornithology
Blake, Otis, 83–85
Blanchot, Maurice, 94
Botany, 44, 229–230, 234, 253, 370, 425n23; and Candolle and Sprengel, 229, 231–232; and ferns, 224–226, 417n4; "of fungi," 131, 225–227; and galls, 231, 240–243; and Harvard vitalists, 128, 135–136, 138, 172, 190, 237, 253; and moss, 44–45, 74–75, 138, 239; and weeds, 171, 177, 200–201, 339, 400. *See also* Trees
Boudreau, Gordon V., 423n20, 425n23
Breitwieser, Mitchell, 325–326, 354
Brennan, William, 419n31
Brown, Bill, 269
Brown, John, 235
Brown, Lucy, 31, 58–59, 78, 89
Buell, Lawrence, 19, 41, 91, 163, 253, 410n1
Buffon, Georges-Louis Leclerc Comte de, 234
Burbick, Joan, 237, 420n38

Cabot, James Eliot, 180–181, 415n36
Cadava, Eduardo, 198
Cafaro, Philip, 182, 410n1, 417n43
Cameron, H. D., 399n16
Cameron, Kenneth Walter, 400n3
Cameron, Sharon, 29–30, 101–104, 254, 400n2, 429n2
Canby, Henry Seidel, 395n16
Candolle, Augustin Pyramus de, 229, 231–232
Canguilhem, Georges, 234–235, 239, 414nn29–30, 426n33
Case, Kristen, 422n39
Cavell, Stanley, 6, 9, 11, 13–14, 251, 253–254, 314, 434n10; and affirmative reading, 15, 18, 390nn21–22; and the *Bhagavad Gita*, 267–268 and critique of Kant, 259–261; on loss, 258–261, 267–268, 326–329
Cemeteries, 192, 201; and Mount Auburn Cemetery, 194–198; and Rural Cemetery Movement, 194; and Sleepy Hollow Cemetery, 385–387; and Walter Benjamin, 177–179. *See also* Graves and graveyards
Chadbourne, P. A., 204
Channing, William Ellery, 120, 346
Christenson, Andrew L., 203–204, 419n30
Cicero, 392n23, 392n24
Clarke, Graham, 423n14
Clarke, James Freeman, 401n4
Coetzee, J. M., 254–255
Colebrooke, Henry, 272–276, 277, 430n2
Coleridge, Samuel Taylor: and harmony, 46–47, 172–173; and theory of life, 129–132, 137; and walking, 282–284
Combes, Muriel, 122–123
Concord, 3, 10, 52, 66–67, 206, 223, 244, 268, 278, 317, 340, 346, 351–352, 419n29; and Concordians, 334, 338. *See also* Concord River
Concord River, 22, 66–68, 70, 74, 171, 342. *See also* Concord; Merrimack River
Contemplation, 262–267, 272, 274–279, 293–294, 305–317, 435n13. *See also* Hinduism; Sitting
Cramer, Jeffrey S., 49, 394n2
Cudworth, Ralph, 409n32
Cuvier, Georges, 163–168, 413n24

Daguerre, Louis Jacques, 214–219
Darwin, Charles, 109–110, 128, 133, 172–175, 177, 179, 412n22, 413n24, 414n31
Davidson, Frank, 437n5
Dayan, Colin (Joan), 2
De Brébeuf, Jean, 357–360
De Certeau, Michel, 283
Deleuze, Gilles, 18, 125, 127, 133, 140, 142, 390n27, 435n12
Deleuze, Gilles, and Guattari, Felix, 312
De Quincey, Thomas, 356
Derrida, Jacques, 199–200, 282, 427n52, 441n9

Descartes, René, 35, 146, 263, 296, 314–316, 429n2, 435nn12–13
Desor, Edouard, 109–110
Detienne, Marcel, 395n10
The Dial, 34, 79–80, 117, 401n4
Diamond, Cora, 254–256
Dickinson, Emily, 101–104
Dimock, Wai Chee, 77–78
Duffy, Charles, F., 433n3
Duffy, John, 236, 427n46

Ecology, 19, 222, 254, 277, 425n23. *See also* Nature
Elements, 25, 37, 44, 45, 119, 136, 137, 157, 159, 166, 169, 171, 187, 264, 308–309, 377, 407n63; and Deleuze and Guattari, 311–312; and Emerson, 126; and relics, 350; and things, 331–332. *See also* Nature; Objects
Emerson, Ralph Waldo, 29, 42, 58–60, 79, 101, 121, 126, 129, 258–259, 286, 307, 326, 346, 375, 428n9, 438n9
Empedocles, 394n4
Empiricism, 35, 44–45, 121, 126, 252, 416n37; and birds, 156; and fish, 168, 180; and radical empiricism, 180; and Thoreau's "intense empiricism," 9, 37, 41–42; and Thoreau's "Kalendar," 208, 210–211, 422n39
Esposito, Roberto, 140–142, 325, 405n26, 426n37, 427n52
Eurydice, 33, 94–95
Euripides, 31, 33, 35, 60

Faraday, Michael, 118
Farrar, John, 73–74, 399n1
Ferns, 44, 224–226, 417n4
Fichte, Johann Gottlieb, 253
Fink, Steven, 82
Fire Island, 340, 346–350, 352, 356. *See also* Fuller, Margaret
Fish, 1, 15, 128–129, 139, 161, 163–186, 203–204, 307, 322, 385–386, 410n1, 411n14, 412n22, 415n36, 416n37, 439n13. *See also* Fossils; Ichthyology
Flaubert, Gustave, 321–322
Fossils, 24–25, 139, 163–171, 187, 191, 211, 335, 353. *See also* Birds; Fish; Paleontology
Foucault, Michel, 121, 140–141, 146, 149, 176–177, 235–236, 241
French, Stanley, 194–195
Freud, Sigmund, 58, 369–379, 441n9
Friedrich, Paul, 267
Frost, Brazillai, 143–144
Fuller, Arthur, 438n10
Fuller, Margaret, 340, 346. *See also* Fire Island
Fuller, Richard, 49
Fungi, 131, 225–229, 234, 239, 425n8

Galls, 231, 240–243
Geology, 26, 104–105, 118, 128, 130, 133–134, 160, 187; and Arnold Guyot, 190–191; and Louis Agassiz, 165–167; and middens, 203–205; and mourning, 370; and Robert Hunt, 214, 219–220, 423n15; and things, 335, 339. *See also* Stones
Giblett, Rodney James, 225, 424n3
Gilloch, Graeme, 442n9
Goethe, Johann Wolfgang von, 128, 148, 413n25
Goldberg, Shari, 420n37
Goldstein, Amanda Jo, 220, 423n16
Goodman, Russell, 254
Gould, Stephen Jay, 406n36, 412n16, 417n4
Graves, Robert, 403n15
Graves and graveyards, 25, 72, 192, 195, 197–198, 200, 226, 288, 290–291, 295–298, 248, 358–361; and John Thoreau, 143–144, 386; and Thoreau's grave, 387. *See also* Cemeteries; Stones
Grice, Steve, 346–347, 439n12
Grief, 20, 22, 23, 26, 29–31, 36, 39, 58–66, 71–76, 78, 79, 87, 93–94, 257, 328–329, 342–343, 369–370, 435n13, 441n7; and Antigone, 32–33, 244–248; and letter to Isaiah T. Williams, 58, 75–76; and letter to Lucy Brown, 58–59, 71, 78, 89; and letter to Ralph Waldo Emerson, 58, 59–60, 65, 70, 71, 101, 290, 375; and paean, 60–66; and perpetual grief, 31–35, 38, 65, 89, 101, 206, 391n3, 396n2; and pure grief, 58–60, 71; and Sophocles, 32–36. *See also* Lamentation; Mourning
Guyot, Arnold, 21, 136–138, 189–191, 407n63, 417n3. *See also* Harvard University

Hadot, Pierre, 311, 313
Hanseen, Beatrice, 442n9
Harding, Walter, 49–50, 55–56, 91, 221, 305, 350, 412n18
Harris, Thaddeus William, 134, 164, 172
Harrison, Robert Pogue, 297
Harvard University, 21, 31, 73, 133–140, 400n3, 409n32; and "Harvard vitalists," 134–139, 147, 164, 194, 236–239, 276, 410n2. *See also* Agassiz, Louis; Bigelow, Jacob; Guyot, Arnold; Holmes, Oliver Wendell; Nuttall, Thomas; Tuckerman, Edward; Ware, John
Hawthorne, Nathaniel, 49–50, 438n3
Hayward, George, 236
Hazlitt, William, 282
Hegel, Georg Wilhelm Friedrich, 200, 244–246, 430n2
Heidegger, Martin, 81, 121–122, 157–158, 270–271, 331
Henry, Michel, 286
Hesiod, 20, 117–118
Hesse, Mary B., 183–184, 417n45
Hildebidle, John, 415n36
Hill, David Octavius, 198–199
Hinduism, 21, 266–267, 272, 274, 277, 293, 430n2, 431n9, 431n14; and the *Bhagavad Gita*, 21, 252, 264–268, 272–275, 429n1, 430n1; and Krishna, 264, 266–267, 274–275, 277, 429n3, 430n1. *See also* Contemplation; Sitting
Holf, Jeff, 408n14
Holmes, Oliver Wendell, 21, 215. *See also* Harvard University
Homer, 20, 31, 34, 51–55, 109, 394n4
Horses, 106, 326–329, 437n5
Hound, 326–329, 437n5
Houses, 9, 65, 74, 268–269, 293–305, 331, 332, 334; and cellars, 227, 296–298, 301–302; and chimneys, 300–302; and estate sales, 22, 334–339, 353; and fireplaces, 302–305, 331–335; and James Collins, 294–305, 433n8; and Thoreau's "dream house," 302–204; and Walden Cabin, 21, 23, 300–302, 306; and windows, 295–296, 302–305, 433n8. *See also* Walden Pond
Howe, Reginald Heber, Jr., 234
Hughes, Ted, 255
Humboldt, Alexander von, 133, 190
Hunt, Martha E., 342–343
Hunt, Robert, 139–140, 222; and light and photography, 213–221, 422n9, 423n19, 424n22; and nature of matter, 118–120, 213–14, 423n15
Huxley, Thomas, 24
Hyde, Lewis, 391n6

Ichthyology, 31, 139, 163, 170, 181, 223, 410n1. *See also* Fish
Impersonality, 79, 101, 253, 264, 276, 280, 288, 289, 375, 376, 378, 379, 383, 404n31, 429n2
Irmscher, Christoph, 145

Jackson, Charles Thomas, 203–204
Jackson, James, Jr., 236
Jackson, Lydian, 394n3
Jacob, François, 165, 411n11
Jacobs, Carol, 247–248, 395n9
James, William, 126, 140
Jankélévitch, Vladimir, 402n5, 402n8
Jarvis, Simon, 403n22
Johnson, Barbara, 8–9, 326–329, 437n5
Johnson, Linck C., 391n30, 412n18
Johnson, Philip, 304
Jones, Anna, 340–346
Jones, Elisha, 350–351
Jones, Ephraim, 353–356
Jones, Russell M., 426n38

Kant, Immanuel, 4, 35, 121, 127–128, 259–260, 282–285, 309, 314–316, 434n11, 435n12
Kazin, Alfred, 437n5
Keats, George, 401n4
Keats, John, 401n4
Kete, Mary Louise, 370
Kohn, Eduardo, 7
Krishna, Isvara, 430n2, 431n9
Krutch, Joseph Wood, 97, 403n24

Lamarck, Jean-Baptiste, 118–120
Lamb, Charles, 88
Lamentation, 26, 31, 59–66, 70, 81, 93–94, 101, 247–248, 328–329, 362, 396n2, 398nn11–12. *See also* Grief; Mourning
Lang, Karl, 411n14
Lear, Jonathan, 17
Lebeaux, Richard, 56–57

Leibniz, Gottfried Wilhelm von, 4, 121–122
Le Jeune, Paul, 357
Le Mercier, François, 357
Linnaeus, Carolus, 128, 225–226, 413n25
Loewer, H. Peter, 240
Loraux, Nicole, 31–33, 60–64, 396nn1–2, 399n14
Lorch, Fred W., 401n2
Lowell, James Russell, 97
Luciano, Dana, 369
Lucretius, 20, 392n23
Lurie, Edward, 412n19
Lyell, Charles, 118, 128, 187–192, 198, 219–220, 417n1, 417n4
Lyotard, Jean-François, 375–379
Lyric, 92–105, 396n2, 398n11; and Chippewa language, 95; and flute, 93–95, 101; and impersonality, 101; and lyre, 92–95, 101; and Orpheus, 92–95, 101; and poetry, 93, 95, 101, 104; and Thoreau's *Journal*, 99, 101–105; and Thoreau's poetics, 95–96, 98–99, 105. *See also* Music

Malabou, Catherine, 433n11
Marder, Elissa, 425n8
Materialism, 130–131, 158, 222–223, 385, 409n32; and Harvard vitalists, 139; and pantheism, 124–125; and Thoreau's vitalism, 121–122, 124
Mather, Cotton, 419n31
Matthiessen, Francis Otto, 401n2
McGill, Meredith, 401n2
McGregor, Robert Kuhn, 267, 429n3, 431n14
Meehan, Sean Ross, 424n22
Melville, Herman, 109–113, 239
Memory, 56, 72, 75, 135, 201, 207, 317, 333, 336, 339, 348, 353, 376–377, 435n13; and birds, 23, 26, 160–161; and dreaming, 106–109
Merleau-Ponty, Maurice, 120, 140
Merrimack River, 67, 69, 70, 192, 207. *See also* Concord River
Metamorphosis, 4–5, 16, 108, 127–128, 156, 165, 219, 289, 292, 294–295, 298, 307, 363–367, 375, 433n11; and birds, 14, 23, 25, 36–37, 160, 364, 367; and vegetal life, 136, 233–234
Meyer, Herman von, 24
Michaels, Walter Benn, 97, 253
Michelet, Jules, 128
Miller, Perry, 97, 252–253
Milton, John, 23, 401n4
Miracle, M. Eulàlia Gassó, 413n28
Moss, 44–45, 74–75, 138, 224–226, 234, 239
Mott, Valentine, 236
Mourning, 29–30, 35–40, 56, 66, 86–87, 91, 93, 132–133, 200, 248, 351–352, 396n1, 438n9; and communal mourning, 22, 329, 338, 346, 353; and epistemology of mourning, 254–257; and natural mourning, 59–60, 289–290, 297, 442n9; and perpetual mourning, 20, 31–33, 35, 105, 379; and Sigmund Freud, 369–379; and Thoreau's "The Funeral Bell," 86–88, 101; and time of mourning, 369, 371; and Walter Benjamin, 379–384. *See also* Grief; Lamentation
Museums, 198, 200–201, 297, 334–339. *See also* Archives
Music, 46–48, 91, 107, 175, 393n9; and Coleridge, 46–47, 172–173; and echoes, 92, 101, 308–309, 399n14; and flute, 62, 91–95, 101, 107–108, 402n7; and harmony, 46–48, 50, 58, 61, 93, 98, 403n22; and lyre, 92–95, 101, 107–108; and music boxes, 49–58, 67–70, 394n2; and poetry, 86–105, 108, 182, 191, 256, 401n2, 401n4
Mythology, 20, 26, 31, 32–33, 35–36, 38–39, 60–66, 81–82, 92–95, 101, 117–118, 128, 244–248, 363–365

Nabokov, Peter, 205–206, 420n36
Native Americans, 90, 330, 357–358, 420n36, 420n38, 439n1; and Chippewa, 95; and history, 205–207; and Huron, 357–367; and middens, 139, 187, 202–208, 419nn30–31; and Mucclasse, 332–333, 335, 337, 338, 343; and Thoreau's Indian Notebooks, 357–362
Nature, 5–9, 15, 23, 26, 31, 35, 41–45, 48, 56, 58–60, 67, 78, 87, 95, 101, 105, 112, 120–121, 125–126, 132, 137, 140, 143, 145, 146, 149, 159, 164, 176, 190, 193, 195, 215–216, 226, 230–231, 237–242, 248, 252–254, 274–276, 277, 285, 311, 313, 336, 353, 363, 370, 379, 382–384,

Nature (*continued*)
410n1, 420n36, 424n22, 434n11. *See also* Ecology
Needham, John Turberville, 230–231
Niépce, Nicéphore, 139, 213–220
Nietzsche, Fridrich, 125–127
Nuttall, Thomas, 21, 138, 144–150, 408nn22–23. *See also* Harvard University

Obituaries, 22, 340–346, 350–355; to Anna Jones, 340–345; and the *Concord Freeman*, 341–343; to elm tree, 351–352
Objects, 6, 8, 11, 43, 99–101, 192, 205, 216, 226, 254, 259–260, 330, 331–332; and archives, 334–337; and Arnold Guyot, 136–137; intimacy with, 318–319; and photography, 139–140, 217–218; as relations, 331; and Sigmund Freud. *See also* Elements; Museums; Things
Ornithology, 25, 138–139; and John James Audubon, 408n13; and John Thoreau, 144–150; and Thomas Nuttall, 148–149, 190. *See also* Birds
Orpheus, 92–95, 101
Ortiz, Alfonso, 206
Ossoli, G. A., 348–349
Ovid, 128, 363–367
Owen, Richard, 24

Paleontology, 24–26, 134, 169, 391n31, 411n14. *See also* Fossils
Pantheism, 123–125
Parrish, Susan Scott, 283
Passivity, 55, 83–84, 88, 89, 272–273, 277, 293, 407n63; and drugs, 278–280; and letters to Otis Blake, 83–85. *See also* Hinduism; Sitting
Patton, Paul, 5, 11
Paul, Sherman, 23, 253, 410n1
Peck, H. Daniel, 208, 212, 254, 395n11, 422n39, 422n41
Peirce, Charles Sanders, 7, 140
Photography, 139–140, 219–220, 416n38, 424n22. *See also* Hunt, Robert; Niépce, Nicéphore
Pierre, Vidal-Naquet, 398n9
Pindar, 20, 31, 60, 82, 400n3; and ontology of life, 80–83; and Thoreau's translation of, 34–35, 79–80
Plato, 4, 36, 38–39, 60, 91, 93–94, 101
Plotinus, 252, 311–314
Political, 21, 22, 142, 148, 168, 180, 235–236, 238, 295, 345, 354, 396n2; and animation, 142; and Antigone, 244–246; and biopolitics, 140–142, 168, 426n37; and John Thoreau, 143; and Thoreau's politics, 20, 129, 140
Ponge, Francis, 227, 425n8
Posnock, Ross, 41
Pratt, Kenneth J., 205
Pratt, Minot, 438n9
Pratt, Theodor Parker, 438n9
Prometheus, 36, 38, 117–118
Pythagoras, 363–367

Raghavan, Ellen M., 272
Redfern, Margaret, 427n49
Rena, S. E., 91
Ricciardi, Alessia, 372
Rice, Charles, 88
Richardson, Robert D., 117, 394n3, 410n2
Rossi, William, 417n4, 425n23; on "Autumnal Tints," 232–234; on mourning, 435n13
Ruwick, Mortin J.S., 410n2, 411n10, 411n14

Sagard, Gabriel, 362
Saint-Hilaire, Étienne Geoffroy, 175
Sanborn, Frank, B., 88, 357
Sauvagnargues, Anne, 123
Sayre, Robert F., 357–358, 439n1
Schelling, Friedrich Wilhelm Joseph, 129
Scott, Clive, 96
Scott, David, 433n1
Seeman, Erik R., 361, 440n8
Sewall, Edmund, 437n5
Sewall, Ellen, 437n5
Seybold, Ethel, 394n5
Shanley, J. Lyndon, 97
Shelley, Percy Bysshe, 60
Simondon, Gilbert, 140, 223, 365–366, 424n2
Sitting, 10, 21, 55, 293–294, 305–308. *See also* Contemplation; Hinduism
Slatkin, Laura, 391n3
Smellie, William, 134
Smith, Asa, 399n1
Socrates, 157

Sophocles, 31, 64, 81; and burial, 246–248; and perpetual grief, 32–36
Spinoza, Baruch de, 125
Sprengel, Kurt Polycarp Joachim, 229, 231–232
Spring, Marcus, 346
Steedman, Carolyn, 336
Stoddard, Elizabeth, 96–97
Stones, 1, 16, 19, 92, 124, 139, 159–161, 171–172, 177–179, 187–188, 191–192, 195, 206, 219–220, 305, 331, 350, 419n31, 420n37. *See also* Graves and graveyards
Sumner, Charles, 346
Sumner, Horace, 346
Sundquist, Eric, 420n38
Swamps, 42, 222, 223–229, 265, 385, 409n30, 424n3

Talbot, Henry Fox, 424n22
Tamen, Miguel, 337
Tauber, Alfred I., 253
Thackeray, William Makepeace, 420n38
Thaker, Eugene, 125
Thales, 20, 156–162, 223, 385
Things, 7, 44, 53–54, 96, 98, 131, 150, 165, 219, 220, 222, 223, 251–252, 259–261, 271, 287–288, 320, 330–337, 338–340, 347–350, 352; and analogy, 185–186; and Aristotle, 158; and contemplation, 277–278, 313; and Fire Island, 340, 346–349; and the flâneur, 320–322; and language of, 5–6, 89; and Leibniz, 122; and the ocean, 109–110, 349–350; and perceptions as, 89, 100–101; and possessions, 269–270, 334–335, 337; and *Prakriti*, 273; and Thales, 157–158, 222–223; and Thoreau's *Journal*, 105, 151, 353–356; and *Walden*, 330, 332, 333, 338; and Walter Benjamin, 7, 319–320, 380–381; and waste book, 353; and William James, 126; and words, 13, 15, 95. *See also* Archives; Elements; Museums; Objects
Thoreau, Helen, 56–57
Thoreau, Henry David: and animation, 7, 60, 67, 122–124, 127, 142, 161–162, 171, 178, 288–289; and Baltimore Oriole, 150–152, 409n29; and the *Bhagavad Gita*, 21, 252, 264–268, 272–275, 429n1, 430n1; and birds, 1, 15, 17, 22–26, 36–38, 123–124, 138–139, 143–162, 167–168, 178, 186, 247–248, 257, 289, 303–307, 321–322, 364–367, 408n11, 410n38; and bittern, 15, 22, 156, 159–161; and cellars, 227, 296–298, 301–302; and cemeteries, 192, 195, 198, 201; and chickadees, 17–18; and chimneys, 300–302; and Concord, 3–4, 10, 52, 66–67, 206, 223, 268, 338, 340, 346, 351–352; and Concord River, 22, 66–67, 70, 171; and contemplation, 262–279, 294, 305–317, 435n13; and critique of idealism, 181–182, 199–200, 256–257, 336–337; and crows, 23, 145, 257; and *The Dial*, 34, 79–80, 117; and "dream house," 302–304; and dreams and dreaming, 82, 84, 106–109, 126, 279, 284, 286–287, 290–292, 303, 343; and drugs, 278–280; and ecology, 19, 222, 254, 277, 425n23; and estate sales, 22, 334–339, 353; and ferns, 44, 224–226, 417n4; and fire and fireplaces, 302–305, 331–335; and Fire Island, 346–352; and fish, 1, 15, 128–129, 139, 161, 163–186, 203–204, 307, 322, 385–386, 410n1, 411n14, 412n22, 415n36, 416n37, 439n13; and fish-hawks, 37–38, 151, 152, 153, 393n9, 420n38; and flute, 91–95, 101, 107–108, 402n7; and fossils, 24–25, 139, 163–171, 187, 191, 211, 335, 353; and fungi, 131, 225–229, 234, 239, 425n8; and galls, 231, 240–243; and geology, 139, 142, 188, 190, 192, 211, 370; and graves and graveyards, 25, 72, 192, 195–198, 200, 290–291, 295–298, 358–361; and Harvard University, 21, 31, 73, 133–140, 147, 164, 194, 400n3, 411n2; and Hinduism, 21, 266–267, 272, 274, 277, 293, 430n1; and horses, 106, 326–329, 437n5; and hound, 326–329, 437n5; and houses, 9, 23, 65, 74, 268–269, 293–305, 331, 332, 334; and individualism, 21–22; and James Collins, 294–305, 433n8; and Krishna, 264, 266–267, 274–275, 277, 429n3, 430n1; and loon, 13–14, 23, 128; and memory, 23, 26, 56, 75, 107–108, 135, 160–161, 201, 207, 317, 333, 348, 377; and Merrimack River, 67, 70, 171, 192;

Thoreau, Henry David *(continued)*
and metamorphosis, 4–5, 14, 16, 23, 25, 36–37, 108, 127–128, 156, 160, 165, 219, 233, 289, 292, 294–295, 298, 307, 363–367, 375; and middens, 139, 187, 202–208, 419nn30–31; and moss, 44–45, 74–75, 138, 224–226, 234, 239; and Mount Auburn Cemetery, 194–198; and museums, 198, 200–201, 297, 334–339; and music boxes, 49–58, 67–70, 394n2; and nature, 5, 7–9, 23, 26, 31, 35, 41–45, 56, 58–60, 87, 101, 125, 140, 193, 206, 216, 230, 238, 246, 252, 254, 274–275, 353, 383; and obituaries, 22, 340–346, 350–355; and owls, 144, 151, 153, 308, 354, 408n13, 409n27, 409n30; and partridges, 152, 154, 156; and photography, 139–140, 219–220, 416n38, 424n22; and poetry and poetics, 86–105, 108, 182, 191, 256, 401n2, 401n4; and sitting, 10, 21, 55, 293–294, 305–306; and Sleepy Hollow Cemetery, 385–387; and stones, 1, 16, 19, 92, 124, 139, 159–161, 171–172, 177–179, 187–188, 191–192, 195, 206, 219–220, 305, 331, 350, 419n31, 420n37; and swamps, 42, 222, 223–229, 265, 385, 409n30, 424n3; and Thoreau's grave, 387; and trains, 307, 344, 393n9; and translation, 31, 33, 63–70, 79, 117, 396n6, 399n13, 430n1; and trees, 1, 10, 16, 74, 92, 172, 175–176, 177, 178, 189, 191, 210, 226, 229, 233, 240–241, 267, 271, 279, 290, 297, 302, 305, 309, 330, 351–352, 416n37; and turtle doves, 23, 326–329, 437n5; and Walden Cabin, 23, 268, 298–306; and Walden Pond, 12, 180, 191, 221, 281, 293–294; and walking, 15–16, 23, 44, 77, 106–108, 150–151, 176, 195, 221–222, 223, 257, 263, 268–269, 281–292, 293–296, 318–322, 432n10; and weeds, 171, 177, 200–201, 339, 400; and whippoorwills, 307; and windows, 295–296, 302–305, 433n8

Thoreau, Henry David, works by: "Autumnal Tints," 229–234; *Cape Cod*, 109–110, 325, 354, 439n13; "The Funeral Bell," 86–88, 101; Indian Notebooks, 357–362; *Journal*, 1, 13, 17–18, 19, 23, 29–30, 37–38, 43–46, 47–48, 50–51, 53–56, 57, 77, 78–79, 86–87, 88, 89–90, 91–92, 95, 98–99, 100–105, 106–108, 123–124, 129, 131–132, 137, 150–156, 161–162, 170–171, 173–176, 178, 180, 181–183, 184, 186, 187–189, 190–191, 194–195, 200, 205, 213, 220, 221–222, 223–230, 232–233, 239–243, 251, 257, 266–267, 272–273, 275–276, 279, 281, 284, 286–287, 290–291, 293, 304, 309–310, 312, 315, 335–336, 339, 340, 346–356, 375, 385–387, 404n31, 409nn29–30, 416nn37–38, 420n38, 423n19, 430n1, 431n1; "Kalendar," 208–212; letter to Isaiah T. Williams, 58, 75–76; letter to Lucy Brown, 58–59, 71, 78, 89; letters to Otis Blake, 83–85; letter to Ralph Waldo Emerson, 58, 59–60, 65, 70, 71, 101, 290, 375; "Natural History of Massachusetts," 123–124, 131–132, 182, 185–186, 204; obituary to Anna Jones, 340–345; obituary to elm tree, 351–352; "The Old Marlborough Road," 285–286; Pindar's Odes, translation of, 31, 34, 35, 79–82; *Prometheus Bound*, translation of, 38, 79, 117–118, 392n23, 396n6; "The Service," 46–47; *Seven against Thebes*, translation of, 62–66, 67, 70, 79; Sophocles, translation of, 33; *Walden*, 1, 3–4, 5–6, 9, 11–12, 13–14, 21, 22, 43, 95, 96–97, 99, 124–125, 152–153, 154, 176, 189, 192, 198–199, 206, 222, 251, 256–257, 262, 268–270, 272, 281, 283–290, 293–310, 315, 318, 326–329, 330–337, 338, 393n9, 433n8; "Walking," 213, 219, 220–221, 222, 272, 283–290, 307, 432n10; *A Week on the Concord and Merrimack Rivers*, 1, 8, 9, 13, 15–16, 21, 22, 25, 32–33, 36–37, 42, 43, 66–70, 71–74, 75, 79, 81–83, 90, 92, 96, 99, 101, 120, 125, 126, 128–129, 139, 156, 159, 163, 167, 168–172, 176, 177–180, 192–193, 195, 201, 206, 207–208, 244–248, 251, 262–265, 272, 277–279, 281, 345–346, 363, 371, 377, 383, 419n31, 432n1

Thoreau, John, 20, 23, 30, 35–36, 49–51, 56–57, 66–70, 83–84, 91–92, 138–139, 289–290, 342, 385–387, 394nn2–3, 395n13, 410n2; on birds, 143–156, 409n27

Thoreau, Sophia, 23, 57, 109
Thorson, Robert M., 139
Time, 50, 51, 66, 74–76, 187, 198, 206, 207, 212, 220, 265, 280, 328, 371, 435n13; and "afflicted time," 369–370; and chronic time, 52–54, 56, 57, 70, 85, 208, 394n8, 395n11; and communal time, 352; and geological time, 211; and human time, 34, 47, 53, 334; and "night of time," 263–265, 272; and "perfect time," 52; and photography, 139–140, 217, 219; and "sacred time," 369; and Thoreau's house, 302; and Thoreau's "Kalendar," 208–212. *See also* Benjamin, Walter; Music; Stones
Torok, Maria, 441n7
Translation, 31, 33, 117, 396n6, 399n13, 430n2. *See also* Aeschylus; Pindar
Trees, 1, 10, 16, 74, 92, 172, 175–176, 177, 178, 189, 191, 210, 226, 229, 233, 240–241, 267, 271, 279, 290, 297, 302, 305, 309, 330, 351–352, 416n37
Tuckerman, Edward, 21, 137–138, 172–175, 240, 413n28. *See also* Harvard University

Van Anglen, Kevin P., 396n6, 397n8, 398n11
Van Doren, Mark, 432n3
Vernant, Jean-Pierre, 36, 52–53, 117–118, 394n5

Walden Pond, 8–9, 180, 191, 221, 281, 293–294, 338; and Flint's Pond, 12; and individualism, 21–22. *See also* Houses
Walking, 15–16, 23, 44, 77, 106–108, 150–151, 176, 195, 223, 257, 268–269, 281–292, 318–332, 432n10, 439n13; and Coleridge, 282–284; and dreaming, 284, 286–287, 290–292; and the flâneur, 319–322; at night, 15–16, 221–222, 263–264, 283–284; and "The Old Marlborough Road," 285–286; and sitting, 293–296; on time, 53–57. *See also* Benjamin, Walter
Walls, Laura Dassow, 19, 133, 180, 183, 186, 225, 410n1, 415n34
Ware, John, 21, 134–138, 164. *See also* Harvard University
Warren, John Collins, 236
Weeds, 171, 177, 200–201, 339, 400
White, Deborah Elise, 100
Wilkins, Charles, 262, 273, 430n1
Williams, David B., 391n31
Williams, Isaiah T., 58, 75–76
Willis, Frederick, L.H., 23
Wilson, Horace, 272–274
Wilson, Leslie Perrin, 438n3
Wood, Barry, 272
Woodard, Ben, 227
Wordsworth, William, 282
Worster, David, 9–10, 37, 41–42
Wyman, Jeffries, 204